COMPREHENSIVE BIOCHEMISTRY

ELSEVIER SCIENTIFIC PUBLISHING COMPANY
335 Jan van Galenstraat, P.O. Box 211, Amsterdam, The Netherlands

AMERICAN ELSEVIER PUBLISHING COMPANY, INC.
52 Vanderbilt Avenue, New York, N.Y. 10017

Library of Congress Card Number 62–10359
ISBN 0-444-41192-5

With 67 illustrations and 20 tables

PRINTED IN THE NETHERLANDS

COMPREHENSIVE BIOCHEMISTRY

COMPREHENSIVE BIOCHEMISTRY

SECTION I (VOLUMES 1–4)
PHYSICO-CHEMICAL AND ORGANIC ASPECTS OF BIOCHEMISTRY

SECTION II (VOLUMES 5–11)
CHEMISTRY OF BIOLOGICAL COMPOUNDS

SECTION III (VOLUMES 12–16)
BIOCHEMICAL REACTION MECHANISMS

SECTION IV (VOLUMES 17–21)
METABOLISM

SECTION V (VOLUMES 22–29)
CHEMICAL BIOLOGY

SECTION VI (VOLUMES 30–33)
A HISTORY OF BIOCHEMISTRY

(VOLUME 34)
GENERAL INDEX

COMPREHENSIVE BIOCHEMISTRY

EDITED BY

MARCEL FLORKIN

Professor of Biochemistry, University of Liège (Belgium)

AND

ELMER H. STOTZ

Professor of Biochemistry, University of Rochester, School of Medicine and Dentistry, Rochester, N.Y. (U.S.A.)

VOLUME 29 PART A

COMPARATIVE BIOCHEMISTRY, MOLECULAR EVOLUTION

ELSEVIER SCIENTIFIC PUBLISHING COMPANY

AMSTERDAM · LONDON · NEW YORK

1974

CONTRIBUTORS TO THIS VOLUME

MARCEL FLORKIN, M. D.

Emeritus Professor of Biochemistry, University of Liège, 17 Place Delcour, Liège (Belgium)

TONY SWAIN, Sc.D., F.L.S.

Director, ARC Biochemical Laboratory, Royal Botanic Gardens, Kew, Richmond, Surrey (Great Britain)

GENERAL PREFACE

The Editors are keenly aware that the literature of Biochemistry is already very large, in fact so widespread that it is increasingly difficult to assemble the most pertinent material in a given area. Beyond the ordinary textbook the subject matter of the rapidly expanding knowledge of biochemistry is spread among innumerable journals, monographs, and series of reviews. The Editors believe that there is a real place for an advanced treatise in biochemistry which assembles the principal areas of the subject in a single set of books.

It would be ideal if an individual or small group of biochemists could produce such an advanced treatise, and within the time to keep reasonably abreast of rapid advances, but this is at least difficult if not impossible. Instead, the Editors with the advice of the Advisory Board, have assembled what they consider the best possible sequence of chapters written by competent authors; they must take the responsibility for inevitable gaps of subject matter and duplication which may result from this procedure.

Most evident to the modern biochemist, apart from the body of knowledge of the chemistry and metabolism of biological substances, is the extent to which he must draw from recent concepts of physical and organic chemistry, and in turn project into the vast field of biology. Thus in the organization of Comprehensive Biochemistry, the middle three sections, Chemistry of Biological Compounds, Biochemical Reaction Mechanisms, and Metabolism may be considered classical biochemistry, while the first and last sections provide selected material on the origins and projections of the subject.

It is hoped that sub-division of the sections into bound volumes will not only be convenient, but will find favour among students concerned with specialized areas, and will permit easier future revisions of the individual volumes. Toward the latter end particularly, the Editors will welcome all comments in their effort to produce a useful and efficient source of biochemical knowledge.

M. Florkin
E. H. Stotz

Liège/Rochester

PREFACE TO SECTION V

(VOLUMES 22–29)

After Section IV (*Metabolism*), Section V is devoted to a number of topics which, in an earlier stage of development, were primarily descriptive and included in the field of Biology, but which have been rapidly brought to study at the molecular level. "*Comprehensive Biochemistry*", with its chemical approach to the understanding of the phenomena of life, started with a first section devoted to certain aspects of organic and physical chemistry, aspects considered pertinent to the interpretation of biochemical techniques and to the chemistry of biological compounds and mechanisms. Section II has dealt with the organic and physical chemistry of the major organic constituents of living material, including a treatment of the important biological high polymers, and including sections on their shape and physical properties. Section III is devoted primarily to selected examples from modern enzymology in which advances in reaction mechanisms have been accomplished. After the treatment of Metabolism in the volumes of Section IV, "*Comprehensive Biochemistry*", in Section V, projects into the vast fields of Biology and deals with a number of aspects which have been attacked by biochemists and biophysicists in their endeavour to bring the whole field of life to a molecular level. Besides the chapters often grouped under the heading of molecular biology, Section V also deals with modern aspects of bioenergetics, immunochemistry, photobiology and finally reaches a consideration of the molecular phenomena that underlie the evolution of organisms.

Liège/Rochester

M. FLORKIN
E. H. STOTZ

CONTENTS

VOLUME 29 A

COMPARATIVE BIOCHEMISTRY, MOLECULAR EVOLUTION

Chapter II. Biochemical Evolution of Plants
by T. Swain

Volume 29. Part B

COMPARATIVE BIOCHEMISTRY, MOLECULAR EVOLUTION *(continued)*

Chapter 3. Biochemical evolution of bacteria
by J. de Ley

Chapter 4. Biochemical evolution of animals
by M. Florkin

Chapter 5. Prebiological evolution
by M. Florkin

COMPREHENSIVE BIOCHEMISTRY

Section I — Physico-Chemical and Organic Aspects of Biochemistry
Volume 1. Atomic and molecular structure
Volume 2. Organic and physical chemistry
Volume 3. Methods for the study of molecules
Volume 4. Separation methods

Section II — Chemistry of Biological Compounds
Volume 5. Carbohydrates
Volume 6. Lipids — Amino acids and related compounds
Volume 7. Proteins (Part 1)
Volume 8. Proteins (Part 2) and nucleic acids
Volume 9. Pyrrole pigments, isoprenoid compounds, phenolic plant constituents
Volume 10. Sterols, bile acids and steroids
Volume 11. Water-soluble vitamins, hormones, antibiotics

Section III — Biochemical Reaction Mechanisms
Volume 12. Enzymes — general considerations
Volume 13. (third edition). Enzyme nomenclature (1972)
Volume 14. Biological oxidations
Volume 15. Group-transfer reactions
Volume 16. Hydrolytic reactions; cobamide and biotin coenzymes

Section IV — Metabolism
Volume 17. Carbohydrate metabolism
Volume 18. Lipid metabolism
Volume 19. Metabolism of amino acids, proteins, purines, and pyrimidines
Volume 20. Metabolism of cyclic compounds
Volume 21. Metabolism of vitamins and trace elements

Section V — Chemical Biology
Volume 22. Bioenergetics
Volume 23. Cytochemistry
Volume 24. Biological information transfer. Immunochemistry
Volume 25. Regulatory functions, membrane phenomena
Volume 26. Part A. Extracellular and supporting structures
Volume 26. Part B. Extracellular and supporting structures (continued)
Volume 26. Part C. Extracellular and supporting structures (continued)
Volume 27. Photobiology, ionizing radiations
Volume 28. Morphogenesis, differentiation and development
Volume 29. Part A. Comparative biochemistry, molecular evolution
Volume 29. Part B. Comparative biochemistry, molecular evolution

Section VI — A History of Biochemistry
Volume 30. *Part I.* Proto-biochemistry
Part II. From proto-biochemistry to biochemistry
Volume 31. *Part III.* The sources of free energy in organisms
Volume 32. *Part IV.* The unravelling of biosynthetic pathways
Volume 33. *Part V.* History of molecular interpretations of physiological and biological concepts, and of the origins of the conception of life as the expression of a molecular order

Volume 34. *General Index*

Chapter I

Concepts of Molecular Biosemiotics and of Molecular Evolution

MARCEL FLORKIN

Department of Biochemistry, University of Liège (Belgium)

I. MOLECULAR EVOLUTION

1. Evolution

By evolution we mean* the divergence in the gene pool of populations of organisms. This results in differential selection, by differential reproduction, of genetic variations in the populations concerned, and also results in random genetic drift. To quote Prakash *et al.*[1],

"The kind and magnitude of genetic divergences between populations is governed mainly by interaction of population size, migration rate and selection intensity."

2. Phylogeny

Along the formalist approach adopted by Linné, naturalists have elaborated a system of classification of organisms and attempted to build a "natural system" taking into account all of the characters identified.

Phylogenesis corresponds to the changes of the organisms in the succession of generations and it has its sources in the process of ontogenesis. The "natural system" became the source of phylogeny in which an abstraction is formulated which corresponds to the supposed succession of living forms in the course of time, as derived from the whole of our knowledge of living species and of the teachings of paleontology.

---

* Definitions for a number of terms are given in the Glossary on p. 123.

References p. 117

In this presentation we designate by "phylogeny" the theory in which is embodied the whole of our knowledge of the "progressive" specialization of organisms, while phylogenesis means the process of the advent of new species and new branches of the phylogenetic tree.

It should be acknowledged that phylogeny must be considered in a critical way. The higher categories which appear on the tree as phyla, classes, orders, families or genera are abstractions defined by the persistence, among the species considered as having successively appeared, of certain common characteristics which do not evolve in the limits of the higher category considered. A scientific theory of evolution can only be a theory of descendance as already stated by Lamarck when he wrote:

"la nature n'a réellement formé ni classes, ni ordres, ni familles, ni espèces constantes, mais seulement des individus qui se succèdent les uns aux autres et qui ressemblent à ceux qui les ont produits" (in ref. 2).

As stated by Teissier[2]:

"We should not be afraid of archetypes, they do not exist" (ref. 2, translation by author).

Ideally, we should therefore be able to compare species along their direct derivation, which is impossible, as the species from which the present species are derived have either disappeared or have themselves changed. All the species we study today have passed through two billion years or more of evolution in descent. In the species which are designated as primitive we find a mixture of specialized and of archaic traits and it is not easy to make the distinction of such traits in the morphological order.

Many branches of phylogeny are of a controversial nature and some sections have multiphyletic origins. But on the other hand, some sections of phylogeny are well documented.

3. The comparative approach to biology

Comparative studies on organisms have been first oriented towards the identification of life's common aspects. Since Darwin, the accent of comparative studies has been laid on the fundamental genetic unity of organisms and the same tendency has been fostered by the definition of general physiology coined by Claude Bernard, who defined it as concerned with phenomena common to all organisms, or as the physics and chemistry of living cells. The general cellular phenomena are of obvious paramount importance and it is not surprising that the physiological aspects common to all cells have aroused such constant interest among those who studied "general physiology".

Comparative physiologists, on the other hand, have considered the traits of the different categories of organisms under their phenomenological aspects, as "functions" defined at the level of the organism as a whole, and within the framework of the relationship of the organism and its environment. Still influenced by the idea of "anatomia animata", the physiologists identified at the organismic level the material substratum of each "function": vascular systems for the circulation of fluids, respiratory organs for ventilation, glands for secretion, etc. They identified the material object of the "function": a gaseous medium for ventilation, a liquid object for currents, etc. Classical physiologists, having identified the mill and the grain, looked for the wind acting on the mill and in each case described a "stimulus" responsible for initiating the reaction of the organ to its material object. Comparative physiology was therefore led to compare systems performing the same "function" (analogous systems) without any reference to homologies, which were left to the realm of comparative anatomy.

It is natural that the same "non-phylogenic" tendency also appeared in the first endeavours of comparative biochemists. When von Fürth published, in 1903, his *Vergleichende Chemische Physiologie der niederen Tiere*[3], he was strongly under the influence of the ways and habits of classical physiology and he grouped his collection of biochemical data in "physiological" divisions: blood, respiration, nutrition, excretion, animal poisons, secretions, muscles, supporting structures, etc. Under these headings, the phenomenological aspects presented by different animal phyla are successively considered. In fact, it is the physiological notion of "function", residing in the relationship of substratum to object, which prevails in von Fürth's book devoted to the comparison of similar phenomena, without reference to phylogeny. The same kind of approach is found in Baldwin's book, *An Introduction to Comparative Biochemistry*[4] the first edition of which appeared in 1937 and which had the merit of arousing a great deal of interest in comparative biochemistry. "The task of the biochemist is, after all, the study of the physico-chemical processes associated with the manifestation of what we call life — not the life of some particular animal or group of animals but life in its most general sense", writes Baldwin, and while insisting on the concept of the biochemical unity of life he makes no reference to the phylogeny of biomolecules or of the systems resulting from the association of these bio-

References p. 117

molecules. The accent had already been placed on the biochemical aspects of the unity of cells as organisms in a paper published by Kluyver and Donker, entitled *Die Einheit in der Biochemie*[5]. Comparative biochemistry is understood by Kluyver as dealing with the study of the unity of biochemical systems corresponding to a metabolic step in different organisms and, reciprocally with the dynamics of known biochemical systems which may serve to explain a similar metabolic step in other, even phylogenetically unrelated, organisms. That there exists a common central pattern of metabolism in all cells is one of the basic concepts on which as we shall see, the evolution of the metabolic pathways is constructed.

4. The concept of molecular evolution

The expression "molecular evolution" covers a number of different topics. One of them is prebiological evolution or "chemical evolution". This domain is based on the study of models and on the experimental development of different hypotheses, a subject which will be dealt with in Chapter 5 (Vol. 29B).

Another aspect of molecular evolution, to which the present chapter is devoted, is the study of the changes which can be identified at the molecular integrative level at different points of the phylogenetic sequence of organisms (biochemical evolution). The concept of descent with change at the molecular level was formulated in 1944 by the present author:

> "The study of biochemical characteristics depends upon techniques which are frequently complicated, and such a study is more difficult to accomplish than direct observation of external morphological characters. Nevertheless, had naturalists started from these, rather than from morphological observations they would have been bound to conceive the idea of evolution of animals". (Florkin[6], translation in ref. 7)

This "confident statement" as it is called by Dessauer[8], has been ratified by molecular evidence. As pointed out by Simpson[9], the concept of evolution at the molecular level was therefore conceived before the advent of molecular biology: "Before the importance of DNA was known Florkin had already discussed the systematics and evolution of various families of molecules." (ref. 9). To quote H. F. Blum[10] :

> "It was brought into particulary clear focus by Florkin in his little book: *L'évolution biochimique* (1944) that we may trace an evolutionary pattern of biochemical compounds which corresponds in its implication to our tracing of the evolution of morphological aspects of the species of living organisms."

Once for all we wish to state that the expression "biochemical evolution" is an elliptic one meaning the pattern of changes of biomolecular epigenesis (diachronic molecular epigenesis) along the phylogeny of organisms, as derived by naturalists.

5. Oppositions to the concept of biochemical evolution

The concept of "biochemical evolution" *i.e.* of the existence of a pattern of evolution at the molecular level[6] has been the object of much adverse criticism. Mayr[11], for instance, has stated that

"Much of the difference among organisms is a matter of difference of systems rather than in unit components."

The same view, current at the time of the publication of Florkin's book, has been formulated again by another veteran biologist, P. Weiss[12], who brings up against molecular evolution the concept of the biochemical unity of organisms as claimed by Baldwin[4] and by Florkin[6].

This is a matter of semantics, as what these authors showed is the biochemical unity of organisms, as expressed in the common "central metabolic pathway" common to all cells, from Protista to insects and vertebrates.

It is abundantly clear that the changes undergone by the metabolic pathways grafted on to the central system, along the evolution of organisms consist, not only of changes of systems by the recombination of existing units, but also of an increasing number of unit components, the biocatalysts of the terminal and lateral extensions in the metabolic pathways as well as of end-products[13]. Another objection is expressed by Mayr[11].

"No case is known to me", — he writes — "in which a change of body chemistry initiated a new evolutionary trend. Invariably it was a change of habits or habitat which created a selection pressure in favor of chemical adjustments."

It will be enough to consider the importance of such biochemical innovations as the system of calcification of the cuticle of a turbellarian ancestor of the molluscs, the sclerotization of the cuticle of the annelid ancestors of Arthropoda, the patterns of ossification in fish, the biochemical patterns of the amniote egg as it appeared in reptiles, or the isosmotic intracellular regulation which allowed marine invertebrates to invade fresh water and land, to escape, while recognizing the rôle of natural selection in its proper perspective, falling in a naive panselectionism.

In a recent book, Dobzhansky[14] states that biological laws

References p. 117

"deal with particular patterns of physical and chemical processes that occur only in living bodies".

They are, he writes, "simply irrelevant, not contrary, to physics and chemistry." This will take aback those who are aware of the accomplishments of the biochemists and biophysicists who have precisely devoted so much successful effort in unravelling the particular patterns of physical and chemical processes that obtain only in cells and organisms.

Comparative biochemistry has also been charged with the sin of reductionism, *i.e.* the methodological fault made when applying to the aspects of a given level of integration, concepts derived at another level.

It might be claimed that the epistemology of evolution is, as we have repeatedly said, situated at the level of organisms, while the changes described by comparative biochemists are situated at the molecular level. This would not be a fair deal, as comparative biochemistry studies molecular changes with recourse to concepts concerned with molecular configurations, reactions and activities, phylogeny being called upon to provide the scheme of descendance, a frame and a situation of the problem.

II. SYNCHRONIC MOLECULAR BIOSEMIOTICS

6. The fitness of the environment

The surface of the earth receives energy almost entirely as solar radiations and this intake of energy is compensated by terrestrial radiation. The biosphere is in a steady state remote from the equilibrium state. The entropy of the biosphere defined as the whole biomass and its environment increases with time.

In an influential book published in 1948, Schrödinger[15], impressed by the new knowledge of the ordered hereditary transmission of relations between nucleic acids and proteins, appearing as a contradiction with the statistical laws of physics and with the tendency of things to go over into disorder, wrote:

"Life seems to be ordered and lawful behaviour of matter, not based exclusively on its tendency to go over from order to disorder, but based partly on existing order that is kept up."

Schrödinger defines the process by which an organism maintains itself at a fairly high level of orderliness (fairly low level of entropy) as "sucking orderliness from its environment" by which he means that while in the whole biosphere entropy increases with time, entropy is kept at a fairly low level at certain points in the system by displacement from randomness to order. Schrödinger states that the organism not only transforms free energy into work and increases the entropy of the whole system, but also feeds on "negative entropy" (abbreviated by Brillouin[16] to negentropy). Entropy is zero at 0 °K and positive above it, while it is impossible to go below. "Negative entropy" in Schrödinger's expression means a displacement from disorder to order. In the consideration of the flow of energy through the biomass, it is "free energy" which is generally referred to in calculations. The molecular free energy is defined as follows

$$G = H - TS$$

in which H is the molar energy content, T the absolute temperature and S the molecular entropy content. As said above, the main free-energy source in organisms is presently considered as being the high-energy bond of ATP. While it is commonplace to describe a chemical reaction in living organisms by defining the reactants, the products and the free-energy changes occurring in the reaction, in a number of reactions in which ATP is hydrolyzed or synthesized, the process to which the chemical reaction is coupled remains obscure; such is the case in muscular contraction, active transport, oxidative phosphorylation or photosynthetic phosphorylation. These appear as aspects of "molecular energy machines" (McClare[17]) the theory of which is not dealt with in classical thermodynamics and remains to be developed as does the energetics of protein folding.

In 1913 appeared the widely read book of L. J. Henderson[18], *The Fitness of the Environment*, which was considered by some as a platitude and by others as deserving endless commentaries which still are a subject of writing nowadays[19,20]. Henderson rejects all sorts of teleological views and quotes Bacon concerning final causes, which are like vestal virgins.

"They are dedicated to God, and are barren".

The core of Henderson's thought is expressed by the sentence in which he considers:

"that peculiar and unsuspected relationships exist between the properties of matter and the phenomena of life; that the process of cosmic evolution is indissolubly linked with the fundamental characteristics of the organism; that logically in some obscure manner, cosmic and biological evolution are one",

References p. 117

and also, more concisely when he writes:

"the whole evolutionary process, both cosmic and organic, is one".

In a lucid comment, Speakman[21] expresses Henderson's thought as follows:

"The environment seems to be well fitted for the sustenance of life in ways over which the organism has no control and which, indeed predate the existence of the organism."

The ways in which "cosmic and biological evolution are one" have been sought in the laws of classical thermodynamics and more particularly in the Carnot–Clausius principle. In the overall balance, cosmic and biological evolution result in an increase of entropy as predicted by the second law. The concept of evolution as an aspect of this law has been formulated by a number of authors and especially by Lotka[22] and by Blum[23]. It has been considered unacceptable by a number of authors on the basis of the idea that evolution, as well as the maintenance of the organisms imply, as stressed by Schrödinger, an increase in order.

That "cosmic and biological evolution are one", as stated by Henderson, could only be understood after the formulation by Prigogine (for a general formulation, and literature, see Prigogine[24]) of the concept of chemical instabilities leading to a spontaneous self-organization of the system both from the point of view of its space order and its function: "there indeed exists a *new state of matter* induced by a flow of free energy far from equilibrium"[24].

Short-time oscillations have been detected in a number of biochemical reactions[24] and arguments have been formulated in favour of instabilities breaking dissipation symmetry, and leading to spatial organization. Chemical instabilities may be associated with biological structures and with their change in descent. These views confer a new interest upon "traffic" pathways in the cybernetic network of the biochemical continuum and upon the system of signification expressed in their branching, control and regulation.

Furthermore it leads to a better definition of the vaguely defined "internal factors" of evolution, possibly by a consideration of the limitations imposed on primordial proteins, with respect to their primary structure, by their mode of synthesis and their persistence in the medium in which they developed, by taking into account the laws of the thermodynamics of irreversible phenomena and particularly the entropic criterion of evolution of energetically open systems (Glandsdorff and Prigogine[25], Prigogine[26]) according to which, as stated by Buvet[27] "in open systems, compounds of internal structures different from equilibrium ones spontaneously arise, which may be considered to have no "raison d'être" other than bending this entropy production in the way of minimalization". In such a system the existence of which involves a source of energy and a sink for the energy flowing through the system, energy is used up and dissipated for the maintenance of the steady state remote from the equilibrium, a steady state depending on cyclic chemical processes and on the chemical and energetic coupling of anabolic and catabolic cycles.

7. Molecular information

The entropy (in the sense it has in thermodynamics) of a DNA molecule is a function of the configuration in which the atoms are ordered. It is expressed in the relation:

$$k \cdot \log D$$

(k, Boltzmann constant; D, measure of atomic disorder). In statistical terms, entropy is defined as

$$S = -k \sum_{i=1}^{i=n} \log p_i$$

($\Sigma_1^n = 1$; p_i, the probability of an idealized physical system in the state i of n possible states). Szilard[28] has pointed out the formal similarity of this equation defining entropy and the equation defining information

$$H = -\sum_{i=1}^{i=n} p_i \log p_i$$

($0 \leqslant p_i \leqslant 1$; ${}_1^n p_i = 1$; p_i = relative probability of the ith symbol generated by a source). The fundamental relations between information and entropy have been developed by Shannon and Weaver[29] and by Brillouin[30], to whose writings the reader is referred.

In our present theory, displacement from randomness, or negentropy is quantified in information, or rather in the quantity of information, expressed in "bits".

This unit of information corresponds to the simplest possible form of message transmission. In the case of a binary symbol, *i.e.* which can take two or only two alternative forms (for instance "yes" or "no", or a dot or dash in the Morse system). The "elementary quantity of infor-

References p. 117

mation" that can be transmitted by a simple "binary" symbol (choice between two alternatives) corresponds to a *bit* (binary digit)

2 bits correspond to a choice between 4 alternatives
3 bits correspond to a choice between 8 alternatives
4 bits correspond to a choice between 16 alternatives

or in general

n bits to a choice between 2^n alternatives

The information content I corresponds to the number

$$I(\text{bit}) = ld\,2^n \qquad (ld = \log_2 = \log \text{ to the base } 2)$$

When we deal with several (n) symbols and with several (m) symbol types, the general equation reads

$$I(\text{bit}) = ld\,m^n = n\,ld\,m = 3.32\; n\; \log_{10} m(\text{bit})$$

As pointed out by Quastler[31] who was the first to understand the biological pertinence of the information theory formulated by Shannon[32] in 1948, an amino acid has about the information content of a word and a protein that of a prose paragraph. He calculated the capacity of a mammalian DNA molecule as $2 \cdot 10^{10}$ bits of information. Morowitz has calculated the information content of a cell of *Escherichia coli* and found it equal to $2 \cdot 10^{11}$ bits. These data illustrate the high content of a mammalian cell in information defined as negentropy *i.e.* in physical terms.

8. Biomolecular *relata*

As said above, negentropy is an extensive property and the information theory quantifies it. As noted by Johnson[34]:

"... in biology, the extensive property, information or negative entropy, can have significance for the feasibility of living process only if it is modulated by an intensive property which indicates the biological relevance or purposefulness of that information".

Brillouin himself has written[30]:

"the present theory of information completely ignores the value (or meaning) of the information handled, transmitted or processed."

The whole mass of living organisms may in the perspective of the present discussion, be considered as a large collection of molecules assembled in a certain order and forming a *biochemical continuum* (Florkin[35,36]) composed of all the molecular aspects of the organisms and of their metabolic extensions. This concept situates the consideration of the whole biomass at the molecular level, *i.e.* a cybernetic molecular network. The sediments at the deepest bottom of the oceans, and the organic matters coating sand and mud particles there, are linked to the surface of the seas by a cloud of molecules, this cloud being of variable density. The same notion applies to soils and to fresh water.

The whole of the biochemical continuum forms a large population of molecules densely associated in parts and more separate in other regions. Some of the dense associations are architectured in the form of the higher level of integration of cell organelles and of cells themselves by recourse to a number of couples of *relata* such as enzyme–substrate, hormone–receptor, multichain regulatory protein, etc., and a mechanism has been developed for the replication of a material including in the frame of the flux of matter and energy, a flux of information. In a cell there is a flow of information from nucleolus to cytoplasm and this is coupled with a flow of material particles, the ribosomes, dependent upon the activity of the nucleolus. Not only are the ribosomal RNAs transferred in that way, but also the RNAs which carry the specificity for protein synthesis. This information depends upon the existence, in the nucleus, of a collection of DNA molecules inherited from the parents, carriers of the molecular order expressed in the structure and nature of the organism.

Monocellular and pluricellular organisms associate in communities whose populations maintain themselves and form, in association with different components of the environment, ecosystems inside which and between which currents, not only of matter and energy, but also of information are flowing[36,37]. The biochemical continuum is the result of a slow evolution starting from the organic abiogenic continuum resulting from the prebiotic (chemical) evolution. The transition from the abiogenic organic continuum[38] to the biochemical continuum is characterized by a decisive step: the emergence of the steady state remote from the equilibrium state and of the prime aspects of nutrition in the form of a flow of matter and energy through the open system of organized aggregates (precells, protocells)

As stated by Eigen[39]

"... General principles of selection and evolution at the molecular level are based on a stability criterion of the (non-linear) thermodynamic theory of steady states. Evolution

References p. 117

appears to be an inevitable event, given by the presence of certain matter with specified autocatalytic properties and under the maintenance of the finite (free) energy flow necessary to compensate for the steady state production of entropy".

The reciprocal relations between the various biosystems and with the medium became more and more elaborate, and it led to the exchange of molecules that were not only of nutritional importance, but acquired regulatory and control properties, acting as transferrers of signs and signals. Many ecological aspects should be classified as *molecular reactions* (liberation of molecular factors by organisms in the habitat) and *molecular actions* (infringements of molecular factors upon organisms). In an ecosystem, besides the contribution of the trophic chains in supplying molecules endowed with nutritive functions and ensuring the flux of matter and of energy one may describe non trophic molecules active in insuring a flux of information as well as the constitution and maintenance of the community (*ecomones*[35,36]). An ecomone may have multiple actions. For instance, the concentration of dissolved carbohydrates in sea water varies from place to place and is a factor of the nature and growth of the phytoplancton. On the other hand the ecological importance and the action as ecomones, of the carbohydrates is expressed in the regulation of the pumping rate of Lamellibranchia.

Some ecomones are recognized as being specifically active in the process of the coaction of organisms upon each other. Such specific substances or *coactones* are determinant in the relationship between the coactor (active and directing organism) and the coactee (passive and receiving organism). A number of coactones are liberated by the coactor in the medium and reach the coactee. These we may call *exocoactones.* Among these are, for instance, the molecular factors of the orientation of animals through the perception of the odour of the animal taken as food. The feeding behaviour of an animal when reaching a plant to which it has been attracted by an exocoactone may be specifically unlocked by plant products which we may call *endocoactones.*

A special category of exocoactones is represented by those substances produced in a coactor and acting on a coactee, both coactor and coactee belonging to the same species. These exocoactones are called *pheromones.* According to Karlson and Lüscher[40]

"pheromones are defined as substances which are secreted to the outside of an individual and received by a second individual *of the same species*, in which they release a specific reaction, for example a definite behaviour or developmental process."

The other exocoactones and endocoactones, transmitting chemical signals to individuals of other species we may call *allomones*, as suggested by Brown and Eisner[41].

The circulation of molecules and macromolecules through the biochemical continuum and the factors controlling their distribution is a field of active biochemical evolution and adaptation in which it is possible to detect the aspects of natural selection shaping both sign and receptor and the evolution of reciprocal interactions between organisms and between groups of organisms.

9. Molecular biosemiotics

The recognition of systems of *relata* at the integrative molecular level indicates that biomolecular order is governed by systems of signification which we may consider in a structuralist (intensive) perspective besides the thermodynamic viewpoint and the quantifying (extensive) viewpoint of the information theory.

When Wiener once wrote that one could consider an organism as a language, he meant that the concepts of information and cybernetics are applicable to organisms. It would be better to resist the temptation of considering biomolecular *relata* along the lines of language studies. Linguistics is dealing with linguistic signs, *i.e.* with psychological entities, relating in the receptor's mind a psychic acoustic image and a concept. Language is a vector of ideas, as are other aspects of semiology, such as the alphabet of the deaf-mutes, symbolic rites, marine signalization, etc. All these languages, framed in the social mass, represent substitutes of human experience which can be transmitted in space and time.

We shall therefore pray for the banishment of the abusive use of the term "language" from the field of molecular biosemiotics, while granting that there is, on the other hand, a language and a graphology of the chemist and of the biochemist, aspects which are not dealt with in this review.

Consequently, we find it advisable to avoid the application of the *specific* concepts of linguistics (word, phrase, etc.) to biosemiotics and, however tempting it may be, to decide not to introduce such expressions into it.

While linguistic semiology deals with human psychological phenomena and the communication of concepts, molecular biosemiotics deals with

References p. 117

existential aspects of cells and the *relata* it studies are not always of the nature of signals. It is on the other hand imperative to keep the concepts of general semiology as understood in the social mass of men, when they go beyond this domain to connote the field of molecular biosemiotics, as we believe that in a future development, linguistic semiology will become based on molecular biosemiotics of the activities of the brain. We shall therefore use in the perspective of this subject several general concepts elaborated by De Saussure[42] such as significant and signified, synchrony and diachrony, syntagm and system with the special meaning they have in molecular biosemiotics. It must be noted that in the mind of F. de Saussure these concepts arose from the consideration of existential (not psychological) aspects of natural science.

To define the concept of significant, Saussure recalls that water, H_2O, is significant in a way which belongs neither to oxygen nor to hydrogen:

"it is a combination of hydrogen and oxygen; taken apart, each of these elements has none of the properties of water"[42] (translation by author)

Synchrony and diachrony are borrowed by De Saussure from the theory of evolution, and diachronic is synonymous with evolutionary.

It is therefore fitting to situate these concepts in the most general context of semiotics, the general science of signification, of which linguistics and biosemiotics are special aspects. However, the concept of the *word* is specifically linguistic as it means the smallest free unit of *denotation.* It is a cultural unit and should consequently not be confused with the material and functional (not denotating) units of molecular biosemiotics. The same holds for the phrase and for discourse, which are the support of connotative developments of values defined by their situation in a semantic field. They belong to dialectic logic and not, as in the case of biosemiotics, to structural logic.

Molecular biosemiotics is an aspect, not of human sciences, but of molecular biology. As stated by De Saussure[42], in linguistics, the sign which he considers as the association of a significant and a signified, is arbitrary with reference to the relation between its two faces. In molecular biosemiotics, on the other hand, significant and signified are in a necessary relation imposed by the natural relations of material realities.

10. Biosemes

Minimal configuration aspects, carriers (significant) of molecular signification (signified), either sequential, structural, functional, protective, connective, motive, signaling, catalytic, processing, regulating, priming, repressing, releasing, etc., we shall call biosemes (units of significant). The sign (theme of signification) is formed by the doublet of significant and signified of the bioseme. The signification system implies a sign (significant + signified) and a receptor, eventually reached through a sequence of intermediary steps.

It must be noted that, while in the terminology of linguistic semiology, the sign is "a double faced psychological entity", the sign of which a bioseme is the minimal significant is a double-faced (significant + signified) material entity (molecular configuration + biological activity). This avoids the controversial nature of the relation significant (symbol) — reference (concept) — referent (the thing signified), as introduced by Richards and Ogden[43] in linguistics. The referent is the object denoted by the significant (in the case of the word *tree*, f.i., the object tree), while the reference is the signification (a concept) carried by the word to the receptor. This, Ullman (1962) has defined as the information carried to the receptor, what Frege (1892) had defined as *Sinn* in contrast to the referent which is *Bedeutung*. To illustrate the distinction, we may write /*Charles Darwin*/and/ the author of the *Origin of Species*/. The *Bedeutung* remains the same but the reference (Sinn) is different. The same difficulty concerning the linguistic signified appears in the opposition of denotatum and designatum, denotation and connotation, denotation and meaning (literature in Eco[44]).

At the level of a bioseme, the significant, an aspect of molecular configuration, and the signified, an aspect of biological activity, compose the sign, and the theme of signification. This is one more reason, if necessary, to avoid referring to "words" and "sentences" when dealing with biomolecular aspects, while it is also a reason for adopting the notion of the sign in a special section (biosemiotics) of general semiotics. This denomination we shall apply to the general science of signs, while applying the name of general semiology[45] to the discipline concerned with systems of signs in relation with the laws of the languages framed in the social mass.

A bioseme carries no *Bedeutung* or *Sinn*. Its significant consists of an aspect of molecular configuration and its signified of an aspect of biological activity.

Biosemes, the units of signification involved in molecular *relata*,

References p. 117

correspond as said above, to intensive aspects of negentropy. The system of signification obtaining in each case of relation with a receptor can be considered, not only from the intensive point of view but also as categories of information. Such is "structural information", defined by Büchel[46] as the information required to construct a system from its parts.

Büchel[46] considers a watchmaker putting together the parts of a watch.

"Whenever a mechanism containing an information of n bits is built, the thermodynamic entropy of that mechanism or its environment must increase by the amount of at least $kn \log 2$ where k is the Boltzmann constant. The quantity $kn \log 2$ is, in fact, the information measured not in bits but in caloric units (cal/grad) [cal/deg. C]. So we may consider the information to be "structural negentropy" contained in the mechanism."

Büchel's structural negentropy or structural information is denoted by Ryan[47] by the symbol I_S.

Another category of information, "bound information", an expression introduced by Wilson[48] is defined by Ryan[47] as the information which is "required to specify the precise microstate of any resonant system". What Ryan[47] calls "functional information" corresponds, according to his definition, to the entropy change corresponding to the order put in (or maintained in) the environment-of-action. We may tentatively recognize this functional information as corresponding to an extensive aspect (information theory) of the intensive concept of the signified as defined above in the context of molecular biosemiotics.

While placing the general (extensive) aspect of organic evolution and its inevitability in a thermodynamic context, we should carefully avoid deriving mechanisms of molecular evolution as a process (intensive aspects of biomolecular changes in the phylogeny of organisms) from thermodynamic concepts, as it has unfortunately sometimes been done, leading to utter confusion.

11. Biosyntagms

In the Macedonian "phalanx", the term "syntagm" designated the compact unit composed of specialized categories of warriors in several rows, pointing their weapons at the periphery. A biosyntagm we shall recognize as being a unit of signification higher than a bioseme, and composed of significant units in a relation of reciprocal solidarity, *i.e.* an associative configuration of biosemes.

A biomolecule may carry several biosyntagms and, on the other hand, a biosyntagm may extend to several biomolecules. The biosyntagm corresponds to the metonymic aspect of molecular biosemiotics. Cistrons, operons, the double helix of DNA, enzymes, biosynthetic assembly chains, are examples of biosyntagms at the integrative molecular level, and are composed of aspects of signification solidary in the course of the information process and presenting a relation of the whole (biosyntagm) to the parts (biosemes)*.

The biosemiotic aspect of the biochemical continuum is the largest biosyntagm conceivable on our planet. It may also be considered as a section of the organic continuum of the cosmos, modified by the emergence of life, this event being defined by the nature of the resulting systems of communication. In order to illustrate the meaning of the bioseme and of the biosyntagm we may take the case of a primary biosyntagm such as ACTH.

ACTH is a polypeptide hormone consisting of 39 amino acid residues and, as pointed out by Harris[49] and by Burgers[50] there exists a close

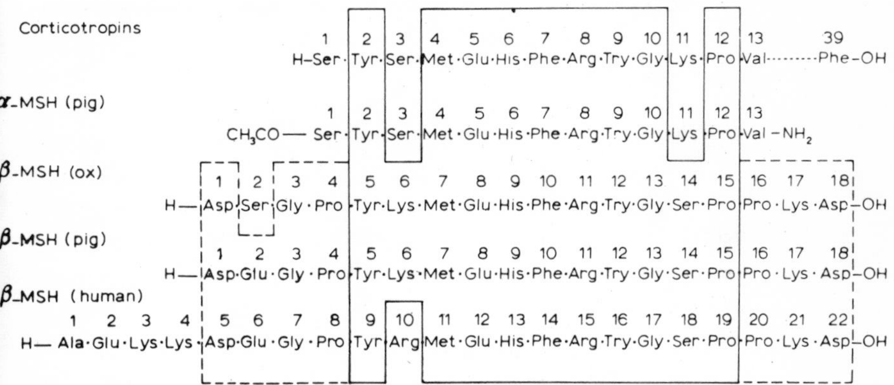

Fig. 1. Amino acid sequences in corticotropins and in the melanocyte-stimulating hormones. (Harris[49]).

structural relationship between ACTH and the melanocyte-stimulating hormone (MSH) which is secreted in the *pars intermedia* of the pituitary gland.

ACTH and MSH have in common the heptapeptide sequence

* The term syntagm has been used by De Saussure[42] in the field of linguistics to designate a succession of semes in time as it is impossible to articulate two different sounds at the same time, an aspect which does not always obtain in molecular biosemiotics, while the time factor is considered in catenary biosyntagms.

References p. 117

Met–Glu–His–Phe–Arg–Try–Gly

To the same family of polypeptides belongs another adenohypophyseal hormone, lipotropin, which is responsible for the mobilization of lipids.

The three hormones are produced by the pituitary gland, embryologically derived from Rathke's pouch and they share the common property, due to bioseme 1 (4–10) and increased in Amphibia by bioseme 2 (1–3), of

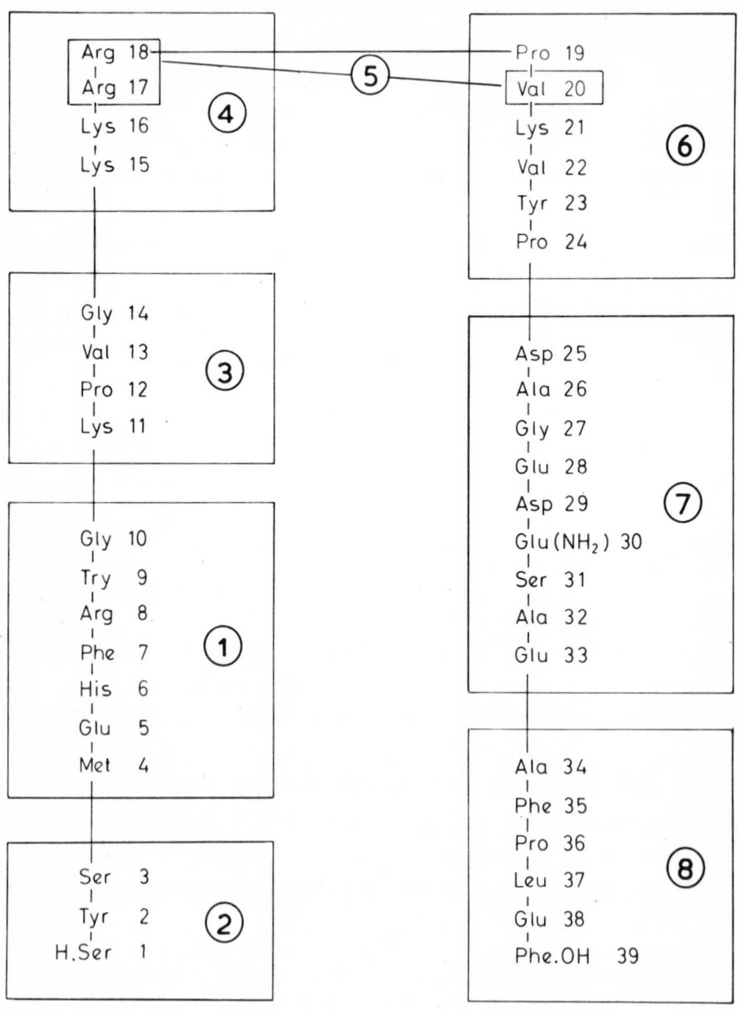

Fig. 2. The primary biosyntagm and the biosemes of ACTH.

causing the dispersion of melanin granules in the melanophores of certain of the lower vertebrates. In mammals, two MSH (α and β) have been isolated.

ACTH increases the production of corticosteroids. This is accomplished through a sequence of steps. It produces an increase in cyclic AMP which activates phosphorylase. The production of G-1-P and of G-6-P from glycogen is increased. The catabolism of G-6-P being increased, there is consequently an increased production of NADPH, which acts at several steps of corticosteroid biosynthesis. With the dominant bioseme 5 (17, 18, 20) of the biosyntagm active at the level of the adrenal cortex is associated the accessory bioseme 4 (15–18) insuring the recognition of a corresponding site on a protein of the adrenal cortex (couple of *relata*). Such action is also attributed, as well as a stabilizing effect of the hormone during transport, to the biosemes 6 (19–24) and 8 (34–39).

Another bioseme can be recognized in ACTH: it is the species-specific label of the peptide, 25–33 (called *specion* by Hechter and Braun[51]) which, in association with bioseme 34–39 is responsible for the production of specific antibodies (according to Schwyzer[52]).

12. Tertiary biosyntagms

While in molecules carrying primary biosyntagms as is the case for ACTH, the biosemes are directly related to amino acid sequences; in tertiary biosyntagms, biosemes are established by spontaneous folding as a consequence of structural relations between remote segments of a polypeptide biosynthesized under the influence of a cistron. The distinctive configuration results from spatial folding bringing a number of specific biosemes together. A cistron (see Vol. 24 of this Treatise) is a biosyntagm composed of a number of biosemes and its sequential signified indirectly commands the alignment of amino acids in a polypeptide chain.

There is another category of biosyntagms arising from a single cistron and a monomeric mRNA, the resulting polypeptide chain of which has no signified, however. It acquires a significant and a signified as a consequence of a postsynthetic cleavage of a polypeptide chain, followed by a spontaneous folding into a significant tertiary compound.

The protein component of myoglobin* is an example of tertiary biosyntagm. Myoglobin has the shape of a shallow box built up from

* For a vivid presentation of the subject and a list of literature references the reader is referred to Dickerson and Geis[53].

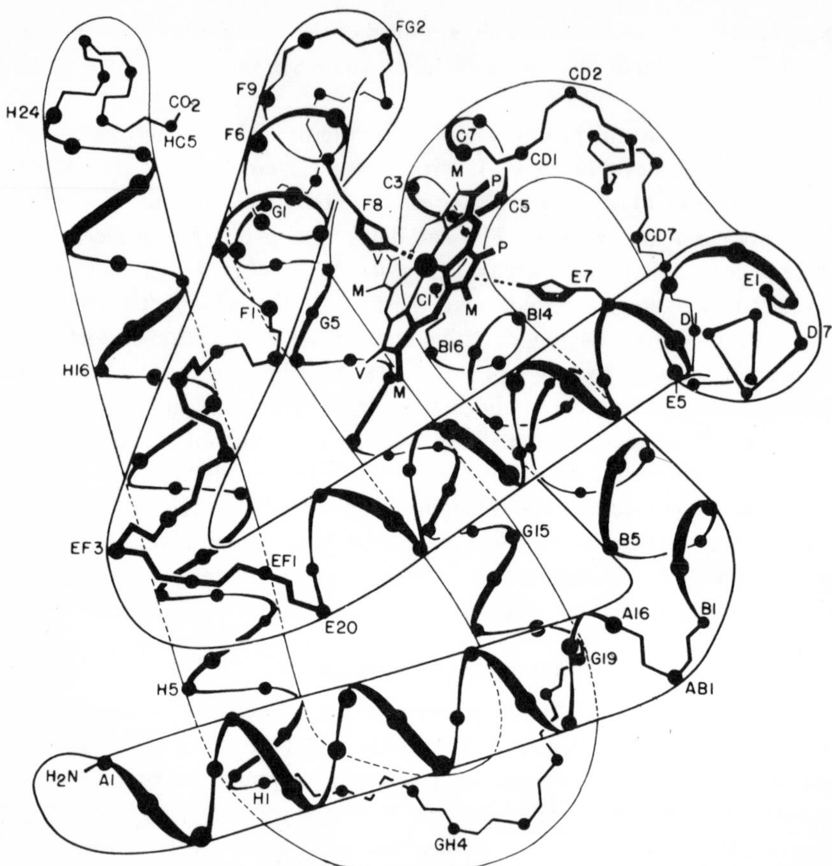

Fig. 3. The three-dimensional conformation of the myoglobin molecule (Dickerson[54]). Large dots represent α-carbon positions. For identification of residues see Table II in Dickerson's text. Heme group framework is sketched in perspective, with side groups identified as: M = methyl, V = vinyl, P = propionic acid.

eight pieces of an α-helix. In this box the heme is buried, except for the edge bearing the two hydrophilic propionic acid groups.

The charged polar groups (hydrophilic) are spread over the surface of the molecule, and the entire pocket for the heme is strongly hydrophobic. The *relata* of the biological activity of the myoglobin are the heme-protein molecules on one hand and the oxygen molecules on the other.

This relation of signification is insured by a definite scaffolding of the heme iron which is octahedrally coordinated. Four of the atoms donating an electron pair to the orbitals around the iron are nitrogen atoms in the porphyrin ring of the heme. A fifth one is one of the ring nitrogens of His F_8 of the globin, locked into its position by a hydrogen bond of its second ring nitrogen with the carbonyl oxygen of Leu F_4. Astonishingly enough, while in metmyoglobin (resulting from the oxidation of the heme iron) the sixth octahedral ligand is water, in reduced (ferrous) myoglobin this sixth ligand site is empty. This is the oxygen-binding site.

The scaffolding of the three-dimensional structure insuring the existence and the maintenance of the oxygen combining site in the heme pocket is due to the eight pieces of α-helix, all right-handed and the length of which ranges from 7 to 26 residues. The helices are folded back to form the pocket for the heme and there are non-helical regions between two of their pairs. The hydrophilic groups are spread over the external surface and the hydrophobic side chains (Val, Leu, Ile, Met, Phe) are either situated at the interior side of the helices packed together, or lining the heme pocket. The dominating bioseme of the myoglobin tertiary biosyntagm is the sixth ligand site of the ferrous iron in reduced myoglobin. Accessory biosemes insure the establishment of this configuration (significant), the signified of which is the oxygen-binding site. These accessory biosemes consist of the structural information introduced by the amino acid side chains and the amino acid sequence, as well as by the scaffolding of the heme iron by its binding to definite N atoms of the protein, as described above. Other forms of tertiary biosyntagms, such as trypsin, chymotrypsin, etc. are described in Chapter 4 (Vol. 29B).

13. Quaternary biosyntagms

These biosyntagms are established through spatial relationships between the biosemes (subunit interaction) carried by the polypeptide chains of a multichain protein, the chains of which are associated by residual valencies. Regulatory proteins belong to this category. The simplest case is represented by *homopolymer* biosyntagms (examples: alkaline phosphatase of *E. coli*, glutamine synthetase). Each of the units is biosynthesized following instructions from a monocistron mRNA, the chains being subsequently polymerized. The signified of a homopolymer quaternary biosyntagm is different from the signified of each protomer (Monod *et al.*[55]) (the identical units associated) as a result of subunit interaction at the significant level. Different polypeptide tertiary biosyntagms can be associated to form a *heteropolymer* quaternary biosyntagm (examples: pyruvate dehydrogenase, tryptophan synthetase, fatty acid synthetase complex, aspartate transcarbamylase of *E. coli*).

The cistrons coding for the subunits are, in Procaryotes, located on an operon (a biosyntagm composed of adjacent cistrons) giving rise to a polycistronic mRNA. In Eucaryotes, the subunits of a heteropolymer biosyntagm are coded by unlinked structural genes (producer genes) coordinated by integrator genes into a battery of genes (see Section 17).

(a) Homopolymer quaternary biosyntagms

An example is provided by glutamine synthetase.

The glutamine synthetase of *E. coli* (mol. wt. 592 000) is composed of 12 structurally identical units, each of mol. wt. 49 000, disposed in two superimposed hexagonal arrays. Each unit carries at least three biosemes: a binding site for the substrate ATP and allosteric binding sites for AMP and tryptophan

$$\text{Glutamic acid} + \text{ATP} + \text{NH}_4^+ \rightarrow \text{glutamine} + \text{ADP} + \text{Pi}$$

The reader is referred to the *Harvey Lecture* devoted by Stadtman[56] to this enzyme. He will find in this review an authoritative presentation and the relevant literature concerning the following data.

There are a number of compounds which have been recognized as causing some inhibition of the enzyme. These substances are, besides alanine and glycine, several end-products of different branches of glutamine

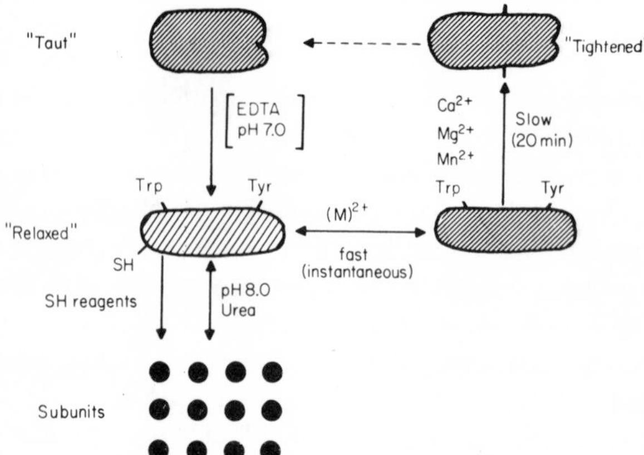

Fig. 4. A schematic representation of the interconversions of taut, relaxed and tightened forms of glutamine synthetase (Stadtman[56]).

metabolism (tryptophan, histidine, CTP, AMP, carbamyl-P and glucosamine-6-P), each producing only a partial lowering of signified and these lowerings are cumulative, indicating the existence, in the biosyntagm of a different separate bioseme for each of the inhibitors mentioned. On the other hand, the sensitivity to feedback inhibition is modulated by the adenylation of a single unique tyrosyl group among the 15 per unit, the degree of enzyme inhibition by a particular end-product depending on the number of adenylated units. Depending on the absence or presence of divalent cations, the enzyme may exist in several states. In the so-called "taut" enzyme, corresponding to nearly spherical particles, the enzyme as isolated contains per unit 3 or 4 equivalents of manganese resulting from the addition of manganese salts during the isolation of the enzyme, to protect it from inactivation.

If the divalent cations are removed, the enzyme is converted into a "relaxed" form which is without signified. This appears to result in a transfer of 12 to 24 residues of tryptophan and tyrosine from a non-polar to a polar environment. In this state about 80 per cent of the SH groups are also exposed. The addition of divalent cations reverts the relaxed enzyme to the active state. This process consists of two phases. In the first one, the SH groups become buried and the hydrophobic groups become exposed. In the second phase, the tryptophanyl and the tyrosyl groups are buried and the activity restored (tightened enzyme, to distinguish it from taut enzyme which is more soluble in dilute salt solutions).

The relation of the signified of the biosyntagm to adenylation depends on the presence of divalent cations. In the presence of Mg^{2+} only the activity decreases with increasing states of adenylation while it increases if the test is accomplished in the presence of Mn^{2+} only. The unadenylated enzyme is completely inactive in the presence of Mn^{2+}. In other words, the unadenylated enzyme has an absolute requirement for Mg^{2+} while the fully adenylated enzyme has an absolute requirement for Mn^{2+}.

From what is said above we may infer that the enzyme, with its 12 identical units, may contain from 0 to 12 adenylated subunits, which means that there are 13 possible forms of the biosyntagm. In fact it appears that forms containing both adenylated and non-adenylated subunits exist and that interactions between adenylated and non-adenylated units in the same molecule affect the catalytic characters (signified) of the biosyntagm.

The biosyntagm of glutamine synthetase is provided with an accessory

References p. 117

complex system of deadenylation. When *E. coli* is grown in the presence of ample nitrogen supply, it is almost completely adenylated, whereas in the condition of nitrogen starvation it is completely non-adenylated.

(b) Heteropolymer quaternary biosyntagms

An example of heteropolymer quaternary biosyntagm is afforded by the tryptophan synthetase of *E. coli.* It is a tetramer composed of two α-chains (mol.wt. 29 500) and two β-chains (mol.wt. 50 000) the structural genes of which are adjacent in the biosyntagm of the tryptophan operon. The conversion of indole-glycerol phosphate into tryptophan is accomplished in two steps, respectively catalyzed by the α and β subunits, along the A and B steps shown in Fig. 5.

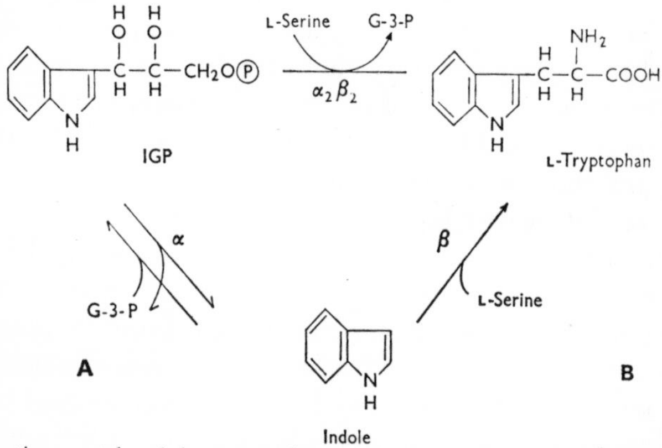

Fig. 5. Reactions catalyzed by tryptophan synthetase and its α and β subunits. IGP, indole-glycerol phosphate; G-3-P, glyceraldehyde 3-phosphate (Williamson[57]).

The signified of the tetramer is different from the sum of the addition of the signifieds of the individual chains. It has a specific activity in reaction A 100 times greater than β_2 subunits[58]. The biosemes for the different reactions are on separate chains but the syntagm of biosemes carried by the tetramer shows a change of signified resulting from a modification of significant.

In this case, enzymes catalyzing separate steps combined in a biosyntagm with a change of significant and of signified.

In the case of the aspartate transcarbamylase of *E. coli*[59], the first enzyme involved in the pathway of pyrimidine biosynthesis, one type of subunit contains all the catalytic activity while another carries all the binding sites for CTP, the normal allosteric inhibitor. The regulatory units differ in different bacteria while the active sites remain the same.

The isolated units may be active, as in the case of tryptophan synthetase, the multichain enzyme showing a change of signified. But in many cases the separate units show no signified, which only emerges from the interaction of subunits. This, we have seen, is true for the glutamine synthetase of *E. coli* and the emergence of the signified depending on the association of the polypeptide chain is the case also in the fatty acid synthetase complex (see Vol. 18 S).

14. Sequencing biosyntagms

The signified of such biosyntagms, the composing biosemes of which are nucleotides or codons, is of a sequencing nature, governing the sequence of nucleotides or of amino acids, in their polymers. To this section belongs the secondary sequencing biosyntagm of DNA, the tertiary biosyntagm of tRNA and the primary biosyntagm of mRNA. The reader is referred to Vol. 24 of this Treatise for a detailed treatment of the biochemistry of these polynucleotides and of the genetic code governing the couples of *relata* between mRNA, tRNA and ribosomes. Limiting our attention to molecular biosemiotics, we may recall that the duplex of DNA consists of two long polynucleotide chains which are wound around each other, the two strands being associated by hydrogen bonds (not covalent bonds) at the level of the composing structural biosemes involved in the purine and pyrimidine complementary couples of *relata*.

In DNA replication, as it takes place at each cell division, the parental duplex unwinds and the two strands act as templates (sequencing biosyntagms) of sequences of base pairs for the synthesis of the complementary strand by hydrogen bonding. In DNA replication, the synthesis is sequential commencing at a fixed bioseme, the origin, and proceeding in both directions in bacteria and viruses[60, 220, 221]. In Eucaryotes, a number of biosemes act as origins in each chromosome and the synthesis also proceeds in both directions[222].

The structural information of DNA is introduced in the course of its synthesis which can be reproduced *in vitro* in the presence of a DNA polymerase, a model in which the information is introduced by hydrogen-

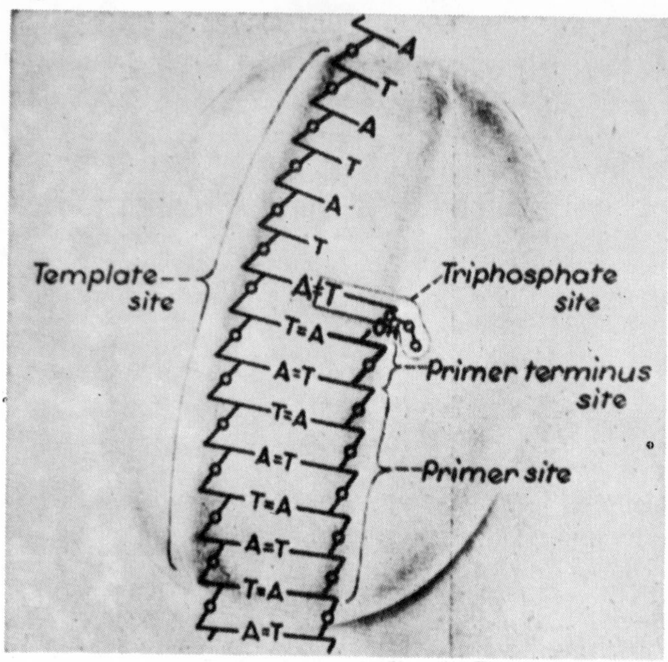

Fig. 6. Catalytic sites in the active center of DNA polymerase (Kornberg[61]).

bonding base pairing but in which base pairs are recognized by the enzyme involved, which also excludes other base pairs. In one of a number of models proposed to account for this property it is considered that, as the polymerase contains a bioseme acting as a site for triphosphate, the template binds a base which is accepted or rejected by the polymerase. When the base satisfies the geometry and is acceptable, the enzyme may undergo a (ligand-induced) configuration change and the synthesis proceeds.

DNA polymerase, consisting of a single polypeptidic chain carries a biosyntagm of several biosemes, coupling for a number of couples of *relata.* Kornberg[61] records at least five of them. The *template site* binds the chain where a base pair is formed. At the *primer site*, the growing chain is located. The *primer terminus* is a site for recognition of 3′-hydroxyl group of the primer's terminal nucleotide. The incoming nucleotide substrates bind at the *triphosphate site*. There is also a site for the hydrolytic cleavage of 5′-monophosphates from this end of the chain.

Heteropolymer tetramers frequently present multiple forms of associations of their constituting units in the same organism or even in the same cell (polymorphism). A familiar example is afforded by lactate dehydrogenase, which is formed by four polypeptide chains, each of mol.wt. 33 500 and which exists in five different forms in rat tissues (isoenzymes). The overall reaction catalyzed is the same in all cases, but they differ in an aspect of the signified, the K_m value for the substrate. The five isoenzymes are five different combinations of the two polypeptides M and H, which are homologous but show differences in amino acid content and sequence. M_4 predominates in muscle and H_4 predominates in heart. The other forms are M_3H, M_2H_2 and MH_3, the relative amounts of which are under genetic control and differ from tissue to tissue[59a].

Many cases of isoenzymes are known and they also are heteropolymer biosyntagms, the different forms of which show differences with respect to kinetic characteristics. Vertebrate hemoglobins are in most cases heteropolymer tetramers composed of chains α and β in equal proportions ($\alpha_2\beta_2$). Polymorphism has been recognized in the hemoglobins of many vertebrate species. The horse has two hemoglobins, with two different α chains and a common β chain (Trujillo *et al.*[59b]). The two different α chains differ by the substitution of a glutamine for a lysine[59c].

The well-known polymorphism of human hemoglobin results from the presence of β and δ chains, a result of gene duplication. Fish have up to twenty hemoglobins. In all these cases, common in animal forms, polymorphism results from the existence of multiple polypeptide chain types and the possibility results in the formation of multiple heteropolymer tetramer types, showing diversification at the level of aspects of the signified.

15. Biosemiotics of the flux of information from DNA to proteins

The origin of this flux resides in the sequencing syntagm of DNA and results in the sequenced syntagms of proteins.

(a) Genomic transcription (biosemes = bases)

The process of genomic transcription, in which, in each form of cellular differentiation, a certain set of genes is active and others are inactive is ruled by sequence-specific binding. Only a section of one strand of the DNA duplex is transcribed in mRNA. It is generally accepted that unwound single-stranded DNA acts as the template. A local unwinding of the DNA duplex uncovers the biosemes (bases) at the head of the RNA under

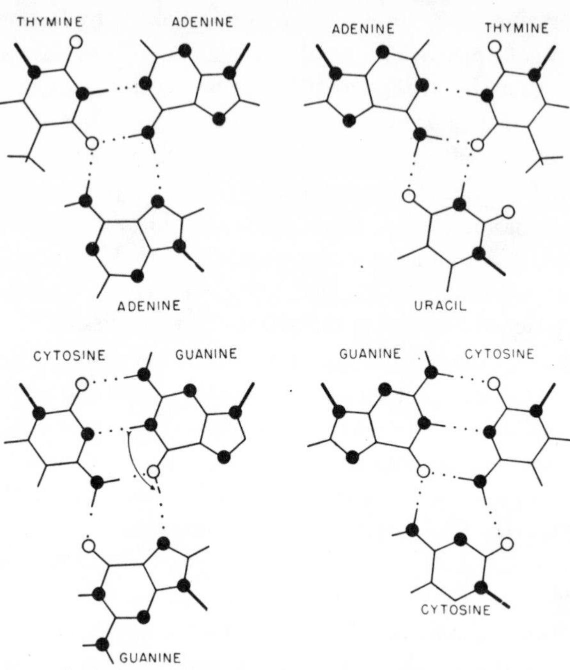

Fig. 7. Messenger synthesis through base triplet-hydrogen bonding between its bases (lower) and the Watson and Crick base pairs (upper) of the duplex. Nitrogen atoms = filled circles; oxygen atoms = open circles. (Stent[62]).

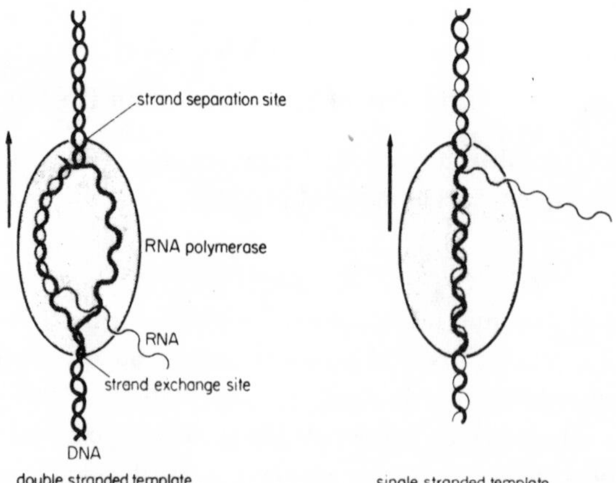

Fig. 8. A model for transcription by local unwinding. (Lewin[63], after Fuchs *et al.*[64]).

synthesis, and hydrogen bonding of complementary bases takes place. This coupling is represented in Fig. 7.

Fuchs *et al.*[64] have proposed a model for transcription by local unwinding. In this model, the authors consider the RNA polymerase, as possessing two sites involved as biosemes in the transcription. These two biosemes (see Fig. 8) are the "strand separation site" and the "strand exchange site", the unwound segment progressing along the DNA duplex in the course of mRNA synthesis.

(b) Translation (biosemes=codons)

The biosynthesis of a polypeptide (see Vol. 24 of this Treatise) begins with the binding to a polysome of the amino-acyl-tRNA bearing the *N*-terminal amino acid, and of the adjacent amino acids to their respective codons (triplets) followed by the peptide-bond synthesis between the carboxyl group of the first and the amino group of the other. This peptide-bond biosynthesis results from a transfer of the polypeptide chain attached to the tRNA in the P site (the ribosome bioseme bearing the peptidyl-tRNA) to the aminoacyl-tRNA on the A site (the ribosome bioseme corresponding to the site entered by the incoming aminoacyl-tRNA). This is catalyzed by an enzyme attached to a specific site on the 50 S subunits of the ribosome.

The first tRNA leaves the ribosome and a polypeptide chain remains attached to the tRNA on the A site. The ribosome moves a triplet (bioseme) further along. The messenger biosyntagm, read in biosemes (triplets) from its 5′ end to its 3′ end bears a series of ribosomes, each successively carrying a greater length of polypeptide chain, which begins before the completion of the translation of the mRNA, to take up, as a consequence of the interaction of structural biosemes (amino acids), its tertiary configuration resulting in the emergence of new biosemes. The implication of the structural information carried by the lateral chains of amino acids is introduced in the course of the translation of the sequence of base triplets of an mRNA into the sequence of amino acids of a polypeptide chain through the action of tRNA, the smallest cellular RNA, containing between 75 and 85 nucleotides. It contains unusual bases, among which are pseudo-uridine, dihydrouridine and dimethylguanine. The flux of information from DNA to tRNA goes nevertheless through the usual channels of sequence-specific transcription but this is followed by other introductions of information mediated by different enzymes which modify

References p. 117

the significant and the signified after the transcription is achieved. Among these enzymes are methylases using *S*-adenylmethionine as methyl-group donor (Baguely and Staehelin[65]) and enzymes transferring sulphur from cysteine to tRNA in *E. coli* (Hayward *et al.*[66]).

The primary structure of about 40 tRNAs has been deciphered (yeast, *E. coli*, rat liver, wheat germ, human tumour KB cells, etc.). These molecules are homologous and a generalized clover leaf secondary structure has been proposed for tRNA by Fuller and Hodgson[67]. Its syntagmic tertiary structure is compact and, though it is made of a single polynucleotidic

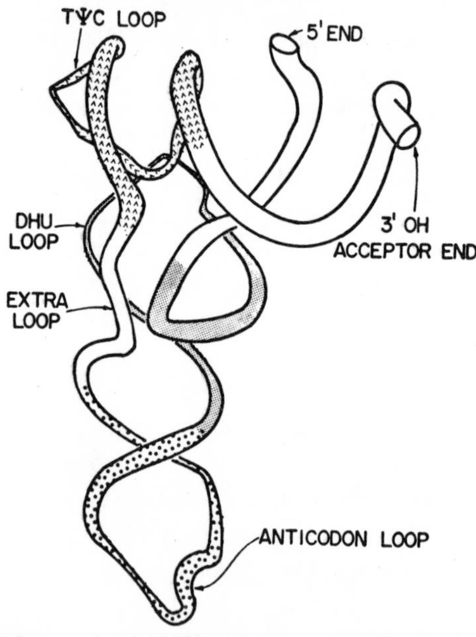

Fig. 9. A perspective diagram in which the polypeptide chain of $tRNA^{phe}_{yeast}$ is represented as a continuous coiled tube. Abbreviations: C, cytidine; ψ, pseudouridine; T, ribothymidine; DHU, dihydrouridine. (Kim *et al.*[68])

chain it carries a number of decisive biosemes. In the model of the tertiary structure of the $tRNA^{phe}_{yeast}$ proposed by Kim *et al.*[68] and shown in Fig. 9, the anticodon sequence coupling to a codon on mRNA is exposed at the end of the biosyntagm opposed to the three loops. The biosemiotic importance of a given tRNA resides in its specific recognition by a specific activating synthetase. The accumulated evidence is in favour of the possibility for one synthetase to charge tRNAs carrying different anticodons. This result is a multiplicity of tRNA molecules representing a single amino acid, due to a redundancy of the several tRNAs responding to the same codon, with differences of structure in other parts, the number of tRNA molecules for each amino acid being between 2 and 5 (no relation to the number of codons for one amino acid).

The question remains of the nature of the system of recognition between the *relata* of the couple amino acid and tRNA through the action of the synthetase. The bioseme relating amino acid and tRNA is common to all of them: it is a three-base sequence situated at the 3′-hydroacyl end. The amino acid is first activated by the formation of an amino acid–enzyme–AMP complex, in the presence of Mg^{2+} and ATP. The same enzyme aminoacyl-tRNA synthetase, or activating enzyme is responsible for this reaction and for the formation of the aminoacyl-tRNA.

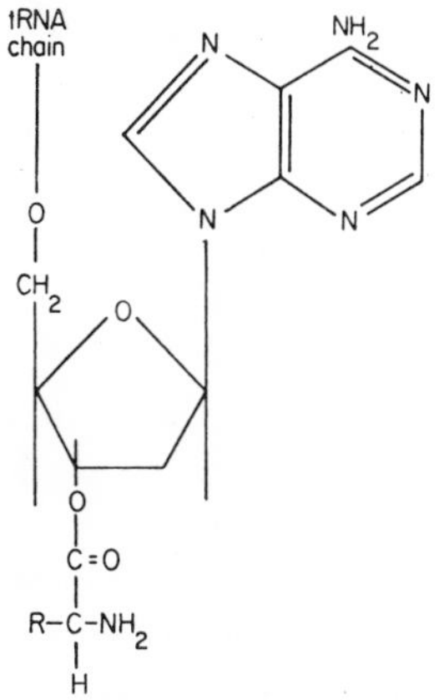

tRNA chain

Fig. 10. The 3′-terminus of aminoacyl-tRNA. (Lewin[63])

As shown in Fig. 9, the TψC and DHU loops come in very close contact and act as structure biosemes. Evidence has been obtained (see Lewin[63]), showing changes in configuration of a given tRNA due to it becoming charged (ligand-induced configuration changes). It appears that more than one factor is involved in the recognition of the correct tRNA by its activating enzyme, and that there are other elements of enzyme recognition in the tertiary structure of this fascinating biosyntagm.

Protein synthesis and its accuracy are highly dependent on the biosemiotic relation of amino acids with the specific synthetases. Each synthetase forms an aminoacyl-tRNA for a single amino acid, though each amino acid may be coupled with a number of tRNAs. On the other hand the two

References p. 117

reactions catalysed by a synthetase (formation of amino acid-AMP and of aminoacyl-tRNA) have different biosemiotic specificity. For instance in *E. coli*, isoleucyl-tRNA synthetase converts either isoleucine or valine to the corresponding adenylate, but it only transfers isoleucine into isoleucyl-tRNA[69].

The couple of *relata* formed by the aminoacyl-tRNA and the ribosome is of a complex nature, well clarified by Lipmann (see his reviews[70]). It is bound to its codon on the ribosome in the form of a complex with GTP and factor T_u. The formation of this transitory complex is catalyzed by another factor, T_s.

The transitory biosyntagm is formed in two stages the first of which does not involve the aminoacyl-tRNA which is specifically involved in the formation of a second complex responsible for the transfer of the aminoacyl-tRNA on the ribosome (Lucas-Lenard and Haenni[71]):

$$\text{GTP} + \text{T}_u \xrightarrow{\text{T}_s} \text{T}_u\text{–GTP complex}$$

$$\text{T}_u\text{–GTP complex} + \text{aminoacyl-tRNA} \rightarrow \text{aminoacyl-tRNA–T}_u\text{–GTP complex}$$

$$\text{aminoacyl-tRNA–T}_u\text{–GTP complex} + \text{ribosome} \rightarrow \text{aminoacyl-tRNA–ribosome complex}$$

The polypeptide chain synthesized so far is transferred from the tRNA in the P site to the aminoacyl-tRNA on the A site and the formation of a peptide bond is catalyzed by the peptidyltransferase of the 50S ribosomal subunit. The translocation on a further codon on mRNA takes place and the tRNA (uncharged) on the P site leaves the polysome, its place on the P site being taken by the polypeptidyl-tRNA previously on the A site. This happens in the presence of factor G and requires the hydrolysis of GTP. The initiation of polypeptide chains requires, in *E. coli* at least, other protein factors than those involved in their elongation.

Met-tRNA, when formylated (fmet-tRNA) acts as initiator, at least at low Mg^{2+} concentrations. In *E. coli*, it is the result of the formylation of met-tRNA in the presence of a specific enzyme (Marcker *et al.*[72]).

$$\text{met-tRNA} \xrightarrow[\text{formyl-THF-met-tRNA transformylase}]{\text{10–formyltetrahydrofolate}} N\text{-formyl-met-tRNA}$$

Clark and Marcker[73] have fractioned the methionine-accepting activity of tRNA of *E. coli* into one form which can be formylated (met-tRNA_f) and one which cannot (met-tRNA_m). Both species responded to codon AUG for methionine but, in supplement, the fmet-tRNA_f responded also to GUG, the codon for valine. But if this codon recognizes fmet-tRNA_f

externally, it inserts valine internally. AUG and GUG both code for the *initiator* bioseme fmet-$tRNA_f$ externally, but both of them internally code for their normal transfer species met-$tRNA_m$ and val-tRNA.

On the other hand, in translation, the couple of *relata* established by mRNA is established between a codon on itself and the anticodon nucleotide sequence of tRNA, the only molecule to be recognized by mRNA. *It is in the process of charging tRNA that the signified of an amino acid expresses itself with respect to the accomplishment of polypeptide sequencing*. This very important biosemiotic aspect has been established in a most elegant way by Chapeville *et al.*[74].

The ribosome-mRNA syntagm is known as a polysome, in which the number of ribosomes associated with mRNA varies.

An initiation codon as bioseme in the reading of a message phrases its translation from triplets into amino acids.

It was difficult to understand how AUG could recognize the initiator transfer externally while recognizing its normal aminoacyl-tRNA internally until it was known that different systems were involved in the two processes.

The initiation is a property of the 30S subunits of the ribosomes. Only fmet-$tRNA_f$ is able to bind to 30S units in response to AUG, while with 70S ribosome, reading is only permitted as met-$tRNA_m$. This is the result of the interaction with aminoacyl-tRNA of the elongation transfer factors (T_u and T_s) which do not form an intermediate complex with fmet-tRNAf, therefore excluding it from internal recognition.

When at the beginning of a messenger, the first AUG codon is encountered by a 30S unit, the initiation complex is formed. Only after the attachment of the 50S subunit, do the AUG codons unambiguously code for met-$tRNA_m$.

It is supposed that after the translation of a polycistronic messenger, the ribosome dissociates into its subunits upon chain termination at the end of a cistron. If the 30S particle remains attached to the messenger, it will recognize the AUG codon at the beginning of the next cistron as an initiator, and commence synthesis of the next protein specified.

Ribosomes bind directly at the initiation site (Bretscher[75,76]).

The formation of the initiation complex, a transitory biosyntagm, takes place between a 30S subunit, mRNA and the *initiator* fmet-$tRNA_f$, in the presence of GTP and initiation factors. These differ from the proteins involved in chain elongation. There are three initiation factors, f and f_2 required for the binding of fmet-$tRNA_f$ to ribosomes in response to AUG

(Anderson *et al.*[77]) and f_3 stimulating the binding of 30S units to mRNA (Brown and Doty[78]). A first reaction binds the initiator, f_2 and GTP. In a second reaction, f is added to form the biosyntagm transferring the fmet-$tRNA_f$ to the 30S subunit. In the course of protein synthesis, polypeptide chains are attached to tRNA by the ester bond between the last added amin acid and the terminal adenosine of its tRNA. The completed protein is released and the tRNA liberated for re-use.

Among the various nonsense codons it is only the *ochre* which is used for chain termination *in vivo.*

When the translation of a messenger is completed, the polypeptide chain synthesized, tRNA and the 70S ribosomes are released (though the presence of free 70S ribosomes remains controversial). When they complex with a dissociation factor, they contribute to the pools of 30S and 50S subunits. When a 30S subunit couples with a messenger in the formation of the syntagm, the dissociation factor is released and the 30S subunit combines with a 50S subunit to form a 70S ribosome undertaking translation.

The sequencing syntagm of DNA is composed of biosemes consisting of nucleotides while in the couples of *relata* involving mRNA the biosemes are the codons (base triplets, at least in general).

To recapitulate the biosemiotic aspects of the flux of information in tein synthesis, we may state that a first couple of *relata* is established between an amino acid and a specific activating enzyme. A second couple of *relata* is established between the enzyme (with amino acid attached) and a specific tRNA. But the point of attachment is non specific, as all tRNAs have the same three terminal nucleotides (see Lewin[63]). In a third couple of *relata, a specific triplet codon* on mRNA is coupled *with its anticodon on tRNA*. This is the domain of the code-translating mechanism.

From the point of view of biosemiotics it is important to recognize the universality of the generality of codon assignments. An important biosemiotic aspect is the fact that, though the code is degenerate (*i.e.* that some amino acids have assigned to them more than one codon) at the stage of translation (and of the formation of a couple of *relata* between tRNA and the codons of mRNA), the couple of *relata* involved is highly specific of the codon and of the specific anticodon bearing tRNA, *i.e.* that for every codon there is a specific tRNA. The degeneracy of the code is not biosemiotically situated at the translation stage, but in the fact that more than one type of tRNA, in a given organism, is associated with some of the amino acids (literature in Lewin[63]).

16. Biosemiotic characters of amino acids

We have met these properties when considering the nature of structural information in the process of the folding of tertiary biosyntagms, and also in the stage of formation of aminoacyl-tRNA. Information belonging to the amino acid units of configuration in polypeptide chains should be classified thermodynamically as structural information. From the point of view of molecular biosemiotics, we may consider the amino acid side groups as endobiosemes. In Fig. 11, the amino acid side groups have been arranged to emphasize their relationship and their biosemiotic content. The side

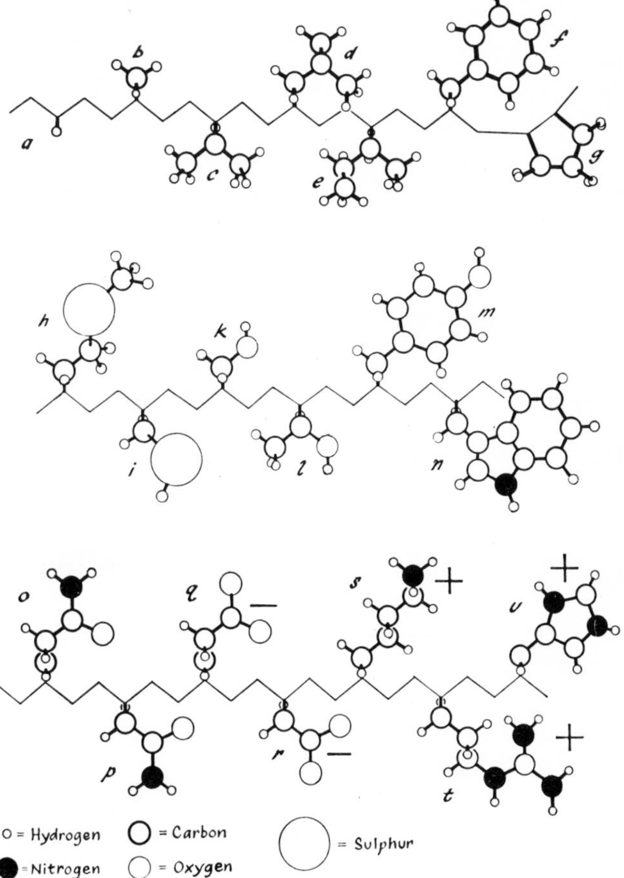

Fig. 11. Amino acid side chains. (Cairns-Smith[79]).

groups b–g are hydrocarbons of progressive bulk and of parallel progressive hydrophobicity. Phenylalanine, besides its great hydrophobicity, contains an aromatic ring and like tyrosine and tryptophan, is able to contract couples of *relata* with other aromatic rings through overlapping of π electron clouds (*e.g.* in the interaction between protein and heme in myoglobin and hemoglobin). Histidine has one pair of electrons on one of its nitrogen atoms, which is used in metal binding. Cysteine is able to form disulphide bridges. In proline the side chain may force a bend on the polypeptide chain and may disrupt an α helix. Some of the side groups contain flat rings (f, g, m, n, u); two carry negative charges (q, r), three carry positive charges (s, t, u). Three are hydroxyl-containing (k, l, m).

"According to the sequence of these groupings on the main chain, the protein molecule may fold into a complex three-dimensional "globular" structure in which a large proportion of the groups fit into a precise but irregular arrangement." (Cairns-Smith[79]).

Dickerson and Geis[53], discussing the nature of the structural information governing the signification of animal structural fibrous proteins record three basic configurations in these protein chains: the α-helix, the collagen triple helix and the β sheet. If a strong and rigid material is used (tendon, fish scales) the collagen helix is adopted, while the α helix is used if a more flexible material is required, and the β sheet (silk) if the material has to be flexible, but strong. The resulting structure is determined (structural information) by the amino acid content and sequences.

"Pro and Hypro are incompatible with an α helix and such a helix occurs in the absence of large quantities of these residues. Polyglycine itself, lacking side groups, has been crystallised in both the β sheet and the polyproline helix, and the two classes with high glycine content are just the silks and collagen. Just as the sequence (–Gly–Ser–Gly–Ala–Gly–Ala–)$_n$ induces a β sheet structure, so the sequences similar to (–Gly–X–Pro–)$_n$ induce a collagen triple helix." (Dickerson and Geis[53]).

We may now return to the couples of *relata* established between tRNA, synthetase and amino acid, a relation in which the properties of the amino acid are involved as specific biosemes. It has been shown that there is a change of signified of the activating enzyme during complex formation, as a result of a configuration change (significant), as a consequence of its association with tRNA. It has been recognized, for instance, that in the cases of proline and valine, the enzyme–tRNA complex is more resistant to heat that the enzyme (Chuang *et al.*[80]) and that the binding of tRNA to the synthetase–amino acid–AMP complex induces in the protein structure, a configuration change resulting in a loss of helical structure (Ohta *et al.*[81]).

It appears that the binding of the amino acid to its synthetase increases the availability of the site to tRNA on this synthetase. A model has been proposed by Yarus and Berg[82], indicating a number of successive ligand-

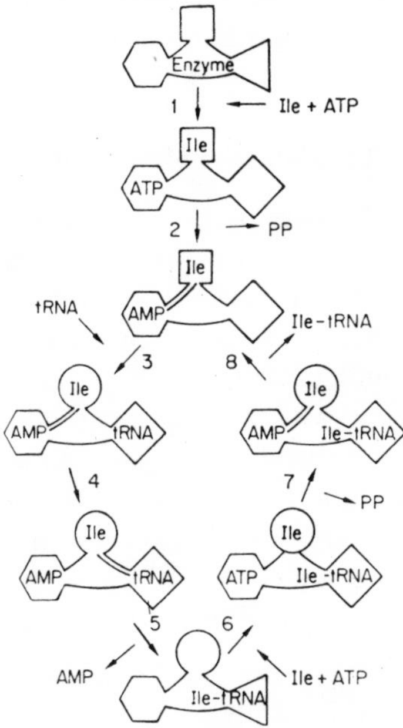

Fig. 12. The catalytic cycle of isoleucyl-tRNA synthetase. (Yarus and Berg[82])

induced configuration changes modifying the significant and the signified of successive biosemes.

As suggested in Fig. 12, the synthetase passes through four stages:

(*1*) isoleucine and ATP bind to the synthetase and there is a configuration change of the bioseme which acts as binding site for tRNA (from closed to open, a state in which rapid entry and exit of tRNA is possible).

(*2*) As a result, isoleucyl-AMP is formed on the enzyme.

(*3*) The tRNA binds and the configuration of the binding site specific for the amino acid is thereby modified (change of significant).

(*4*) Isoleucine is transferred on the tRNA.

(*5*) The amino acid leaves its site, with a change of the tRNA-binding site into the "closed" form.

References p. 117

(*6*) Isoleucine and AMP bind, reconverting the tRNA-binding site into the "open" form.

(*7*) With a loss of inorganic phosphate, isoleucyl-AMP is formed.

(*8*) Isoleucyl-tRNA is released and the conformation of the amino acid binding site is changed.

As demonstrated by Baldwin and Berg[83] the transfer of a wrong amino acid to tRNA is avoided by the destruction of any "wrong" complex.

17. Multigene biosyntagms

Chromosomes are mixed biosyntagms composed of molecules of DNA, of RNA and of proteins. Crick[84] has proposed a model for the chromosomes of higher organisms in which the DNA coding for the biosynthesis of polypeptide chains, as related in Section 15 (p. 27) is only a small part of the amount of DNA in the chromosome. This part is contained in the interbands while the major part of DNA, used for control purposes, is contained in the bands. For control purposes the configuration used in higher organisms for the formation of couples of *relata* with regulatory molecules is represented by unpaired stranded stretches of double stranded DNA (Fig. 13).

The biosynthesis of proteins, either structural proteins or enzymatic

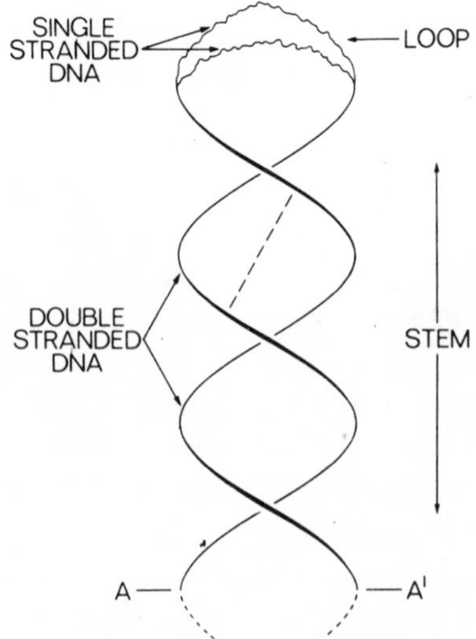

Fig. 13. An example of a possible structural motif within the globular DNA. (Crick[84])

proteins is operated and regulated by batteries of genes, the association of which is of a different nature in Procaryotes and in Eucaryotes.

In Procaryotes, this association is insured by operons, while in Eucaryotes the cistrons cannot be considered as contiguous in the genome. In both cases the catenary biosyntagm is formed of a genetic biosyntagm (operon or batteries of activated non-contiguous operator genes) insuring the biosynthesis of specific enzymes, and of a metabolic catenary biosyntagm involving metabolic intermediates as substrates.

(a) Procaryotes

In Procaryotes, it has repeatedly been observed that the genes governing the biosynthesis of the enzymes of a catenary metabolic syntagm are closely linked, forming a cluster on the genome.

Induction and repression of the whole operon happen coordinately and the definition of the operon involves both structural genes and regulatory functions (Jacob and Monod[85]). An example is provided by the lactose (*lac*) operon of *E. coli* in which three contiguous genes are simultaneously induced or repressed: those for β-galactosidase, β-galactoside permease and β-galactoside transacetylase.

In the *lac* operon, the repressor is a multichain protein (mol.wt. 150000) made of four identical units interacting with a specific repressor site on the DNA. By complexing the repressor with an inducer, derepression occurs. Repression, or specific deletion of enzyme synthesis, was discovered by Monod and Cohen-Bazire[86], who found that tryptophan selectively inhibits the biosynthesis of tryptophan synthetase of *E. coli*. This is not a variation of activity (signified) but a deletion of enzyme synthesis (Yates and Pardee[87]). Certain enzymes of yeast are formed only in the presence of their substrate. This has been shown to be a synthesis *de novo* and not an activation (literature in Lewin[63]).

Jacob and Monod[85] have first proposed the concept of two types of genes in the genome: *structural genes* and *regulator genes*, the activity of which being according to this theory exercised through the biosynthesis of a regulator protein. In the *lac* operon, the regulator gene i^+ governs the synthesis of an *apo-repressor* which binds to the lac *operator*.

(b) Eucaryotes

The genome of a differentiated Eucaryote cell is considered as being

References p. 117

composed of small proportions of DNA coding for proteins and of a much larger portion of genes exerting a regulation *at the level of genomic transcription* (Britten and Davidson[88]; Crick[84]).

As summarized by Britten and Davidson[88]

"Batteries of producer genes are regulated by activator RNA molecules synthesized on integrator genes. The effect of the integrator genes is to induce transcription of many producer genes in response to a single molecular event."

In the concept of gene expression involved in this model, a number of elements are concerned:

Producer gene: a region of the genome transcribed to yield a template RNA (for example for a hemoglobin subunit). It corresponds in Eucaryotes to structural genes as defined by Jacob and Monod[85] in Procaryotes. The products of the producer genes are the RNAs except those exclusively performing genomic regulation.

Receptor gene: a DNA sequence linked to a producer gene and causing the active transcription to happen when a complex (sequence-specific) is formed between the receptor sequence and the RNA molecule called "activator RNA".

Activator RNA: Forms a sequence specific complex with receptor genes linked to producer genes. This complex is established between native (double-stranded) DNA and a single-stranded RNA molecule. The authors recognize that it is feasible to accept that the role attributed above to activator RNA may be carried out by protein molecules coded by these RNAs. In this case, the DNA sequence coding for such proteins should not be called producer genes, a term which should be reserved for sequences of DNA coding for RNAs other than those translated in proteins recognizing the receptor sequence. In the model proposed by Britten and Davidson, the simpler alternative of the activator RNA has been chosen.

Integrator gene: for synthesis of activator RNA (a link of integrator genes are activated together).

Sensor gene: a DNA sequence which is a binding site for agents inducing the occurrence of specific patterns of activities *in the genome.* The binding of these inducing agents (active in intercellular or intra-cellular control) is of sequence-specific nature depending on the sensor gene sequence. The result is an activation of the integrator gene (or genes) linked to the sensor gene. Certain inducing agents will not bind directly to the DNA of the sensor gene but rather to an intermediary structure such as a specific protein complexing with the inducing agent and binding to the DNA of a sensor gene in a sequence specific way.

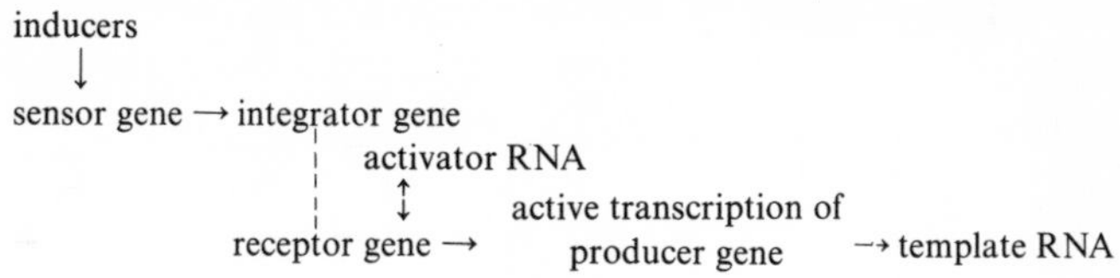

A set of producer genes is activated when a particular sensor gene activates its set of producer genes. Many of these sets may be required in the accomplishment of a particular cell stage.

States of cell differentiation stem from the concerted activation of one or more batteries of producer genes.

Fig. 14 (Britten and Davidson[88]) shows two different basic aspects of integrative systems within the model. In Fig. 14A the producer genes are integrated in three very small gene batteries. Sensor gene S_1 and its integrator gene specify the activation of producer genes P_A, P_B and P_C; while S_2 acts in the same way in the case of P_A and P_B and S_3 for P_A and P_C. The control pattern is different in Figs. 14A and 14B. In Fig. 14A existence of *redundant receptor sequences* is shown and there is only one integrator per sensor. There will be as many copies of a receptor gene as there are producer genes in the biosyntagm. Such biosyntagm may play a role in the control of metabolic catenary biosyntagms (formed by enzymes involved in the same metabolic pathway) specific of a cell category. On the other hand, in Fig. 14B, redundancy obtains between *the integrator genes* of different integrator sets. There will be as many copies of an integrator as there are biosyntagms calling on its producer gene. The system, governing a large diversity of producer genes such as obtains in the primary catenary biosyntagms, is common to all cells.

In the model proposed by Britten and Davidson[88] the regulation is accomplished by sequence-specific binding of an activator RNA, rather than by histone accomplishing specific sequence recognition. According to the authors, histones are general inhibitors of transcription of the genome and the regulation of genome transcription in a biosyntagm submitted to general histone inhibition, depends on specific activation of otherwise repressed sites, rather than a repression of otherwise active sites.

In Fig. 15 (Britten and Davidson[88]) a minimal model is presented in a much simplified way. In such an event as development, massive changes of differentiation take place. Britten and Davidson visualize such phenomena

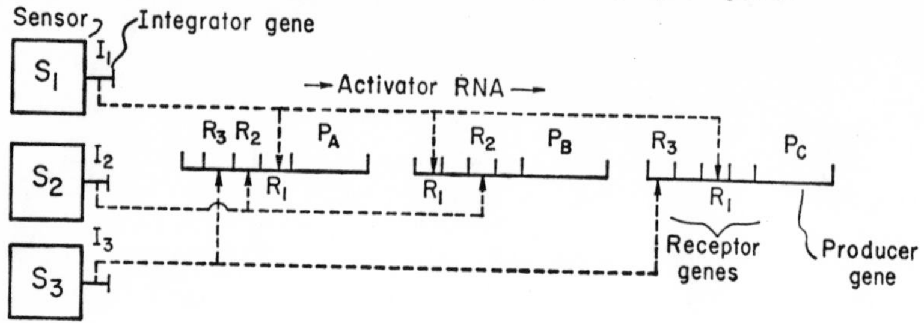

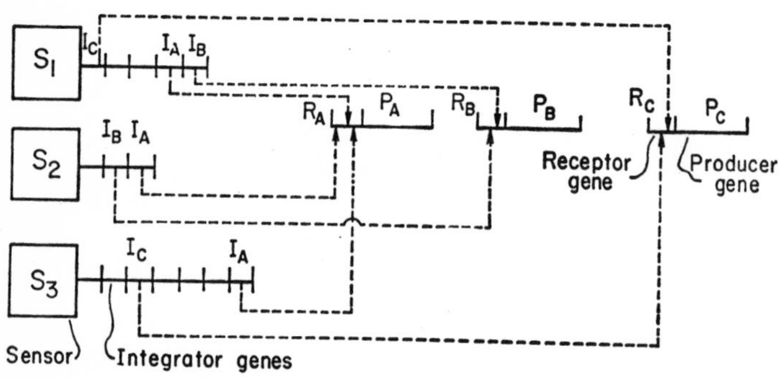

Fig. 14. Types of integrative system within the model. (Britten and Davidson[88]). (A) Integrative system depending on redundancy among the regulator genes. (B) Integrative system depending on redundancy among the integrator genes. These diagrams schematize the events that occur after the three sensor genes have initiated transcription of their integrator genes. Activator RNAs diffuse (symbolized by dotted lines) from their sites of synthesis — the integrator genes — to receptor genes. The formation of a complex between them leads to active transcription of the producer genes P_A, P_B and P_C.

"as being mediated by sensor genes sensitive to the products of integrator genes in *other integrative sets.*" They consider that in the course of synchronic epigenesis, a single inducing agent can lead to the activation of a number of sensor-integrator sets and thereby lead to the activation of a large number of producer genes involved in the new differentiation. Leaving the molecular level of integration for the cellular level, sequential patterns of gene activation could take place if certain sensors respond to the *products* of producers. "In addition, the protein of a newly effective sensor assembly is, in the model, a product of a previously activated producer gene."[88]

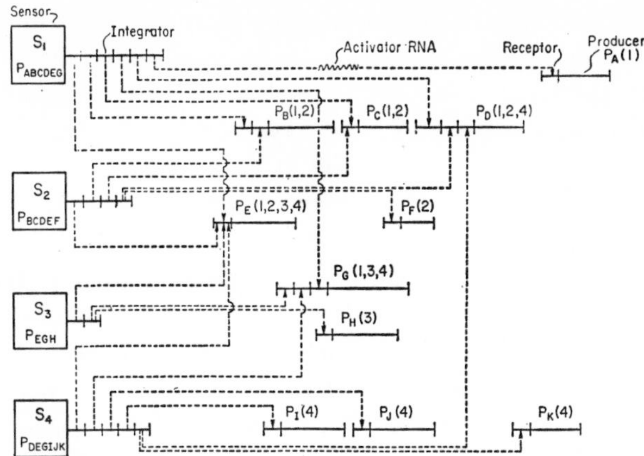

Fig. 15. This diagram is intended to suggest the existence of overlapping batteries of genes and to show how, according to the model, control of their transcription might occur. The dotted lines symbolize the diffusion of activator RNA from its sites of synthesis, the integrator genes, to the receptor genes. The numbers in parentheses show which sensor genes control the transcription of the producer genes. At each sensor the battery of producer genes activated by that sensor is listed. In reality many batteries will be much larger than those shown and some genes will be part of hundreds of batteries. (Britten and Davidson[88]).

In Britten and Davidson's model a given state of differentiation involves the coordinated activity of a number of multi-gene biosyntagms. Producer genes active in a given tissue need not be physically linked in the genome of Eucaryotes, while (polycistronic) operons, in Procaryotes, consist of a biosyntagm of physically adjacents involving direct contiguity of active producer genes.

18. The central catenary metabolic biosyntagm of cells

Catenary metabolic biosyntagms are composed of sequences of enzymes involved in a catabolic or an anabolic metabolic pathway.

Whatever point an organism occupies in phylogeny, its cells are provided with a catenary metabolic biosyntagm which may be qualified as central, because it does not only allow for the acquisition of free energy in the form of ATP "high-energy" bonds, through a traffic regulation of the ways of the flow of matter and energy but because it is also the starting point of the metabolic pathways of biosynthesis which insure the composition and maintenance of the organism and its maintenance in a state remote from the equilibrium. The central metabolic biosyntagm is derived from the glycolytic pathway (from triosephosphate to pyruvate) linked with the Krebs cycle at the level of acetyl-CoA, which can also be derived from the catabolic pathway of fatty acid metabolism. The pathway of the catabolism of amino acids enters the central metabolic pathway by way of aspartate or glutamate, through their connection with the Krebs cycle.

Not only do the organisms feeding on carbohydrates, fats and proteins use the system described for obtaining ATP, but the bacteria feeding on nutrients such as benzoic acid, itaconic acid, uric acid, etc. take advantage of the same central biosyntagm in which free energy can be made available either anaerobically or aerobically by the extension consisting of the respiratory chains.

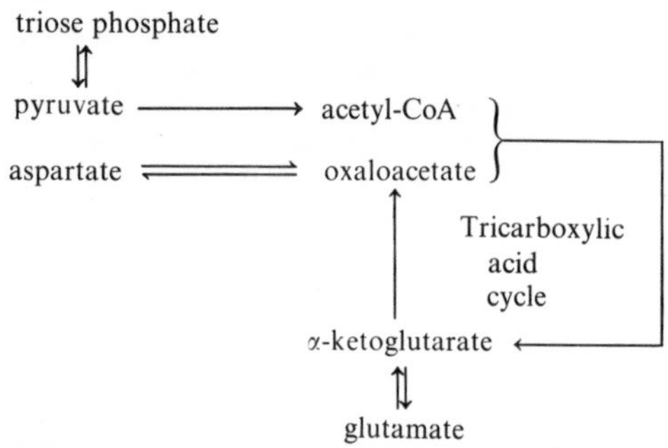

Fig. 16. The central metabolic biosyntagm.

The catenary biosyntagm represented in Fig. 16 may be considered as a frozen catenary biosyntagm underlying the unity of biochemistry.

19. The integrated pattern of primary and secondary catenary biosyntagms

The intermediary steps of the central metabolic biosyntagm common to all cells function as starting informers for biosynthetic primary metabolic pathways (biosynthetic pathways of general occurrence) carrying themselves a number of derivations in the form of secondary biosynthetic pathways (limited to special categories of organisms), the starting points of which are situated on the sequence of a primary pathway. Examples of such relations are found in Fig. 17. The secondary biosyntagms mentioned are but a few of the existing ones and the diagram as a whole does not

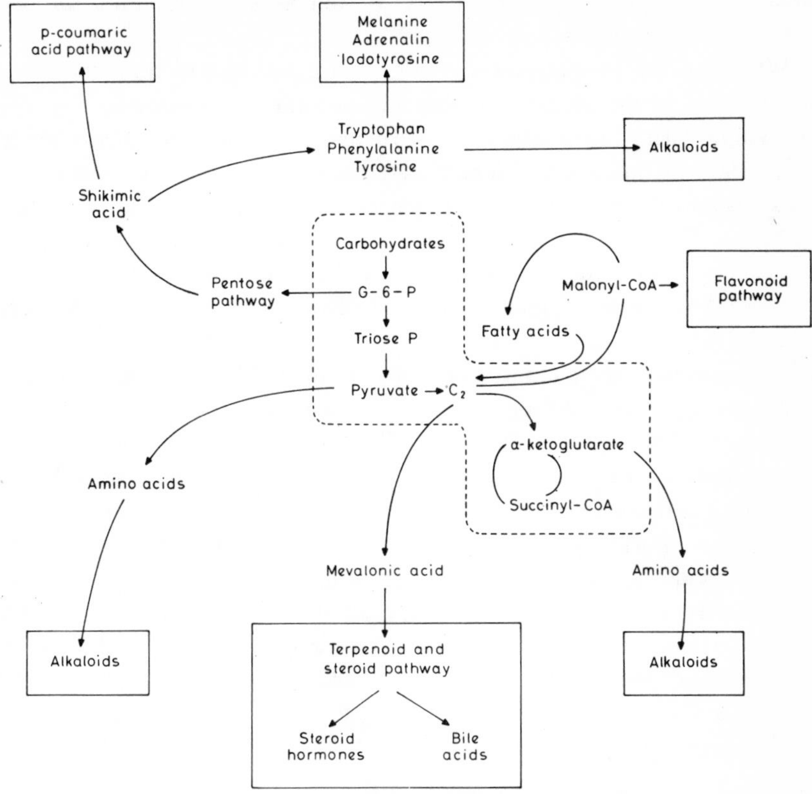

Fig. 17. Examples of the connections of a few secondary metabolic pathways obtaining either in plants or animals and the primary biosynthetic pathways, themselves communicating with the central metabolic biosyntagm.

refer to any particular category of organisms. The flavonoid pathway is for instance mostly found in a category of plants, the Angiosperms. The coumarins and the corresponding secondary catenary pathway belong to Dicotyledons. Steroid hormones and bile acids are found in vertebrates.

A number of heterotrophs degrade the products of photosynthesis and the dissipation of energy from the solar source to the interplanetary sink is accomplished by many species. Biosynthesis starts from small units to produce either body constituents themselves introduced into the pool submitted to energy dissipation in the organism, or acting as ecomones or as nutrients for other organisms, as well as starting points for co-evolutionary developments.

When considering a given ecosystem, we are dealing with an association of segments of the general primary biosyntagm abbreviated in Fig. 17, combining energy dissipation, the permanence of the constituents for a certain lapse of time and the biosemiotic significants informing the metabolic biosyntagms along the catenary sequences of their intermediary articulations.

Recurring metabolic patterns which emerged from the extension of biochemical knowledge are much in line with Kluyver's concepts (see Section 3, p. 2).

The limitation of the number of forms of catenary biosyntagm pathways is an aspect of the molecular logic of organisms which deserves consideration. Pullman and Pullman[89,90] have stressed that the essential components of these catenary biosyntagms are constituted of completely or at least partially, conjugated (resonating) systems, rich in delocalized π electrons. Purine and pyrimidine bases are conjugated heterocycles. Nucleic acids, proteins and ATP are conjugated compounds. Highly conjugated compounds are the current basis of the living system. Electronic delocalization, as stressed by Pullman and Pullman, is the most characteristic property of these compounds. Electronic delocalization confers stability on these compounds. Practically all coenzymes are conjugated organic molecules.

"The choice and utilization of conjugated heterocycles as structural components of life appears thus as one of the most important quantum effects in biochemical evolution"[89] (see also Vol. 22 of this Treatise).

It should be remembered that a given intermediate in a primary metabolic pathway may have a different origin. For instance, if such biomolecules as oxaloacetate, fumarate and succinate involved in free-energy production as well as starting points for biosynthesis, are situated on the catabolic general primary biosyntagm, they are also biosynthesized by the pathways called anaplerotic by Kornberg[91].

20. Ligand-induced modulations of signified

The old controversy between mechanists and vitalists may be considered to be prolonged in our time in the form of diverging opinions of molecular biologists who consider life as the expression of a complex molecular order and organicists who consider that any level of organization above the molecular one is determined not only by the latter but also by the higher levels of organization. One of the arguments of the organicists consists in claiming that the cooperative play of biomolecules depends on a principle which organizes the parts as terms of the whole. This recurrence of the concept of life "as imposed" is another (and unjustified) way of formulating the recognized complexity of the suprachemical systems of cells and organisms. In the field of the integrative molecular level, molecular biology recognizes a biosemiotic aspect of integration and regulation consisting of modulations of the signified of biosemes and biosyntagms as a result of ligand-induced modifications of the significant (an aspect of molecular configurations). Modulation refers to quantitative changes introduced in the context of an existing signified. The traffic on the catenary metabolic biosyntagms is governed by extensive modulations of the signified of the relevant biosemes.

Catenary metabolic biosyntagms involve effectors and regulators, those commanding at the level of the genome the degree of activity of producer genes. Another biosemiotic aspect is the modulation of the signified of effector enzymes. This is generally the result of a ligand-induced configuration change, though in the present situation these aspects are mainly grasped at the signified level. Feedback in metabolic catenary syntagms is a common example of such modulations of the signified. It is well known that in many cases the end-product of a metabolic catenary sequence may exert an inhibitory effect on the first enzyme of the pathway (feedback inhibition, retroinhibition). The enzyme whose activity is modulated is called a regulatory or allosteric enzyme (a multichain protein) and the inducer of the change of signified is the modulator or effector. Feedback may also take place at a branching point of a catenary metabolic biosyntagm. When the regulatory enzyme has only one modulator (effector) it is said to be monovalent while it is said to be polyvalent if it is influenced by several modulators. When the substrate is itself a modulator, increasing or decreasing the signified, the enzyme is called homotropic. There are positive homotropic binding effects and negative ones. Homotropic enzymes carry two or more biosemes for the substrates, one of which is the catalytic site. When the enzymes are stimulated or inhibited by specific naturally occurring modulators (*i.e.* end-products) different from the substrate, the regulatory enzyme is called heterotropic. There are enzymes which are both homotropic and heterotropic, as their substrate is

References p. 117

one of the different modulators to which the enzyme responds positively or negatively. There are regulatory enzymes in which the catalytic site and the receptor site for the modulator are situated on different polypeptide chains. A number of examples of allosteric regulation in enzyme activity have been collected and analyzed by Stadtman[92] and the reader is referred to his excellent review as well as to the reviews published in *Current Topics in Cellular Regulation* (ed. by B. L. Horecker and E. R. Stadtman, Academic Press, New York) and in *Advances in Enzyme Regulation* (ed. by G. Weber, Pergamon, Oxford).

From the viewpoint of biosemiotics, allosteric effects are only one aspect of ligand-induced modulations of the signified which appear as taking part in couples of *relata* of many kinds of proteins, the enzymes being an example among them.

Fig. 18 (Koshland[93]) illustrates the potential relationship of inter-subunit and intra-subunit effects. The units of the model in the upper left-hand corner are represented as spheres, four identical subunits associated at the level of binding domains indicated by the letters p and q indicating that the contact of one sphere with two others implies biosemes composed of different amino acid residues which is imposed by the identity of amino acid sequence in the four spheres. If in one of the spheres, a ligand-induced change takes place, a change of relationship of the subunit contacts may occur and "affect the energy of the interaction and the geometric relationships with neighbouring units"[93].

The multichain protein of the upper right-hand corner in Fig. 18 is composed of polypeptide chains which are not identical (as in the case of hemoglobin).

The amino acid sequences are not identical as in the first case but the subunits derive from a common ancestor and the binding domains should be represented by pp′ and qq′. In the lower left-hand corner of Fig. 18 the case presented corresponds to a protein composed of catalytic and regulatory subunits. In this case in spite of their different activities the ligand-induced distortion will, as in the case of $\alpha^2\beta^2$, affect the subunits similarly. While in the model shown in the lower right-hand corner of Fig. 18 the catalytic and regulatory sites are situated on the same polypeptide chain the induced deformations induced either at the catalytic or at the regulatory site will also have an impact on the neighbouring subunits, in addition to the direct effect of the regulatory sites on the catalytic site. From the data analyzed by Koshland it appears that the most

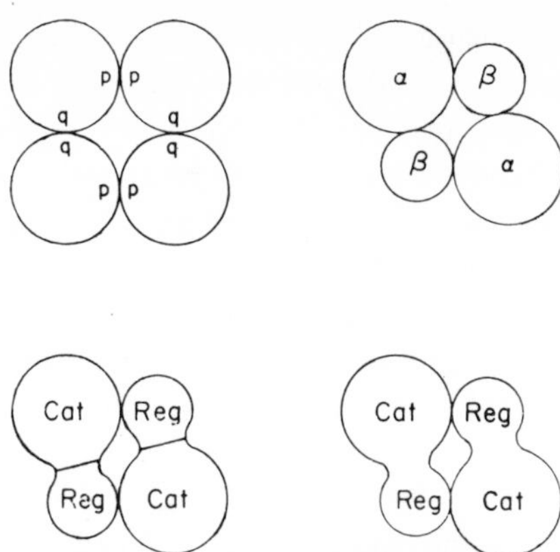

Fig. 18. Schematic illustration of the relation between intra- and inter-subunit effects. (Koshland[93])

frequent ligand-induced model corresponds to a distortion not only of the subunit to which the ligand is attached, but also of the whole molecule and of its subunits. This obtains for instance in the case of glutamic dehydrogenase or CTP synthetase.

In other cases, as in the example of the combination of oxygen with hemoglobin, the attachment of the ligand does not appear to change the molecule as a whole (symmetry model) but only the interaction between subunits.

In the case of phosphofructokinase there are indications that the model proposed by Monod, Wyman and Changeux[55] involving a simple subunit structure with transitions from one stage to the other involving simultaneous changes in all the identical subunits is relevant.

Examples of ligand-induced changes of signified

(i) Hemoglobin and oxygen

The hemoglobin molecule is approximately spherical and the four heme pockets are exposed at its surface. The difference between myoglobin and hemoglobin consists in the fact that the latter is a multichain syntagm in which a supplementary aspect of signification is acquired, the fixation of oxygen at the level of one heme pocket inducing a conformation change, the ligand fixation at one pocket influencing the ligand fixation at another pocket. The plot of percentage oxygenation *versus* P_{O_2}, a rectangular hyperbole in the case of myoglobin becomes

References p. 117

a sigmoid curve in hemoglobin. The result is that, while myoglobin remains loaded at relatively low partial pressures of oxygen, hemoglobin unloads its oxygen at these low pressures. The biosemiotic aspect of chain interactions, the oxygen fixation on a first heme resulting in a conformation change and interaction with the other chains, remains to be unravelled. The first oxygen attaches itself slowly, to a heme pocket, but the second and third bind more rapidly and the fourth several hundred times more rapidly than the first. A change is therefore significant for the other and this controls the sigmoid curve and its slope (Shulman *et al.*[94]).

(ii) CTP synthetase

This enzyme catalyzes the conversion of UTP to CTP with the utilization of ammonia and ATP. It is a dimer of two probably identical units. L-Glutamine can substitute for ammonia (Chakraborty and Hurlbert[95]). The purified enzyme can use glutamine instead of ammonia as nitrogen donor, if GTP is present as an allosteric effector (Long and Pardee[96]).

The enzyme shows positive cooperativity for the two ligands ATP and UTP while it shows negative cooperativity for glutamine and GTP. This interesting case has been analyzed by Koshland in a *Harvey Lecture*[97] and in a contribution to a more recent symposium[93].

To quote Koshland[93]

"Apparently the GTP, which activates the protein by accelerating the glutamine reaction, causes a conformation change which makes it more difficult for the next GTP to bind. Thus GTP has a positive effect on the active site within its subunit and a negative effect on the other GTP sites on neighbouring subunits. Moreover, the conformation induced by ATP is different from that induced by GTP since it causes positive cooperativity. Other evidence shows that the conformation change induced by UTP is different from ATP and GTP. Glutamine, in turn, causes other conformational states".

Different ligands therefore appear to introduce different changes of signified as a result of changes of significant, *i.e.* of conformational states. Fig. 19 illustrates the GTP induced significant changes.

(iii) Messenger hormones

Hormones regulate *existing functions*, they modulate the signified of a bioseme by increasing or decreasing the action already in existence. We may take cyclic AMP as the receptor in a couple of *relata* of which the other partner is a bioseme of a circulating nature, *i.e.* carried by a hormone accomplishing direct information (messenger hormones), such as epinephrine, glucagon, insulin, gastrin, secretin, parathyroid hormones,

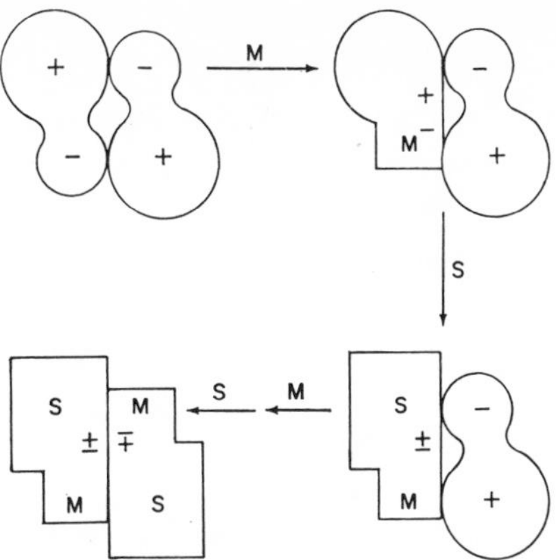

Fig. 19. Binding of substrate and modifier to a dimeric protein. Binding of modifier changes the shape of its subunit in such a way to make the binding of substrate occur more readily by altering the position of the attractive groups. It will also change the strength of subunit interactions with neighbouring chains. It can readily be seen that the ligand-induced conformational chains can vary with the ligand bound and the site to which it is bound. Thus, the modifier on one site may enhance binding of S but decrease binding of M as was observed in the case of CTP synthetase. (Koshland[93])

calcitonin, and the hormones of the anterior and posterior lobes of the hypophysis (in contrast to steroid hormones, thyroid hormones, growth hormone, whose receptor is in the genome. These hormones are sometimes called permissive hormones, developmental hormones, or maintenance hormones).

The mode of action of hormones is dealt with in Vol. 25 of this Treatise to which the reader is referred. From the biosemiotic viewpoint, the couple of *relata* exists between the hormone and the enzyme adenylcyclase and is followed by another couple between adenylcyclase and 5-AMP in the presence of ATP and Mg^{2+} and by a third coupling between cyclic AMP and a protein kinase. Modulations of the levels of cyclic AMP may take place by modulation of the activity of adenylcyclase, of phosphodiesterase and of both enzymes simultaneously (see ref. 98).

But the system may also lead to a second biosyntagm in which after impinging on the target cell at the adenylcyclase system specifically, a first messenger (hormone or neurohormone) affects the activity of adenylcyclase, bringing a change of concentration (increase or decrease) in the intracellular cyclic AMP. The information is thereby translated into cyclic

AMP levels and it is this level which modifies the metabolism of the cell and it acts as a second messenger. In white adipose tissue, or in liver, insulin decreases the level of cyclic AMP while glucagon increases it. This second messenger increases, in such cells as those of thyroid or of adrenal cortex or gonads, the rate at which they secrete their own hormones and liberate them in the circulation and may act in their target cells as "first messenger", a reason for giving up the denomination of "third messenger", in use for some time (see Robison *et al.*[98]).

The nature of the action of a second messenger such as cyclic AMP depends on the biochemical differentiation of the cell and of the couple of *relata* it contracts in this category of cells. It has been demonstrated that not all the effects of cyclic AMP occur in all tissues; on the other hand the response of a given metabolic syntagm to cyclic AMP may be different in different tissues, just as it may be affected in different ways by other forms of regulation.

In the case of the phosphorylase activation in muscle (a process influenced by a number of other factors) a cascade of events has been identified (literature in Robison *et al.*[98]). In skeletal muscle the activity of a protein kinase (kinase II, phosphorylase kinase kinase) is allosterically increased by cyclic AMP. In the presence of ATP and Mg^{2+}, kinase II catalyses the phosphorylation of phosphorylase *b* kinase which becomes activated. In the presence of ATP and Mg^{2+} and Ca^{2+}, this enzyme catalyses the phosphorylation of phosphorylase *b* into phosphorylase *a*, the most active form. In the adrenal cortex, cyclic AMP influences phosphorylase activation in a different way: it inhibits the activation of phosphorylase phosphatase. On the other hand, cyclic AMP exerts no effect on the phosphorylase system of brain.

The effect of cyclic AMP on glycogen synthetase seems also to involve the activation of a protein kinase, but in this case the phosphorylation, at least in some categories of cells, leads to a less active form of an enzyme.

Cyclic AMP stimulates protein kinases of a number of different tissues and the phosphorylation of a protein may be important in general in the effects of cyclic AMP.

In respect to the increase of protein kinase activity, it is now known that the protein kinases responding to cyclic AMP are composed of catalytic and regulatory subunits, the regulatory ones tending to inhibit. As cyclic AMP binds only to the regulatory subunit, it acts as a deinhibitor (see Robison *et al.*[98]).

21. The configuration of biosemes

We know enough of the biosemes involved at the integrative molecular level to define their configurations.

It must be pointed out that in the context of transient information, the transfer is insured by labile couplings. These couplings may be ionic interactions, hydrogen bonding or hydrophobic interactions. On the other hand the forces involved "must be sufficiently localized so that they can build up a defined geometric pattern" (Eigen and De Maeyer[99]). Very specific binding may be due, for instance, to specific types of H bonding as is the case in complementary base recognition[100–102], the pairs AU and GC being the strongest and most stable, probably due to one strongly polarized hydrogen bond (Eigen[39]).

Tertiary biosyntagms often present a configuration situating the non-polar groups (hydrophobic) inside and the polar ones outside. Non polar groups such as Val, Leu, Ile, Pro and Phe show the distribution and as stated above are active in the folding up of tertiary protein biosyntagms and structure stabilization. Non-polar biosemes are sometimes found at the surface of the biosyntagm where they play a role for instance in the formation of quaternary biosyntagms or in the formation of couples of *relata*. Charged side chains of amino acids are generally situated at the outside.

As stated above, the bonds involved in information processing are labile non-covalent bonds. However, the permanent storage of information for hereditary transfer does require the formation of covalent bonds.

For the purpose of processing in synchronic epigenesis, the permanent form of DNA is transferred into a transient form, mRNA, transient with respect to a time lapse greater than the duration of a generation. while it is relatively permanent compared with the time of incorporation of an amino acid into a peptide chain (Eigen and De Maeyer[99]). An example for readout of information is the code transfer from mRNA to an amino acid sequence, in which the transient form is represented

> "by the specific interaction between complementary codons somehow checked by the ribosomal enzyme, the processing of the read information is the storage into the new permanent form of amino acid sequence in the polypeptide chain." (Eigen and De Maeyer[99])

While the information is in the labile form, errors may be introduced during storage or readout for instance at the DNA level, producing point mutations.

References p. 117

Biosemes should not be considered as rigid configurations. Indeed a number of aspects of ligand-induced configuration changes may be identified, accompanied by changes of the signified which are determinant in integration and regulation at the molecular level.

III. DIACHRONIC MOLECULAR BIOSEMIOTICS

22. Diachronic molecular epigenesis

It is important to insist once more on the elliptic character of the expression "biochemical evolution", meaning changes *at the biomolecular level* along the phylogeny of *organisms*. The existence of this kind of change has first been documented by the present author[6,7] who has also used the expression "heteromorphic evolution" to designate, in the phylogenetic sequence of homologous biomolecules, "the acquisition of a modified component, with a lesser degree of isology"[13,103]. But, as noted by Dobhzansky[104]:

"We must remember that heredity, development and evolution are essentially epigenetic and not preformistic. We do not inherit from our ancestors, close or remote, separate characters, functional or vestigial. What we do inherit is instead, genes which determine the patterns of developmental processes."

Given a set of DNA (sequencing) biosyntagms and a given set of environmental conditions, the development is determined. This is a synchronous process, a derivation of order from the order provided by the parental genes, while the "heteromorphic" changes of biomolecules along the evolution of organisms is a diachronic process, resulting from changes at the level of the genome or of the regulation of the genome.

While the derivation of the phenotype from the genome is determined (in definite environmental conditions) "biochemical evolution" is epigenetic as a process as well, as Dobhzansky[104] remarks, as heredity and development.

In evolution at the biomolecular level (molecular changes in the phylogeny of organisms), what we consider is the diachronic (evolutionary) epigenesis of a synchronic epigenesis.

Therefore the present author[105] has suggested using the phrase "molecular epigenesis" (in evolution) or "diachronic (molecular) epigenesis" rather than "molecular evolution" or "heteromorphic evolution". As was stated above, while the information is shunting the metabolic traffic of synchronic molecular epigenesis and flows in the labile form, errors may be introduced during storage and readout, for instance at the DNA level with the production of mutations[106–109]. This mutagenic aspect is lacking in the case of amino acid sequences, unable to reproduce the error copy and subject as stressed in a previous section, to conformation changes involved in metabolic regulation.

When we analyze the simplest type of cell known to us, of a *Mycoplasma*, we find that it is composed of a set of 30 different kinds of biomolecules: twenty amino acids, five nitrogenous aromatic bases (adenine, guanine, uracil, thymine, cytosine), a fatty acid (palmitic acid), two sugars (α-D-glucose and α-D-ribose), glycerol and choline (after Lehninger[110]). If we now consider the molecular composition of a cell of *Escherichia coli*, we find that it is composed of 70 per cent of water and 30 per cent of dry matter. The latter is made of approximately the following numbers of molecules: 3000 proteins, 1 DNA, 1000 RNA, 50 carbohydrates, 40 lipids, 500 intermediates and building block molecules, 12 inorganic ions (percentage in weight: proteins, 15; nucleic acids, 7; carbohydrates, 3; lipids, 2; metabolites 2; inorganic ions, 1)[110].

In a way, the many forms of biomolecules found in mammals or in insects are the descendants by an epigenetic process which is repeated in all individuals in the course of their ontogeny, of those of their ancestors. To say that they are descendants cannot however be taken without a grain of salt. When we refer to DNA the diachronic molecular epigenesis is situated at the level of hereditary transfer and we are dealing with real descendants. When we refer to other biomolecules the epigenetic process consists of a molecular change at the level of the transient molecules (RNA, enzymes, regulators, etc.) involved in the biosynthetic process which is repeated in each of the descendants. Such changes concerned in modifications of biosynthetic pathways involve not only a configuration change of the biomolecules, but a radical change of biological activity. This change of configuration and of activity, which has been designated as "biochemical innovation" (Cohen[111]) as "functional epigenesis", as "functional radiation" (Florkin[105]) we may, in order to connote a biosemiotic aspect, designate as "*commutation*".

Innovation of molecular activities is, as we shall see, one of the ways through which the patterns of biosynthetic pathways are modified in the evolution of organisms.

Examples of biochemical specialization of a new type, due to molecular

References p. 117

commutation are easily identified in the course of animal evolution. Snakes, for instance, do not mix digestive secretions with their prey by a process of mastication. They swallow their prey, as coelenterates also do, after having injected it with a secretion initiating hydrolysis. In the least specialized forms, such as *Colubridae opisthoglyphae* a simple secretory tooth appears at the rear of the upper jaw and serves for the injection of a secretion whose function is purely digestive. In more specialized forms, this organ, following a decrease in length of the maxilla, approaches the anterior part of the buccal cavity and becomes an aggressive and defensive organ, as is the case in *Colubridae proteroglyphae* and even more so in *Viperidae*. The digestive origin of the secretion is further borne out by the presence in snake venom of such hydrolytic enzymes as proteases, peptidases, phosphatases, etc. The new specialization is expressed, biochemically, by the presence of molecular innovations, in comparison with the less specialized forms, *e.g.* by the presence of substances of high toxicity (Zeller[112]).

When we consider the nature and the frequency in evolutionary time, of point mutations at the level of the primary protein structures, we are obviously not studying evolution proper, any more than when we study morphogenesis, for instance. Diachronic (molecular) epigenesis (*biomolecular* changes along the branches of the phylogeny of *organisms*) is one of the many components of the origin of the modification appearing at the level of the population of organisms. They play a role in evolution, for instance on account of the more or less pronounced polymorphism of molecular structures in a species, one of the substrates of the effects of population dynamics. (For literature on the polymorphism of enzymes and other proteins, and its relations with point mutations, see pp. 80–170 of Manwell and Baker[108].) Diachronic molecular epigenesis at the nucleic acid and protein levels is the source of changes at the tertiary structure level and consequently of the resulting signified. It is integrated in the whole complex of multigenic systems along with all the other aspects, either morphological, physiological or biochemical, of the organism. The natural reality, when we consider animals or plants is the population within a species. Modifications appear from time to time at the population level by divergence in the gene pools of populations of organisms. This is evolution proper. We should bear in mind that the higher categories mentioned on the tree of phylogeny are abstractions, defined by the persistence, among the species that appeared successively, of certain common characteristics which do not evolve.

Wald[113] writes "living organisms are the greatly magnified expressions of the molecules that compose them." For the time being, in order to avoid the sin of naive reductionism as well as unnecessary misunderstandings, and in order to designate the biomolecular changes accompanying the evolution of organisms, we prefer to speak of *diachronic molecular epigenesis*, and even at the risk of repetition to state that this is admittedly an elliptic expression, meaning the diachronic epigenesis, in the phylogeny of organisms, of the synchronic metabolic systems insuring the biosynthesis of their biomolecules. In that way we leave open for the time being the nature of the relations which may obtain between molecular changes in phylogeny and the evolution of organisms.

23. Biochemical homology

The concept of biochemical homology was first introduced by Alfred Redfield[114]. He imagined the case of a naive biochemist unaware of the concept. Reviewing the material on the composition of organisms, he would undoubtedly set up two great kingdoms — the chlorophylophores and the chlorophylophages. The former he would characterize by the presence of chlorophyll, and he would associate with it the ability to perform the biosynthesis of a number of carbohydrates and nitrogenous compounds from simple inorganic substances. All the other organisms he would classify among chlorophyllophages, coming close to distinguishing animals from plants. Being beholden to the medical profession, the biochemist would pay most of his attention to the chlorophyllophages and he would separate them into those able to synthesize their full complement of amino acids, the aminoacidogens, and the others. The sub-kingdom of the non-aminoacidogens would be a source of worry, as such different creatures as pathogenic bacteria and the laboratory rat would be members of it. The biochemist would try to separate major groups and in line with his methods, he would pay much attention to body fluids. In one of these fluids, the blood, he would note the presence of a group of coloured substances united to oxygen in a way serving the respiratory transport of oxygen. The biochemist described by Redfield would distinguish one of the main phyla of his classification as Chromosanguinea among which he would have individualized the classes Cuproproteinata and Ferroproteinata. The

References p. 117

class Ferroproteinata would be divided up in three orders, one characterized by chlorocruorin, the second by hemerythrin and the third, the Protoporphyrina, by the particular porphyrin contained in all hemoglobins. As *Planorbis* contains hemoglobin in its blood, our biochemist would decide that it has erroneously been classified with the hemocyanin-bearers gastropods, he would consider the morphological basis of this classification as a matter of convergence in development and he would conclude that the study of morphological arrangements has led biologists to an erroneous classification of organisms. The case of the naive biochemist acutely portrayed by Redfield[114] remains of interest today, first because it embodies a number of traits which are still used as arguments against the concept of biochemical evolution, and, second, because there are still today a number of chemists who have not been informed of the nature of the concept of homology and who still behave like Redfield's naive biochemist*.

To quote Redfield[114], the naive biochemist, in reaching his foolish conclusions,

"... has failed to distinguish between the analogies and homologies in biochemical mechanisms, and as a result he has produced a classification as naive as that of Aristotle. It is well known that quite distinct chemical substances serve analogous functions in different groups of organisms. For example, phosphocreatine, which appears to play an indispensable role in the contraction of vertebrate muscle, is replaced by phosphoarginine in the muscles of many invertebrates. Hemocyanin, hemoglobin and hemerythrin serve a common function. They do so because they have certain physicochemical properties in common; not because there is any generic relationship between the chemical configurations which happen to give them these properties. The resemblance between the respiratory pigments of the cephalopods and the vertebrates has no more evolutionary significance than do the similarities of the eyes in these groups. Both are remarkable cases of convergence to serve a common function.

"If the distribution of chemical peculiarities among the natural groups of organisms is to be given an intelligent interpretation, we must first develop some satisfactory criteria by which to judge what resemblances are significant in an evolutionary sense and what are not. We need some body of chemical doctrine, similar to that which embryology has given to the morphologist, by which to judge our findings ...

"... Comparative physiology and biochemistry alike can contribute to an understanding of organic evolution only in so far as they observe old rules of the game — the distinction between analogies and homologies ...

* It may be noted that the analogous nature of phosphagens already clearly indicated by Redfield, has not reached the mind of a number of authors who still derive phylogenetic conclusions from their consideration, following the pathway of the naive biochemist described above.

"... much of physiology is by nature analogous — being the fortuitous combination of factors to serve a complex end. The morphological factors are the province of comparative anatomy; the chemical factors deserve treatment by a similar discipline. Before they can aid us in understanding the evolutionary problem, we must develop criteria for judging their true homologies."

Such a problem, in the formal order has been present in the history of the classification of organisms, in which it has long been recognized that resemblance and affinity are two different aspects. An earthworm and a lamprey are vermiform, show a metamerized segmentation of muscles, a sucking mouth, nephridia, a skin provided with many glands, a passage of genital products in the coelomic cavity. An earthworm and a butterfly show a similar situation of brain and ventral nervous chain, they both have a ventral ganglionary chain, a segmented body and a dorsal blood vessel. This is no reason to classify earthworms and lampreys, or earthworms and butterflies in the same systematic categories, any more than to situate in the same category all the organisms provided with hemoglobin.

The present author, who had the privilege of being a collaborator of Redfield, has endeavoured to derive biochemical homologies from the chemical properties of biomolecules when taking into account the descendance sequence embodied in phylogeny. He has distinguished the concepts of *isology* or chemical kinship and of *phylogenetic isolog, homology*, *i.e.* common ancestry (Florkin[103]). An example of this method of defining molecular homologies and identifying molecular epigenesis is found in the comparison of the chlorocruorin and hemoglobin in annelids.

Chlorocruorin, an oxygen carrier, is present in the blood of three families of polychete annelids: Flabelligeridae (chlorhemian spioniforms), Sabellidae and Serpulidae. The chlorhemians are sedentary polychete annelids, which are descended from the errant polychete annelids: these forms whose preoral lobe is not sunk into the first segment of the metasome, and which feed on floating plankton gathered by means of posterior antennae in the form of long palps bearing a ciliated gutter. They live in sand or mud and secrete a membranous tube covered with a fine layer of slime. The chlorhemians are spioniforms which have lost the dissepiments and even the external segmentation. Their blood is green and their palps are folded forward. Related to the spioniforms are the cryptocephalic annelids, having a preoral lobe sunk into the first segment of the trunk, but possessing furrowed appendages like those of the spioniforms. Sedentary and tubicolous, the Cryptocephalae comprise two

References p. 117

subdivisions: the Sabellariides, which although sedentary and microphagic, have retained an uneven number of antennae, and the sabelliforms which have an even number of antennae, and palps forming a multicoloured corolla. The sabelliforms are divided into the Sabellidae, having a mucous tube, membranous or cornified, and the Serpulidae, possessing a calcareous tube.

The blood of spioniforms other than the chlorhemians is coloured red by hemoglobin, as in *Sabellaria.* In the sabelliforms, chlorocruorin is the characteristic green blood pigment. All the Sabellidae so far studied contain it. Among the Serpulidae, blood of species in the genus *Serpula* contains both chlorocruorin and hemoglobin, whilst in the genus *Spirorbis*, one species, *S. borealis*, has blood coloured by chlorocruorin, another, *S. corrugatus* contains hemoglobin, and a third, *S. miliaris* has colourless blood. H. M. Fox[115] did not find chlorocruorin in the tissues or in the coelomic fluid of the forms having chlorocruorin in the blood. No doubt, in those forms which contain it, the synthesis of chlorocruorin is a variant of hemoglobin synthesis as it was present in their annelid ancestors possessing this synthetic mechanism. The isology of chlorocruorin and hemoglobin in Annelida was at the time concluded from the fact that the heme of chlorocruorin, or chlorocruoroheme, differs from protoporphyrin heme in only one small detail, the oxidation of vinyl group 2. As for the protein component of *Spirographis* chlorocruorin, its isology with that of other Annelida such as *Lombricus* or *Arenicola* was considered at the time as attested by the high molecular weight ($192 \times 17\,000$) the isoelectric point lower than that of horse hemoglobin and the amino acid composition. Compared to the globin of horse hemoglobin, that of the three polychetes was characterized by common traits: higher in cystine and in arginine; lower in histidine and lysine.

In the case of chlorocruorin it was therefore concluded that the chemical entity of chlorocruorin is very isologous to annelid hemoglobins and that it is present in divisions of systematics in phylogenetic relations with other divisions in which the blood contains hemoglobin. In this context it was possible to conclude that there is homology between annelid hemoglobin and chlorocruorin. It is an example of molecular epigenesis, at the level of the heme as well as of the globin (Florkin[116,117,13]).

This example of the recognition of homology has been retraced in detail in order to emphasize the dramatic change introduced in the field by the advent of molecular biology and of the development of configuration and structure studies on macromolecules.

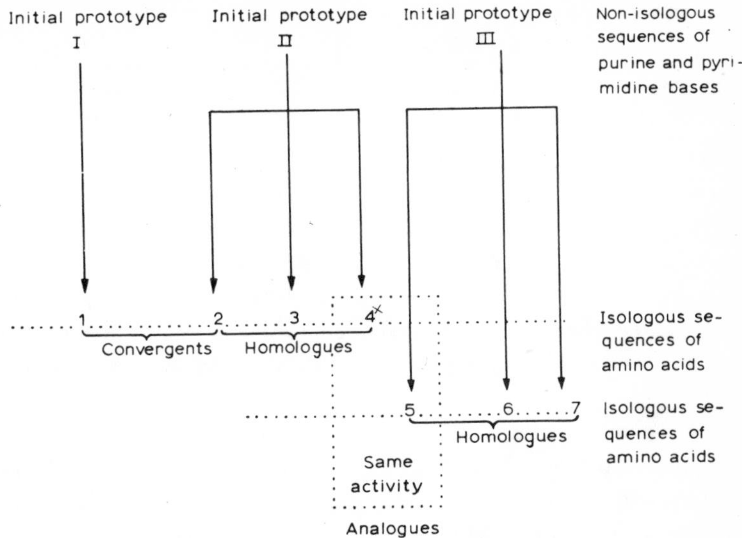

Fig. 20. Isology, homology, analogy and convergence. The roman figures designate sequences of purine and pyrimidine bases. The arabic figures designate sequences of amino acids. (Florkin[117])

In 1958, Crick[118] acutely remarked:

"Biologists should realize that before long we shall have a subject which may be called "protein taxonomy" the study of the amino acid sequences of the proteins of an organism and the comparison of them between species. It can be argued that these sequences are the most delicate expression possible of the phenotype of an organism and that vast amounts of evolutionary information may be hidden away with them."

The following year an influential book of Anfinsen, *The Molecular Basis of Evolution*[119], traced the results and promises of contemporary protein chemistry and genetics and "the possibilities of integration of these fields toward a greater understanding of the fundamental forces underlying the evolutionary process."

Anfinsen refers to "homologous proteins" but with the meaning unfortunately current among enzymologists, of a similarity of function (what we call analogy):

"The principal aim of this book has been to examine the basic principles underlying another possible method for the study of evolution. This method is based on the hypothesis that the individual proteins which characterize a particular species are unique reflections of the genes which control their synthesis. The examination of the chemistry of a series of homologous [meaning *analogous* as defined below] proteins is, of course, a purely phenotypic approach to the problem. Nevertheless, the evidence available to us, even at this early date, suggests that the structure of proteins may be a relatively direct expression of gene structure and that comparative protein chemistry may be a relatively direct expression of gene structure and that comparative protein biochemistry may furnish a qualitative view of genotypic differences and similarities."[119]

Anfinsen does not refer to the concept of molecular homology and he mainly emphasizes the concept of the persistence of genes through long periods of time. In 1962, the present author (Florkin[120]) proposed to accept *that a high degree of isology* in primary structure *permitted the recognition of homology, i.e. common ancestry.* This concept was linked to that of the persistence of genes through long periods of time as emphasized by Anfinsen.

In this terminology, the molecules or macromolecules showing chemical kinship are designated as *isologous.* Cytochrome, peroxidase, catalase, hemoglobin and chlorocruorin show isology as they all are heme derivatives. In the case of the hemoglobins of two men who are identical twins, the maximum degree of isology obtains. It is less pronounced when we consider the hemoglobins of a dog and a jackal, and still less if we consider those of a dog and of a horse. In these cases the protoheme is identical, but the degree of isology depends on the configuration of globin, the protein moiety of the hemoglobin macromolecule. If we compare a hemoglobin and a cytochrome *c*, the isology obtains only at the level of the heme moiety, the sequence of amino acids in the protein parts being non isologous. Isology is a chemical concept, while homology is a genetic one, implying a community of evolutionary origin.

Such a degree of isology as found in cytochromes of different organisms is incompatible with chance effects and points to the persistence noted by Anfinsen[119] of very ancient sequences of DNA. These sequences, reproduced through the ages with seemingly unwavering constancy, may be called *homologues* in the sense used by the biologist, *i.e.* connoting a *common origin and a common line of descent.* Homology at the molecular level is thus a genetic concept, *the emphasis being put on the common descent*, by replication of a DNA sequence of bases, of a common ancestor.

"The whole of our present knowledge suggests the great probability that *the very isologous primary structures of proteins* are replicas of very isologous base sequences of bases, they may be qualified as homologous[120]" (translation by author[117])

In his classical book on hemoglobins, Ingram[121] uses the term homology as we have just used isology.

"When examining the complete amino acid sequences of the α and β peptide chains of adult hemoglobin, one is immediately struck by the great degree of homology which exists between these two peptide chains, since almost 40 amino acid residues (out of 141) are identical"[117].

In a figure in which the sequence of amino acids in human α and β hemoglobin chains is represented, he indicates that "the chains are arranged for maximum homology" (what we call isology). He adds that it is this chemical evidence which leads to postulate that the two chains are perhaps derived from a common ancestor (what we call homology).

A degree of isology in primary structure greater than possible by probability is evidence for common origin in the case of nucleic acids and proteins (homology).

The slow accumulation of knowledge concerning the primary structure of a number of polypeptide chains has led to the development of several methods for ascertaining whether the degree of isology is an indication of homology. Fitch's method[122,123] is based on mutation distances among proteins and requires no preassignment of sequences. The drawback is the necessity of comparing either very long similar undeleted sequences or more closely related short sequences. It may happen that, in a long stretch of diachronic molecular epigenesis, homologous proteins, though of common ancestry, have lost the degree of isology which would have allowed, in previous stages of diachronic molecular epigenesis to recognize the homology, or may present only a short identical section. To deal with these possible aspects of homology, Dayhoff and Eck[124] have imagined a useful empirical test. They have collected data on the frequency of mutations in clearly homologous proteins and have derived a "mutation probability matrix" for amino acid sequences of 100 amino acids by imposing a large number (256) of point mutations on a protein of average amino acid composition. When homology is suspected but not absolutely clear from data on amino acid sequences, the sequences are tested by the Chi square method for correspondence to the matrix with less than 256 point mutations per 100 residues.

A number of examples show that homology takes its own full meaning when derived from complete amino acid sequences of polypeptide chains. When segments are used, comparison has more meaning if the sequences

References p. 117

compared represent identical areas of linear structures, or if they are disposed in corresponding regions in relation to extremities. While it appears hazardous to draw conclusions on the homology of whole polypeptide chains from primary sequences of fragments of them, it should be kept in mind that homology may in fact obtain only at he level of a section of such a chain. Partial internal duplications may have occurred during the evolution of the corresponding structural gene, as a consequence of unequal crossing over within one gene. This is the case in the example of the first 26 residues and the last 26 residues of ferredoxin, displaying remarkable similarity (literature in ref. 125). This also appears to be the case with the two segments of the light chains and the four segments of the heavy chains of γ-immunoglobulins (literature in ref. 125). On the other hand equal crossing-over may take place between two alleles in a heterozygote. This must have happened in the case of the non-α-chains of human *Lepore* abnormal hemoglobin, which are hybrids of δ- and β-chains (literature in ref. 125). Another example is provided by the 2-α chain of human haptoglobin (see Chapter 4).

Neurath[126] recognizes two categories of homologies: intraspecial homology and interspecial homology. As the tertiary structure of a protein is derived from its primary structure, conformities in tertiary structure may become useful as aspects of "conformational homology" (Neurath[126]).

Margoliash, who has devoted to the implications of the homology concept much penetrating thought, distinguishes *orthologous* homology from *paralogous* homology:

"Following a gene duplication, both genes may evolve different functions while descending site by site in the same phyletic lineage. Such genes may be called *paralogous* (para = in parallel) and an example might be human myoglobin and α-hemoglobin. Clearly, although myoglobin and α-hemoglobin may indeed be homologous (*i.e.* have a common ancestor) a species phylogeny of birds and mammals based upon myoglobin sequences from some species, and α-hemoglobin sequences from other species, would be an absurdity because the major dichotomy would not be between avian and mammalian species, but between those species whose myoglobin sequence was utilized. One must compare directly only those sequences whose genes have a lineage that precisely reflects, in a one-to-one fashion, the lineage of the species in which they are found. Such genes may be called *orthologous* (ortho = exact)" (Fitch and Margoliash[123]).

In the case of orthologous proteins, such as cytochromes, as the protein is an expression of the structure of a gene, when we know the primary structure of a number of proteins of a number of species, it may be possible "to read directly the record of the evolutionary history" (Margoliash and Fitch[125]) of these species, encoded in the proteins they synthesize.

If we had a complete knowledge of the primary structures of all proteins we would be in a position to derive by identifying homologies (common ancestry identified by a high degree of isology), the genealogic tree of these macromolecules and consequently the *system* of their diachronic interrelations. What we shall call the system is in fact as we shall see a genetic and biosemiotic system and the relations involved are situated in the ancestry, in the evolution of organisms. The system expresses diachronic relations, *i.e.* relations connected with evolution and situated in the past history of the organisms. Synchronic aspects shall in our terminology relate to the phenotype and genotype considered from the viewpoint of the living organisms.

If a primary structure, a sequence of amino acids for instance, is common to different cells and if it can be shown that the initial prototypes of these sequences are different (see Fig. 20) the common primary structure must be considered as convergent.

As noted by Margoliash and Fitch[125].

..."Common ancestry is not the only possible basis for similarity in amino acid sequence. Indeed, two proteins may be similar at the time they are examined not only because they diverged from a common origin relatively recently in their evolutionary history or because having diverged a long time ago they have followed largely parallel pathways, but also because having arisen from different ancestral origins they have tended to evolve to similar or identical functions in different lines of evolutionary descent, and have therefore acquired the degree of similarity of structure required by this similarity of functions."

As emphasized by Margoliash and Fitch[125] it is necessary, before assuming that apparently similar amino acid sequence means homology, to show that the similarities of primary structure (isology) *are greater than could occur by chance.* In the case of cytochrome *c*, the distribution observed could occur by a chance of less than 10^{-80}.

In order to define whether the significant similarities of primary structure are due to common ancestry or to functional convergence, statistical phylogenetic trees of the macromolecule concerned must be constructed.

There is a tendency among certain workers to derive the homology of enzymatic proteins from the homology of the active site. Nevertheless it is now clear, for instance, that the same active site structure has evolved independently in subtilisin BPN as in the homologous proteins chymo-

References p. 117

trypsin, trypsin and elastase, though subtilisin BPN is not homologous with them and the case is one of convergence of the active site (Wright *et al.*[127]).

Sometimes authors comparing the configuration of active sites in what they call homologous (in fact analogous) proteins do not pay attention to homology as meaning common descent and they do not go beyond the consideration of the active site and the consideration of its implications. A plea should be made in favour of a consistent reference to the terms isology, homology and analogy.

24. DNA in phylogeny

As mentioned earlier, the base sequence of DNA represents the permanent form of genetic information.

Replication of DNA, as described in a previous section of this review, takes place in all cell divisions, but its critical biosemiotic impact is recognized at the level of meiosis, preparatory to gamete formation. At the start of meiosis, homologous chromosomes are synapsed (synapsis, haploid number of chromosome pairs).

Each chromosome, in the next step splits longitudinally into two chromatids, each of the resulting four strands containing double-stranded DNA. The homologous chromatids break at one or several places and the partner strands exchange segments. In this succession of breakage and reunion, new chromatids are formed, containing parts of both parental types.

Afterwards two cell divisions take place, once to produce dyads containing the diploid set and once to segregate the chromosomes into four tetrads. These mature into the gametes. In this process the chromosome set has replicated once but divided twice and the final gametes are formed of sections of the chromosomes of both parents.

The replication which takes place and which was described is a very accurate biosemiotic accomplishment, an aspect which has been emphasized by Monod[128] as the "principle of invariance". This is not an induction but a reality recognized by factual evidence (for literature see Lewin[63]) which is also expressed in Ohno's phrase[129]:

"self-replicating DNA is for the preservation and transmission of genetic messages to the progeny ...".

In the process of replication, shortenings or deletions of portions of strands of the DNA may take place, and "point mutations" may occur at the resulting level.

Point mutations resulting from inaccuracies in the course of recombinations, crossing-over, etc. (literature in Dixon[106], Jukes[107], Manwell and Baker[108], Watts[136], etc.) are considered as a random process (Watson and Crick[130]).

We may repeat here the statement of Eigen[39], cited above

"Evolution appears to be an inevitable event, given by the presence of certain matter with specified autocatalytic properties and under the maintenance of the finite (free) energy flow necessary to compensate for the steady-state production of entropy."

The occasional tautomer of the bases would pair with the "wrong" complementary base. The tautomerically-induced mispairing is not the only possible mechanism of point mutation. A deamination may also be at the origin of mispairing. In the two occurrences, the result is the substitution of one purine by another, or of one pyrimidine by another, an interchange called *transition* by Freese[131] as distinct from *transversion*[132] in which a purine would be inserted in place of a pyrimidine, or *vice versa.*

The constant flux of point mutations represents the background noise of diachronic molecular epigenesis.

In conformity with the principle of invariance, natural selection is active in preserving in descent the base sequence of each gene, and forbids any mutation affecting the biosemes in the resulting protein. Tolerable mutations may affect secondary properties of an enzyme but not its biosemiotic nature. A new gene locus can only be acquired by the accumulation of forbidden mutations at the sequencing bioseme governing the synthesis of an active site, with the emergence of a "commutation" involving a radical change in activity as well as in structure. A commutation corresponds to a new gene (paralogous proteins) while point mutations correspond to the same gene (orthologous proteins).

"By duplication a redundant copy of a locus is created. Natural selection often ignores such a redundant copy and, while being ignored, it accumulates formerly *forbidden* mutations and is reborn as a new gene locus with a hitherto non-existent function." (Ohno[129])

Mechanisms for gene duplication are listed by Ohno[129] and by Watts and Watts[133]. The latter authors recognize eight mechanisms of duplicate gene formation classified in four classes with regard to the relative position of duplicate genes (literature in ref. 133).

References p. 117

Class I. Same chromosome arm; may be adjacent.
1. Chromatid interchange in ring chromosome (reciprocal)
2. Chromatid interchange in rod chromosome (non-reciprocal)
3. Interchange between homologous chromosomes (non-reciprocal)
4. Breakage–fusion-bridge cycle

Class II.
5. Single cross-over with an inversion in an inversion heterozygote
   (a) paracentric inversion
   (b) pericentric inversion
   (c) "terminal" inversion
6. Single cross-over in the common inverted region of heterozygote for partially overlapping inversions

Class III. Same chromosome, different arm.
7. Ring-rod heterozygote

Class IV.
8. Adjacent-I disjunction in translocation heterozygote

Cistron duplication (cistron = portion of nucleotide sequence providing the coding information for a single polypeptide chain) represents an information gain.

Bacteria possess much less DNA than any cell of a higher organism.

In phylogeny there is an increase in amount of DNA concurrently with the increase in information involved in the increased number of cell differentiations.

Evolution as a whole decreases entropy at the level of the biomass and increases order and the regulation of order. The more an organism is adapted to an ecological niche or the more it is liberated from the necessity of a niche, the more value it has in its environment.

Fig. 21 shows a great increase in DNA content as a concomitant to increased complexity of organization. This tendency is not universal.

For instance the cell nuclei of birds contain about one half the amount of DNA found in mammals or reptiles. Among Amphibia, the Urodeles present a relatively high DNA content.

It is implied in the model of Britten and Davidson[88] (Section 17) that the occurrence of sequence repetitions in the genome is of great evolutionary importance. The quantity of DNA, the number of times a given sequence is repeated and the precision of the repetition is variable from species to species. The families of repeated DNA show different degrees of similarity, "from perfect matching to matching of perhaps two thirds of the nucleotides" (Britten and Davidson[88]). In this model as well as in that proposed by Crick[84] the repeated sequences are scattered throughout the DNA and interspersed with non-repeated sequences. In the model of Britten and

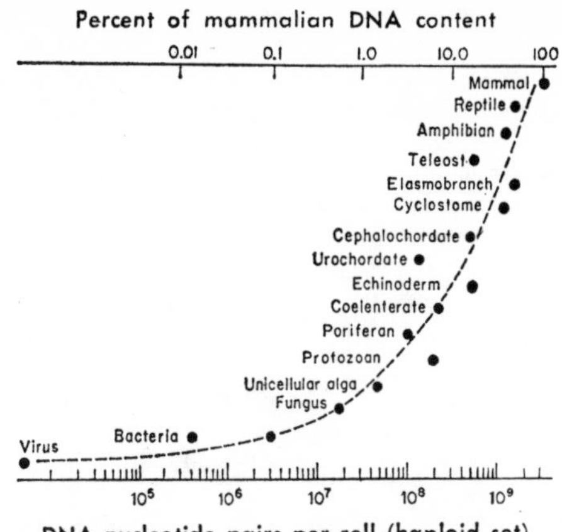

Fig. 21. The minimum amount of DNA that has been observed for species at various grades of organization. Each point represents the measured DNA content per cell for a haploid set of chromosomes. In the case of mammals, amphibians, teleosts, bacteria and viruses enough measurements exist to give the minimum value meaning. However for the intermediate grades few measurements are available, and the values shown may not be truly minimal. No measurements were unearthed for Acoela, Pseudocoela and Mesozoa. The ordinate is not a numerical scale, and the exact shape of the curve has little significance. The figure shows that a great increase in DNA content is a necessary concomitant to increased complexity of organization. (Britten and Davidson[88])

Davidson[88], DNA sequences must be specifically activated in order to be active in transcription (emergence of the signified).

Saltatory replication (Britten and Kohne[134]) can be the source of new regulatory DNA. Saltatory replication means the production of families of repeated DNA in relatively sudden evolutionary events which are initiated and terminated and do not continue through the course of evolution. With respect to the possible mechanism of saltatory replication, Britten and Davidson[135] consider (though no evidence is available) four possibilities.

"(*1*) erratic behavior of a DNA polymerase, perhaps caught in a closed short loop without adequate termination controls;

(*2*) geometric growth as a result of a series of duplications of an initially short duplicated sequence due possibly to unequal crossing-over:

(*3*) the excessive replication of some nuclear elements analogous to the episomes of bacteria (not yet observed in higher cells);

(*4*) the integration into the genome of many copies of viral genome or viral-borne sequence".

In saltatory replications, Britten and Davidson[135] consider the following (hypothetical) processes as necessary:

References p. 117

"(*1*) Many copies are made of a DNA sequence and appear in the germ cells of certain individuals.

(*2*) The copies are somehow integrated into the genome so that they are duplicated.

(*3*) Over a sufficient period of time they are disseminated throughout the population and its evolutionary descendants. Dissemination could result from their association with a favorable genetic element or simply because of their multiplicity.

(*4*) Individual sequences become scattered among many chromosomes and interspersed intimately along the length of the DNA of the genome.

(*5*) The growth of the family of repeated DNA is eventually terminated or controlled. Subsequently individual members of the sequence families would diverge from each other through base substitutions".

"All that is required for regulatory function in this model is sequence complementarity (translocation of members of the same repetitive sequence family to integrator and receptor positions). Almost any set of nucleotide sequences would suffice. *The likelihood of utilization of new DNA for regulation is thus far greater than the likelihood of invention of a new and useful amino acid sequence*, since for the latter case great restrictions on the nucleotide sequence exist." (Britten and Davidson[88]; italics as in original text).

It is seen that Britten and Davidson, in the model they propose, consider that the integrator–receptor matching is implemented through activator RNA and not by a protein, and that the binding of activator RNA to the receptor genes is insured by a degree of sequence isology which is not perfectly complementary. In this situation, the divergence of regulatory sequences may be reversible.

"If the degree of complementarity required for binding between activator RNA and receptor sequence is fairly low then a reasonably good probability would exist for a subsequent base change to restore the complementarity lost by an earlier change" (Britten and Davidson[88]).

In the model, any DNA could be used in the construction of a new control system and natural selection could reversibly affect the integration of individual producer genes into batteries of smoothly changing patterns of integration among many sets of producer genes and allowing for direct adjustment by natural selection. To quote Britten and Davidson,

"The properties of the model regulatory system suggest that *both the rate and the direction of evolution (for example towards greater or lesser complexity) may be subject to control by natural selection*."[88] (italics in original)

R. L. Watts[136] raises an objection to the view according to which any base sequence can provide the material of a new control gene. She stresses that the base sequence must be "sufficiently different from those of preexisting control genes to avoid confusion with them".

As noted by Ohta and Kinamura[137] biochemical evolution consists of a sequence of events in which originally rare molecular mutants (DNA changes) spread into the species.

This concept is often expressed in the phrase "accepted mutations". The incidence of polymorphism of enzymes and other proteins, as detected by electrophoresis, in relation with population dynamics, is one of the origins of this "acceptation" (see Manwell and Baker[108]).

If the information transmitted in descendance in the sequence of phylogeny increases as a result of the multiplication of DNAs in commutated forms or in redundant or polymorphic aspects, the number of possible alternatives or choices is also increased. What determines the acceptance of some among the many mutations is natural selection, extracting more order from order by differential reproduction. Natural selection has the signification of an antichance factor and it should not be considered as a simple sieve, retaining useful mutants and disposing of harmful ones. To quote Dobzhansky[138]:

"Mutation and mendelian recombination, the latter being a corollary of sexual reproduction brings forth an immense variety of different genotypes, the number of which are in higher organisms almost as great as the numbers of individuals conceived. Mutation and recombination link with selection to form a cybernetic system that maintains or enhances the internal teleology, that is the harmony between a living species and its environment. Disruption of this system results in eclipse or extinction of the species."

This effect of natural selection considered at the molecular level commands the meaning of the choice made in this information transmitted and is related to its intensive aspect corresponding to the biochemical relevance of the selected molecular character. It must be emphasized here that, as noted by Blum[23] selection must be considered

"on the basis of the relationship of *all* the characteristics which constitute the phenotype to multiple aspects of the environment."

Natural selection has a biosemiotic aspect and is a form of information, as by reproductive differentiation of organisms it modifies the process of order-to-order transfer by routing a choice between several possible ways: *i.e.* by accepting one among a number of possibilities (yes or no) which could be expressed in *bits*. The choice of selected mutations is a choice of molecular patterns accepted in preference to others. If we could be fully informed of the course of microevolution as ruled by natural selection, and above all of the mutations which were not accepted we would be in a position to express in *bits* the course of accepted mutations. Natural selection may therefore be considered as introducing information at the molecular level. In this presentation, as said at the beginning of the Chapter, the essential aspect of biochemical evolution is considered as the diachronic molecular epigenesis, *i.e.* as changes with descent, of the information described in its intensive aspects, as they modify the control, regulation and modulation of the effectors active in the traffic of catenary metabolic biosyntagms.

25. tRNA in phylogeny

(literature in refs. 107, 124, 139–141)

As described above, tRNA is a highly significant biosyntagm, involved in different couples of *relata* of a sequential biosemiotic nature (amino acid-synthetase–tRNA; tRNA–mRNA; tRNA–ribosome).

While we only have indirect knowledge concerning the sequence of nucleotides in DNA, we know the structure of a number of tRNAs.

tRNA is transcribed from a very small part of the genome (0.02 per cent of the DNA in *E. coli*) which corresponds to the presence of approximately 40 types of tRNA in a cell of *E. coli.* If we compare the composition of the total tRNA in bacteria, plants and animals, we find that the content of G+C does not vary, remaining around 60 per cent. But if we compare specific molecules of tRNA in a species, we see that their composition varies.

It is now possible to compare the primary structures of different tRNAs of the same organism. This can be done (as in the case of proteins) by the insertion of gaps, which show that deletions of genetic material have taken place in phylogeny. The existence of a common origin for the three different tRNAs of yeast has been stated by Jukes[140] who suggests that a common archetype of the cistrons for serine and tyrosine tRNAs probably underwent gene duplication at a time subsequent to the gene duplication which separated the archetypes of alanine tRNA from the common archetype of serine and tyrosine tRNAs.

The tRNAs of *E. coli* differ more from the tRNAs of Eucaryotes than these differ one from another. This indicates that the phylogenic branch leading to bacteria has separated very anciently to the one leading to Eucaryotes.

We have described, in Procaryotes, the transitory biosyntagm fmet–tRNA recognizing the triplet AUG and putting methionine in the first place on the sequence of bacterial proteins. As such a biosyntagm has not been identified in Eucaryotes, it is an indication of its emergence after the separation of the bacterial branch (Jukes[140]).

The fact that fmet–tRNA is found in mitochondria is in favour of the theory of a colonisation of the cytoplasm of Eucaryotes, after aerobic life began, by bacteria which became mitochondria.

26. Primary structures of orthologous proteins as records of evolutionary history

As we still are unable to determine base sequences in DNA, the primary structure of proteins remains our access, indirectly, to these base sequences. It appears utopian to believe that we shall often be able to recognize homology by isology, at long phylogenetic distances, any more than we can recognize morphological homologies at the same remote distances.

Some biomolecules may have lost most of the traces of their isology, while being homologous, and it appears that it is safe to establish comparisons between not too far away sections of phylogeny. A table of primary sequences of the α- or β-chains of primates (as established by Hill and Buettner-Janusch[141]) clearly shows the homology of these chains in the series, while if we compare hemoglobins of very distant forms of phylogeny, recourse to isology to establish homology becomes useless.

In closely related species, homologous proteins show greater similarity than in more distantly related species.

Phylogeny itself becomes hazy when we compare very distantly related species and loses a great deal of its content. It is easy to retrace the phylogeny of vertebrates, but the greatest uncertainty persists, for instance, concerning the derivation of Chordata. There are exceptional cases in which the homology of biomolecules can be identified at very long phylogenic distances and such a case obtains with such an orthologous molecule as cytochrome *c*.

Cytochrome c. (literature in refs. 107, 124, 142–145, 185; on cytochrome *c* in plants, see Chapter 2 of this volume)

In vertebrates, its protein constituent is always composed of 104 amino acids. To this chain, the heme component is attached by two thioester bridges joined to residues of cysteine separated by two amino acids. The cytochromes *c* of vertebrates are acylated at the N-terminal end. This is not the case for the cytochrome *c* of yeast whose polypeptide chain is made up of 108 amino acids while we find 107 residues in the cytochrome *c* chain of *Neurospora.*

Cytochromes *c* can be compared by vertically aligning them at the level of their connection point with heme. The degree of concordance in primary sequence is greater if the species are less distant in phylogeny.

The cytochromes *c* of pig, sheep and cow are identical. Those of man and *Macaca* differ by one single residue, those of man and horse by 12, those of man and dog or ox by 10; of man and rabbit by 11. The cytochrome *c* of chicken differs from that of a number of mammals by 10–15 amino acids and those of fishes differ by 17–21 residues from mammals and birds. But to derive from these figures a knowledge of the number of mutations which have taken place since the separation of the species concerned from a common ancestor is illusory if no correction is allowed for the fact that several substitutions may have taken place at each point.

From the data obtained with a number of cytochromes *c*, as well as with proteins which perform a function which remains established through the phylogenic series concerned, it results that the rate of evolution by point mutations is constant (Dickerson[145]).

This is what we may call a constant flux (constant for each protein species) of diachronic molecular epigenesis.

It appears that the rate at which amino acid substitutions have occurred in different orthologous proteins studied so far, is consistent with a neutral allele-random fixation model utilizing reasonable estimates of neutral mutation rates.

There also appears to exist a significant correlation between the number of synonymous codons for each amino acid and its respective occurrence frequency in the proteins so far studied (Kimura[146]; King and Jukes[147]).

The album compiled by Dayhoff[124], to which the reader is referred, provides an admirably ordered collection of data on protein molecules, most of which have functions which have not changed in phylogeny (orthologous proteins).

These chemical units nevertheless present the picture drawn at the fringes of duplication and replication process, the form of molecular epigenesis which now appears to us as the background noise of commutations, which constitute the essential matter of which biochemical evolution is made in the sense of unleashing new evolutionary trends.

It is a most important concept that those functionally not-evolving molecules allow us to recognize the pattern of the phylogeny of organisms. The phylogenetic trees derived by Dayhoff and her collaborators are essentially based on the nature of the genetic code, a knowledge which, it is important to stress, is not derived from the consideration of phylogenetic aspects but from experimental work with enzymatic protein systems derived from procaryotic cells. Admittedly in some stages of the work, the "biological affinities" *i.e.* classical phylogeny has been taken into account. This is no exception to the general practice of establishing the phylogenetic tree, except if we derive a confirmation of a phyletic derivation when the chemical aspect was derived from this very phyletic derivation.

27. Indirect homology

Given a set of genes and a given environment, the development is determined. This is a synchronic epigenetic process derived from hereditary factors while evolutionary epigenesis is a diachronic process, situated in the flow of time. We have designated as diachronic molecular epigenesis (meaning evolutionary diachronic epigenesis at the level of biomolecules) the changes of informational macromolecules in the course of the phylogeny of organisms. When we consider the micromolecular constituents we are confronted with a more complicated situation.

In DNA, RNA and proteins, the sequence of building blocks carries the biological information corresponding to a segment of the genome, and colinear in each of this kind of informational molecule while micromolecules derive from a more complex interplay of sources of biological information. Each of the micromolecular substances derives from whole assembly lines of enzymes. Each of these is derived from one or several structural genes and these are likely to be controlled by genetic influences. This implies the participation of a large number of genes, even if smaller than in the case of a morphological trait.

The whole of our knowledge suggests the great probability that the very isologous primary structures of proteins are replicas of very isologous sequences in nucleic acids and that, like these sequences of bases, they may be qualified as homologous. The term may likewise apply to chains

References p. 117

of homologous protein biocatalysts, and also to the results of a biosynthesis catalyzed by a homologous enzyme chain. This usage makes clear the distinction between isology and analogy. ATP is isologous in all cells, but it is not always homologous, being, for instance, the product of the action of one chain of biocatalysts in glycolysis, and of another in oxidative phosphorylation. Bile acids, on the other hand, are homologous in all vertebrates, as they are biosynthesized by pathways catalyzed by homologous enzyme chains.

In substances displaying indirect homology, called *episemantides* by Zuckerkandl and Pauling[148], this kind of homology can only be recognized if we know that the biosynthetic pathways are the same (Florkin's *biochemical orthogenesis*, see p. 78) and that the enzymes involved are homologous. There are, of course, many examples of convergence in the case of isologous molecules other than nucleic acids and proteins in which a high isology reveals homology, *i.e.* common ancestry. Isologous microbiomolecules can eventually result from different biosynthetic pathways, not catalyzed by homologous enzyme chains and consequently non homologous. It is true, as Wald[113] wrote, that if the probability is remote of evolution producing the same organ twice (and we may add the same primary protein structure) the probability is greater for the convergent production of non-nucleic acid and non-protein molecular structures. This is certainly true and attention is drawn to these difficulties by the distinction between direct (nucleic acids, proteins) and indirect (other molecules) homology. When indirectly homologous biomolecules are considered, what is really homologous is the sequence of proteins and the proteins (enzymes) composing the biosynthetic assembly chains.

Benzoic acid is not homologous in all its biological localization. It may be derived from shikimic acid, but in other cases it is the endpoint of another enzymatic assembly chain, by cyclisation. Lysine, in certain microorganisms is derived from α,α'-diaminopimelic acid while in others it derives from α-aminoadipic acid. Porphyrins and prodigiosin, though isologous, are not homologous. In the case of micromolecular constituents, isology is not, even if complete, a sign of homology, which can only reside in the homology of the enzymatic assembly chains of the molecules concerned.

28. Analogy

The term *analogous*, as proposed by Redfield[114] is applied to biomolecules which show the same biochemical activity. The luciferins are analogous, though they are not always homologous (McElroy and Seliger[149]). The oxygen carriers (hemoglobins, chlorocruorins, hemocyanins, hemerythrins) are analogous though not always homologous.

A most interesting case of analogy is displayed by the proteins of a parasite and its host, which while not homologous, are not distinguished by the mechanism of elimination of the *non self*. If we consider, in glycolysis, the enzymatic step producing two triosephosphates by scission of a molecule of FDP, we find that two types of aldolases exist, which are not homologous (Rutter[150]). They are analogous biomolecules.

In bacteria, yeasts, mycetes and blue-green algae, we find type II, while green algae, Protozoa, plants and animals have type I. The two types coexist in *Euglena* and in *Chlamydomonas*. Harris (cited by Lai and Horecker[150a]) has isolated and sequenced the cysteine peptides and has found no isology between yeast and rabbit muscle aldolase, the two classes of aldolases appearing as having arisen by convergent evolution as suggested by Rutter[150].

Phosphagens are analogous as a reserve of labile phosphate used to regenerate ATP under conditions of heavy demand. Phosphoarginine, phosphocreatine, phospho-opheline, phosphotaurocyamine for instance have no general relation of common ancestry and are not homologous as defined above, which explains the misgivings which have in the past, accompanied phylogenetic theories derived from the consideration of phosphagens. On the other hand, some phosphagens are homologous: this is the case for phosphohypotaurocyamine and phosphotaurocyamine or for phospho-opheline and phospholombricine. The extracellular proteinases of mammals are non-homologous with the proteinases of bacteria and moulds, the presence of serine and of histidine at the active sites representing a case of convergence (literature in Dixon[106]).

29. Free-energy sources in phylogeny

The energy transformations in organisms have been shown to take place at the molecular level in a very complex network of molecular events resulting in the generation of free-energy sources, mainly in the form of

References p. 117

molecules of ATP. Free energy thus provided is used in endergonic biosynthesis, in setting up and maintaining concentration differences, etc.

To quote Broda who has written review papers on the subject[151,152], the order in phylogeny of the ATP-generating processes may be retraced as follows.

"... (1) fermentation → (2) photo-organotrophy [production of free energy from organic substances] → (3) photolithotrophy [production of free energy from inorganic substances] → (4) phytotrophy (plant photosynthesis) → (5) oxidative phosphorylation (essentially respiration) ...

..."The transition (1) → (2) was forced by shortage of substrates for fermentation, (2) → (3) by shortage of organic carbon, and (3) → (4) by shortage of inorganic reductants other than water. All energy used up to now by organisms is ultimately of solar origin."[152]

30. Colinearity of the phylogeny of biomolecules and of their biosynthesis

The concept we can now express as the diachronic epigenesis of catenary metabolic biosyntagms by terminal or lateral extensions, or terminal or lateral deletions, at the level of the limited number of catenary biosyntagms recognized in organisms, was defined by the present author[6,117,153] under the term "biochemical orthogenesis".

The concept was based on a consideration of the examples of amino acid catabolism and of purine catabolism pathways, showing in phylogeny, successive terminal deletions on a common pathway in the latter and terminal extensions on the common pathway in the former.

In the vocabulary of the present review, the term "biochemical orthogenesis" means that the diachronic epigenesis of catenary metabolic pathways resulted from variations on the recurring lines of what we have called the primary metabolic biosyntagms (not to be confused with the central metabolic biosyntagm), the variations consisting of terminal extensions or deletions or in lateral extensions or deletions. The idea of "orthogenesis" was not to imply an evolution in a straight line but to recognize an evolution of metabolism situated at the level of variations on the lines.

The idea was that, in the emergence of new biomolecules, recourse was commonly made to variations of existing catenary biosyntagms rather than to the emergence of entirely new catenary biosyntagms starting from the small molecular material of biosynthesis (*e.g.*: C_2, C_1, CO_2).

The term "biochemical orthogenesis", was the subject of a discussion covering several pages of the influential book by Blum, *Time's Arrow and*

Evolution[23]. Blum considered the term as "unfortunately chosen". With respect to the notion of "recurring patterns" (which had been expressed by Kluyver), Blum notes that "only certain ones are reproduced by living organisms and these appear to be built up according to a strictly limited number of structural types. If we examine the limitations and search for the origin of chemical types, we may find, as Florkin has pointed out, that biochemical homology is more fundamental than morphological homology, and may require a more basic approach. The fact that biochemical compounds can be arranged in relatively few series of homologous structural categories, seems evidence enough that there is a general limitation to certain basic patterns."

... "once living organisms acquire a basic reproducible pattern, they are likely to repeat it."[23]

The progressive unravelling of an increasing number of catenary metabolic syntagms has afforded a wealth of evidence for the concept of the recurrence

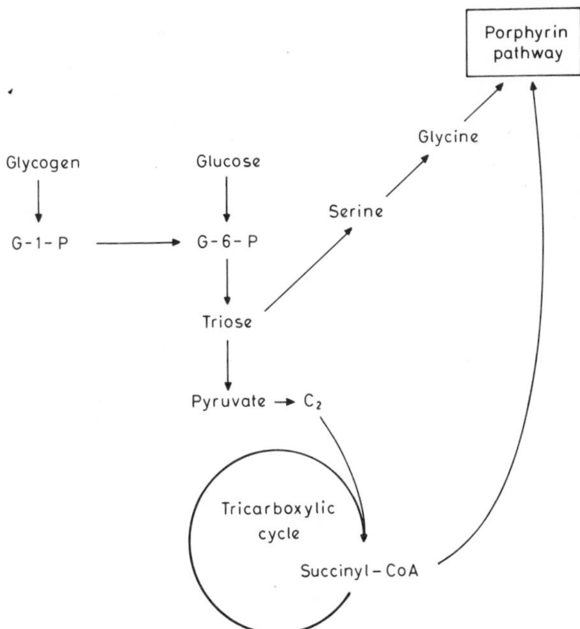

Fig. 22. Pathway of porphyrin biosynthesis.

of a restricted number of pathways as of their diachronic epigenesis by terminal or lateral extensions, or by terminal or lateral deletions. This, as pointed out by Florkin[117] is at the same time, on the one hand an aspect of the unity of organisms, and on the other, the rationale for their diversifications:

"... the organisms in the course of their diversification and of the realization of their adaptation have not indefinitely multiplied the number of biosynthetic pathways, many of which have grown according to the system of extensions on common biosynthetic lines. This aspect, which of course displays many exceptions, is an aspect of "biosynthetic economy" which has a definite impact on the phylogeny of many biosynthetic systems and of their products."

This formulation corresponds to the concept of a colinearity (expressed with the usual reservations) of the biosynthesis of the biomolecules and of their phylogeny. This colinearity was strikingly formulated by Granick in his *Harvey Lecture*[154] when he wrote, on the basis of a study of porphyrin biosynthesis:

"Biosynthesis recapitulates biogenesis."

(a) Porphyrin pathway (literature in Bogorad[155] and in Vol. 19 of this Treatise)

The biosynthesis of porphyrins starts at the level of glycine and of succinyl-CoA derived from the Krebs cycle (see Fig. 22). The δ-amino levulinic acid and the porphobilinogen are situated on this pathway.

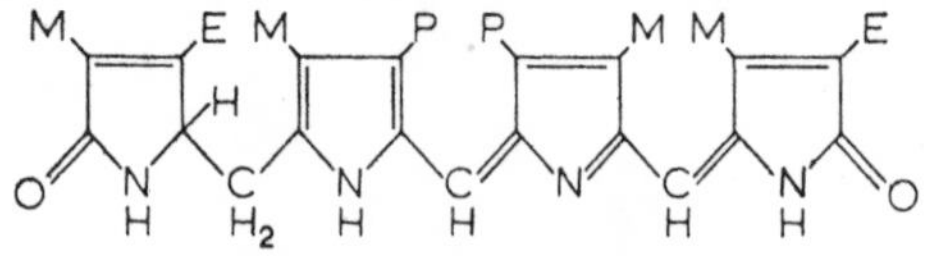

Mesobilierythrin

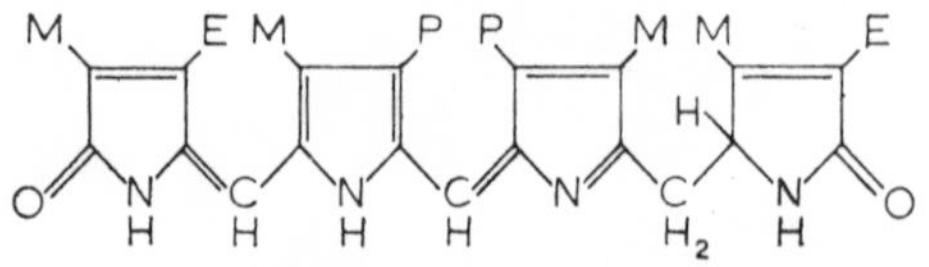

Mesobiliviolin

Fig. 23. Mesobilierythrin and mesobiliviolin (M = methyl group; E = ethyl group; P = propionyl group).

By insertion of iron into the protoporphyrin nucleus, we obtain what Granick[154] calls the "iron branch" of the biosynthetic pathway. Protoporphyrin is also the starting point for the biosynthesis of chlorophyll (the "magnesium branch" of Granick) as well as for the biosynthesis of heme.

A cell capable of photosynthesis contains at least one chlorophyll and at least one yellow pigment. In addition it often contains a phycobilin. The chief pigment in photosynthesis both in algae and in the higher plants, is chlorophyll *a*. In the photosynthetic bacteria, on the other hand, we find a different chlorophyll, bacteriochlorophyll. Whereas in green plants the chloroplasts contain chlorophyll *a* and chlorophyll *b*, in the algae we find a number of combinations:

$$a+b,\ a+c,\ a+d,\ a+e$$

In addition we sometimes find a phycobilin. The phycobilins, which are soluble in water, are proteins combined with a chromophore belonging to the class of bile pigments. The phycoerythrins are predominant in the red algae and the phycocyanins in the blue-green algae.

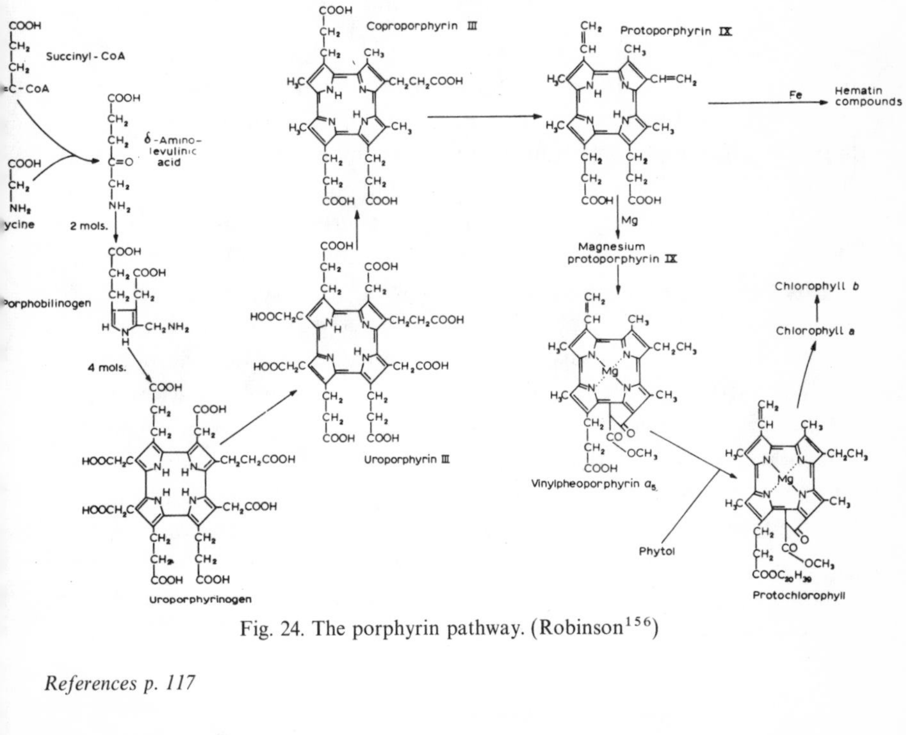

Fig. 24. The porphyrin pathway. (Robinson[156])

References p. 117

The chromophore of the phycoerythrins, phycoerythrobilin, is identical to mesobilierythrin; the chromophore of the phycocyanins is mesobiliviolin. Various phycoerythrins are found in algae, which differ in the structure of the protein moiety; *R*-phycoerythrin is the most common and is found in the *Rhodophyceae*, whilst *C*-phycoerythrin is present in the *Myxophyceae*. Among the phycocyanins, *R*-phycocyanin is present in the *Rhodophyceae* and *C*-phycocyanin in the *Myxophyceae*. The phycobilins serve to absorb light and transmit energy to other systems, notably by the chlorophyll system. Phycoerythrins and phycobilins appear to result from a lateral chain of the porphyrin pathway starting at the level of hematin compounds (see Bogorad[155]).

Plants are able to effect the synthesis of porphyrins along the "iron branch" and along the "magnesium branch" whilst in animals the latter is lacking. However, animals have particularly developed the "iron branch" as far as the biosynthesis of the compound of heme and globin, hemoglobin, is concerned. The biosynthesis of hemoglobin is sometimes observed in plants, for example in the root nodules of legumes. In animals, the presence of hemoglobin other than in the blood has often been demonstrated.

(b) Terpenoid and steroid metabolic biosyntagm

This path (Fig. 25) starting with acetyl-CoA leads to mevalonic acid, the starting point of the biosynthesis of the different terpenes which are constituents of the essential oils of plants. These constituents play an important role in the relationships between plants and insects. They are compounds composed of isopentane units and contain 5, 10, 15, 20 or more carbon atoms and are called respectively hemiterpenes, mono-, sesqui-, and di-polyterpenes. From the material which cannot be distilled in steam, it is possible to obtain by solvent extraction a series of other substances containing 20, 30, 40 or more carbon atoms, and belonging to the group of diterpenes (*e.g.* the resins), triterpenes (*e.g.* the saponins), tetraterpenes (*e.g.* the carotenoids) or to the polyterpenes (*e.g.* rubber). Moreover, a whole series of organic compounds synthesized by plants are related to isopentane, of which they contain varying numbers of units in their structure. Among the isoprenoids are the irones. Many monoterpenes are found in plants and, in general, but not always, one can consider their formula as being based on two isopentane units joined in head-and-tail union. The cyclic monoterpenes and sesquiterpenes can be considered to result from the rolling up of the same chains.

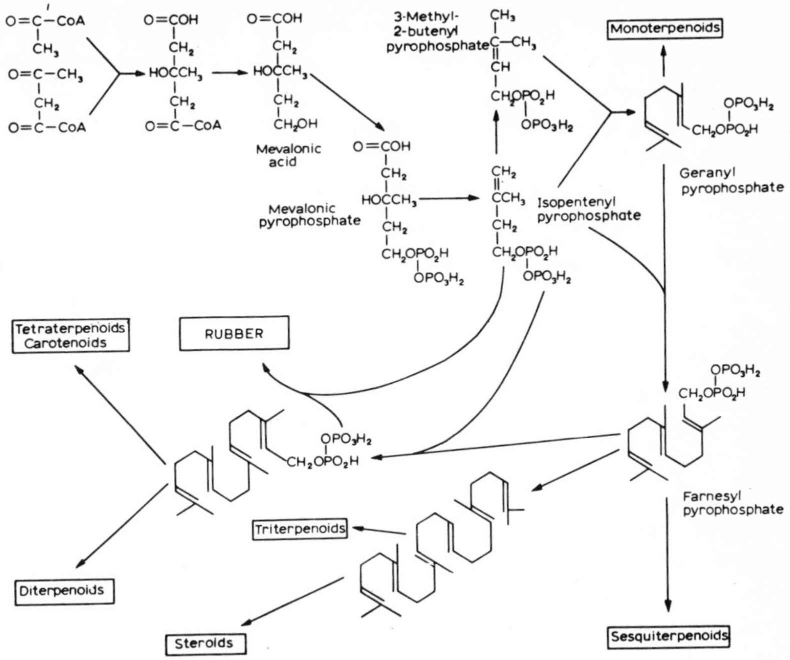

Fig. 25. The terpenoid and steroid pathway.

Certain diterpenes may be regarded as containing four isopentane units in head-to-tail union. This is the case with phytol and vitamin A; others have an irregular arrangement. Regarding tetraterpenes, plants are able to synthesize carotenoid molecules, while animals are only able to modify them, for example, by oxidation. Astaxanthin, a carotenoid usually found in crustaceans, is one such oxidation product.

In mammals, birds and certain amphibians, the ingestion of carotenoids in the food results in the absorption of carotene in the intestine, the extent of absorption depending on the greater or lesser activity of the intestinal carotenase, which converts carotene to vitamin A. As a result, the reserve fats become more or less saturated with carotene. Man and other primates absorb carotenoids in general, as does the frog; other animals exercise a selective absorption. For example, the horse and the cow selectively absorb carotenoids and store them without alteration; birds and fish show a preference for xanthophylls. However, birds and fish modify one of the ingested xanthophylls, lutein, and the products of the oxidation are deposited

References p. 117

in the feathers in the case of birds, and in the skin in the case of fish. The carotenoid structure appears to be connected in a general way with the function of photoreception.

The most primitive type of photoreception, lacking any differentiated photoreceptors, is the type called dermatoptic, which is found in primitive types of organisms, up to the amphibians, and also in plants. The maximum sensitivity of this dermatoptic function is in the ultraviolet part of the spectrum, around 365 mμ; it is operative in photokinetic processes involving tropisms towards light in the above-named primitive types. Now, in a number of cases, photoreceptors have evolved secondarily and developed new kinds of receptor molecules adapted to the light from the sun and the sky.

All these substances belong to the carotenoid group. In plants, phototropic bending depends on the properties of carotenoids such as xanthophyll in *Avena*, or β-carotene in the sporangiophores of *Phycomyces*. The orientation of an animal depends on visual photoreception and requires the presence of other carotenoids showing the same kind of adaptation to sunlight and having a maximum absorption at around 500 mμ. This development is due to the ability of plants, mentioned above, to change some of the carotenoids into vitamin A. There are two types of vitamin A: vitamin A_1 and vitamin A_2.

Let us briefly consider the general system of photoreception in the eye of land vertebrates including birds. The pigment of their retina is rhodopsin, a rose-coloured carotenoid–protein complex. In aqueous solution, its absorption spectrum consists of a single broad band with a maximum at 500 mμ. In light, it is bleached to orange and yellow pigments, and in the process the carotenoid retinene I is liberated. The latter substance has never been found anywhere except in the retina. Its spectrum in chloroform consists of a single band with a maximum at about 387 mμ. In the retina, the mixture of retinene I and protein reverts to rhodopsin and, in addition, retinene I is converted to vitamin A_1, which in the intact eye also reunites with protein to form rhodopsin. This system is not only present in the eye of mammals and birds but also in that of some invertebrates, such as the squid *Loligo* and the crayfish *Cambarus*.

If we consider the system in the eye of marine fish, we find the rhodopsin system as in birds and mammals and invertebrates, but this is not the system to be found in the eye of freshwater fish which utilize another system, the porphyropsin system. Porphyropsin, like rhodopsin, is a carotenoid–protein complex and is purple in colour. Its spectrum resembles that of rhodopsin, but has a maximum at 522 mμ. On exposure to light, a substance having properties similar to rhodopsin is liberated; it is called retinene II. In chloroform it has an absorption maximum at 405

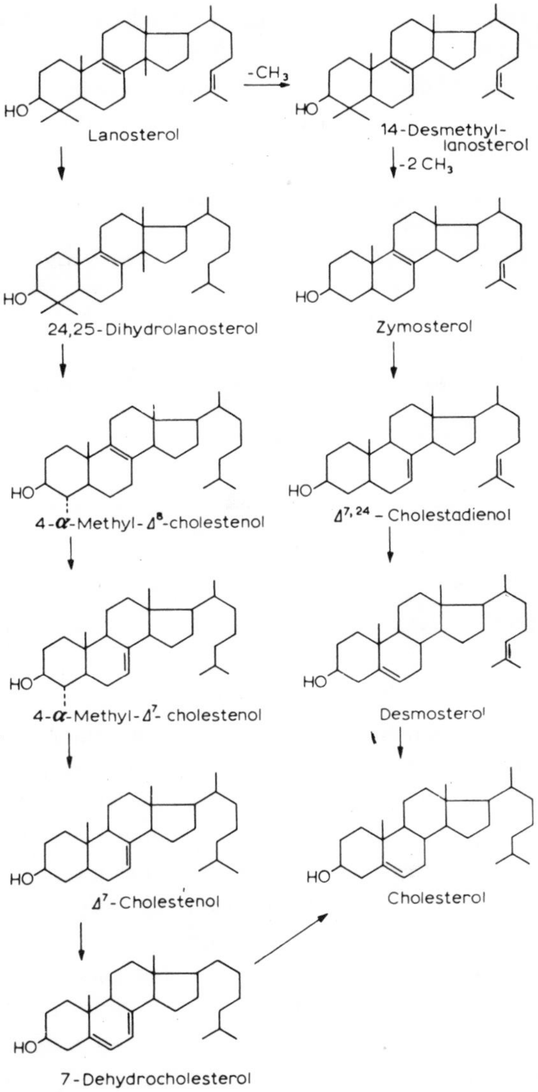

Lanosterol
$-CH_3$
14-Desmethyl-lanosterol
$-2\,CH_3$
24,25-Dihydrolanosterol
Zymosterol
4-α-Methyl-Δ^8-cholestenol
$\Delta^{7,24}$-Cholestadienol
4-α-Methyl-Δ^7-cholestenol
Desmosterol
Δ^7-Cholestenol
Cholesterol
7-Dehydrocholesterol

Fig. 26. Pathways of cholesterol biosynthesis. (Richards and Hendrickson[157])

References p. 117

mμ. In the retina it is converted simultaneously to porphyropsin and to vitamin A_2.

Squalene derives from mevalonate as shown in Fig. 27.

From squalene lanosterol is derived, and from the latter two pathways lead to cholesterol (Fig. 26). Starting from steroids, specialized pathways lead, in vertebrates, to steroid hormones or to bile acids. The example of the terpenoid and steroid pathway (Fig. 25) shows how in biosynthetic pathways new syntheses may result from lateral or terminal extensions on a definite path, which becomes longer or more branched.

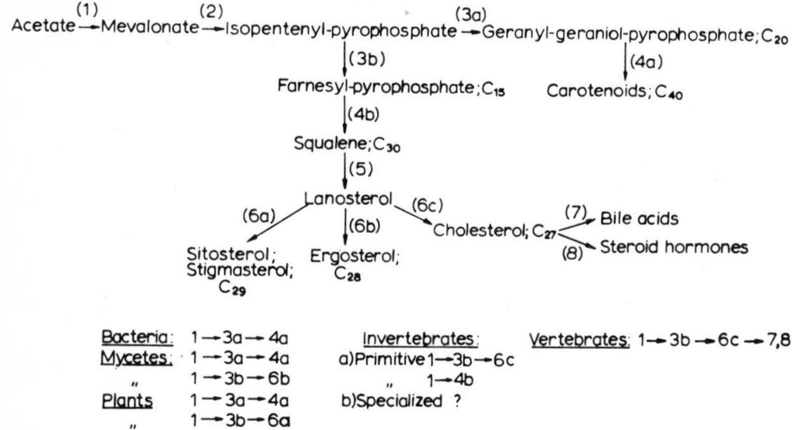

Fig. 27. Phylogeny of metabolic sequences in the mevalonic and steroid pathway. (Bloch[158])

If we consider the terpenoid and steroid metabolic syntagm, no longer from the point of view of the biosynthesis of natural products, but from that of phylogeny, *i.e.* guided by the principle we have already stated, we start from the phylogeny of organisms and follow along the branches of the phylogenic trees, the possible changes at the level of molecular units (phylogeny of biomolecules). These changes, as was emphasized by Bloch[158] (Fig. 27), show that the metabolic pathways are not identical in all groups of classification, but that in each case biosynthesis repeats the phylogeny of the molecule, *i.e.* the changes of structure it has undergone along the phylogenic branches. Along the catenary biosyntagm $1 \rightarrow 3b \rightarrow 6c \rightarrow 7$ which is a character of vertebrates the terminal extension $6c \rightarrow 7$ has been the subject of very penetrating and exhaustive studies by Haslewood and his school. Haslewood has found that the intermediates in the biosynthetic process of the C_{24} bile acids from cholesterol in mammals are serving, as their conjugates, as bile salts in lower vertebrate forms. Haslewood has recently reported on the colinearity of bile salt biosynthesis and their evolution and we shall heavily draw on this paper in the following paragraphs (ref., see Haslewood[159]).

The relations of bile alcohols and bile acids to biosynthesis are shown in Fig. 28.

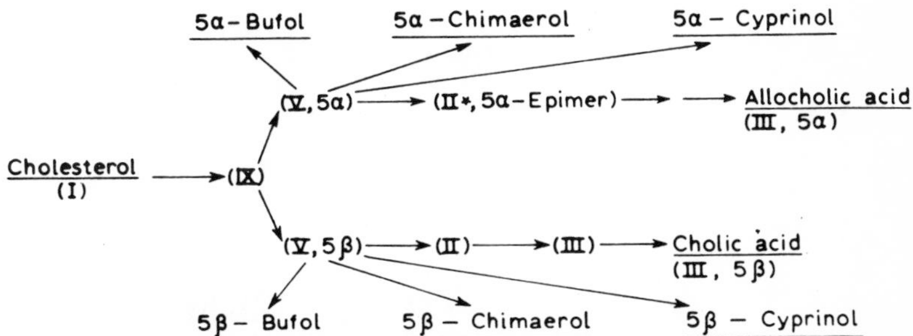

Fig. 28. Relationship of bile alcohols and bile acids to biosynthesis. Shown by Amimoto[160] to be made from IV, 5α in the salamander *Megalobatrachus japonicus*. (Haslewood[159])

Myxinol disulfate (X) is the chief bile salt of the hagfish, the most primitive vertebrate. The bile salt is also primitive, still having the C_{27} skeleton of cholesterol and its 3β-hydroxyl group. It is also specialized, as are the myxins, by having a 16α-OH group.

The lampreys, the other family of cyclostomes, have petromyzonol sulfate (XI). It is primitive as it is a bile alcohol sulphate and also highly specialized as it is a 24-carbon compound.

I. Cholesterol.

References p. 117

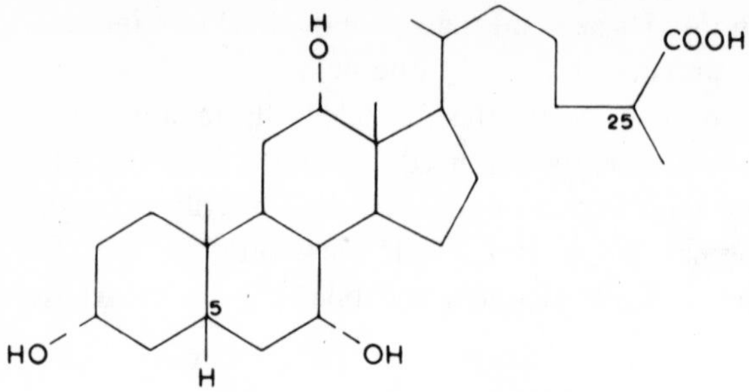

II. Trihydroxycoprostanic acid.

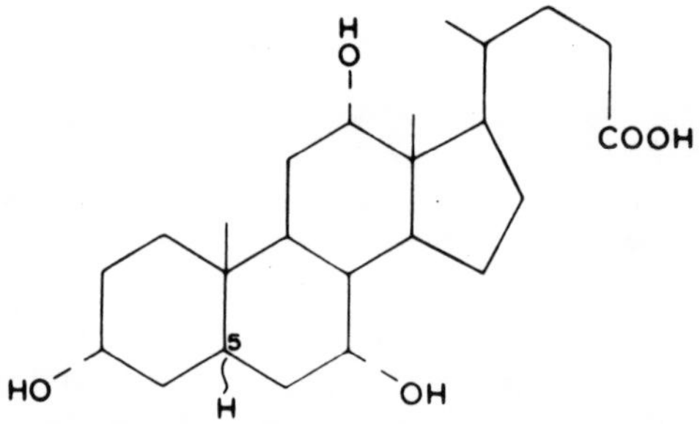

III. Cholic (5β) and allocholic (5α) acids.

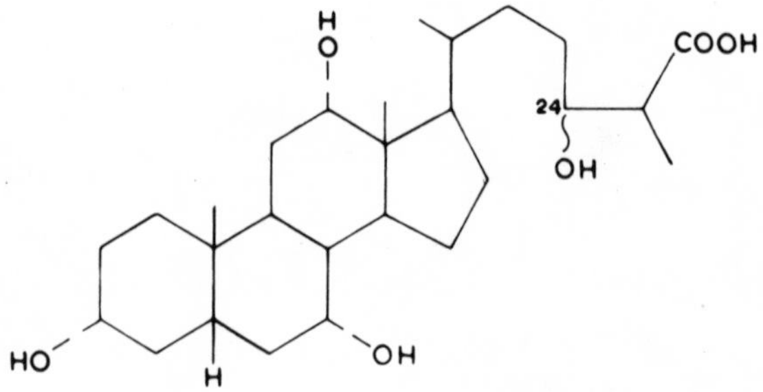

IV. 3α,7α,12α,24ξ-Tetrahydroxy-5β,25ξ-cholestan-26-oic acid.

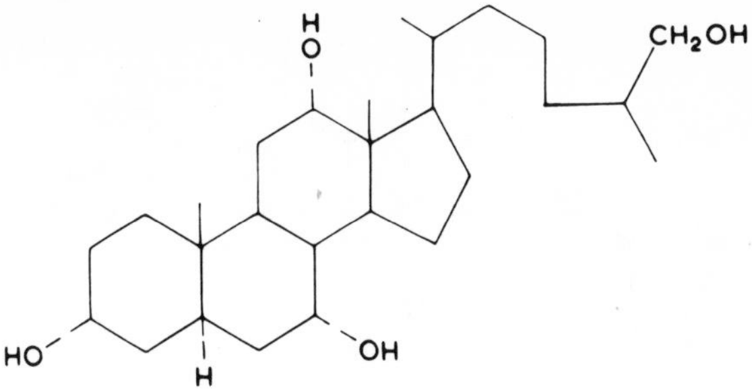

V. 5α,25ξ-Cholestane-3α,7α,12α,26-tetrol.

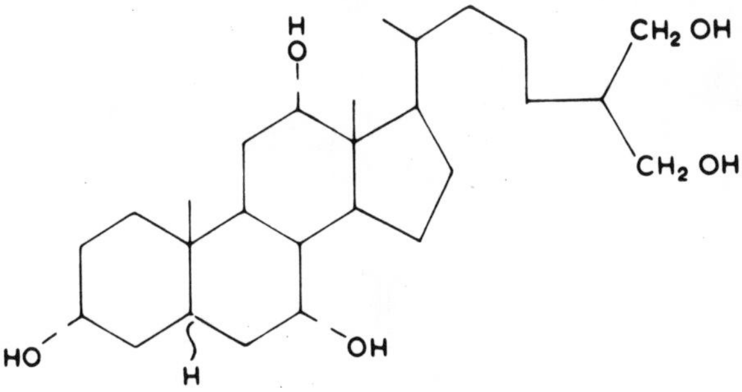

VI. Cyprinols.

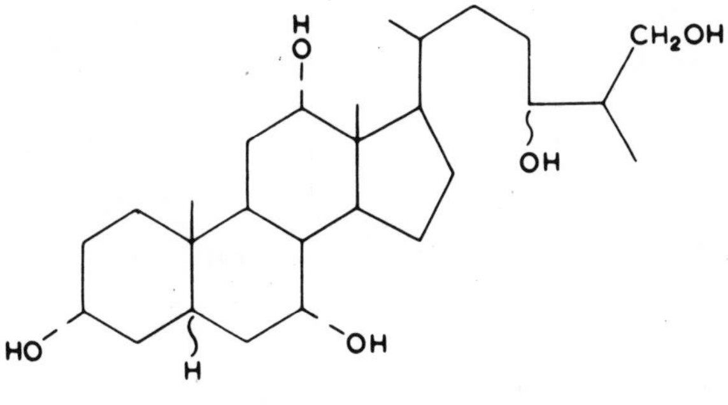

VII. Chimaerols.

References p. 117

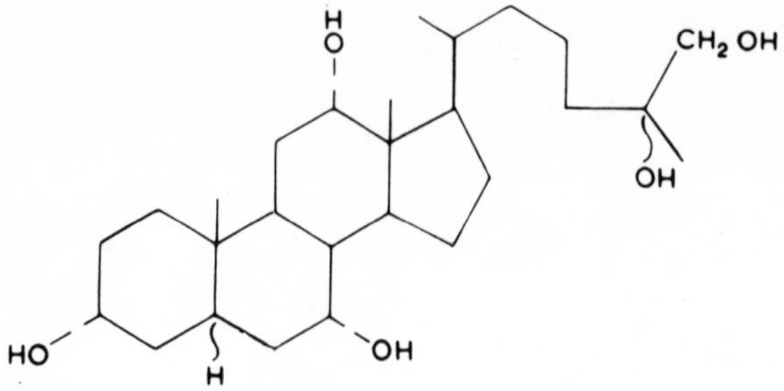

VIII. Bufols.

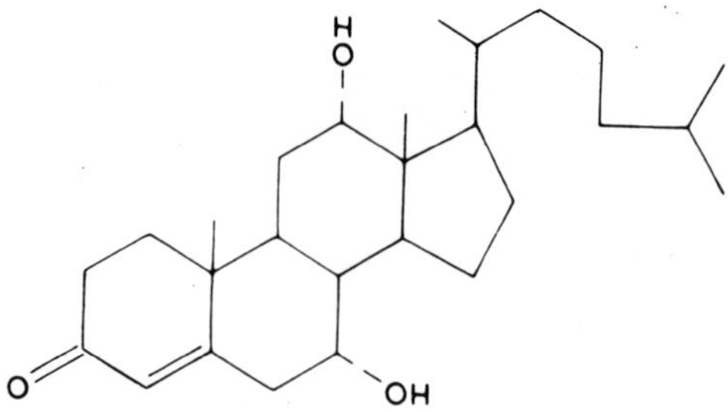

IX. 7α,12α-Dihydroxy-cholest-4-ene-3-one.

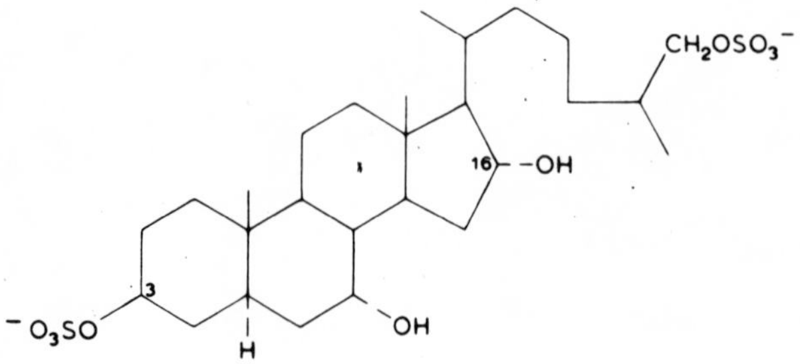

X. Myxinol disulphate.

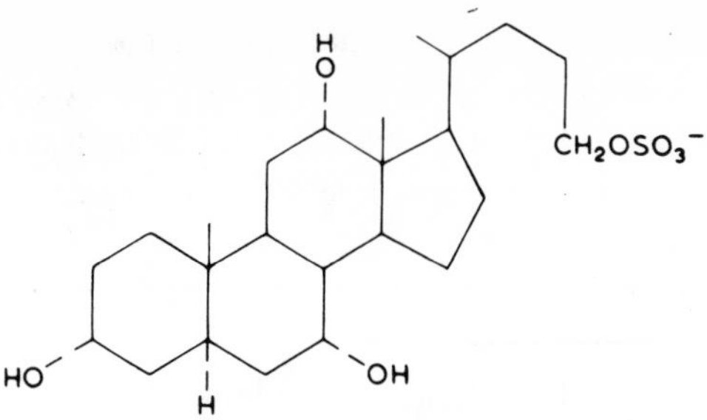

XI. Petromyzonol sulphate.

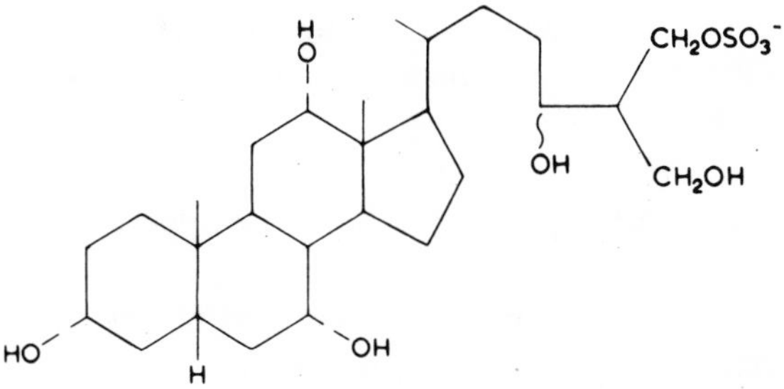

XII. Scymnol sulphate.

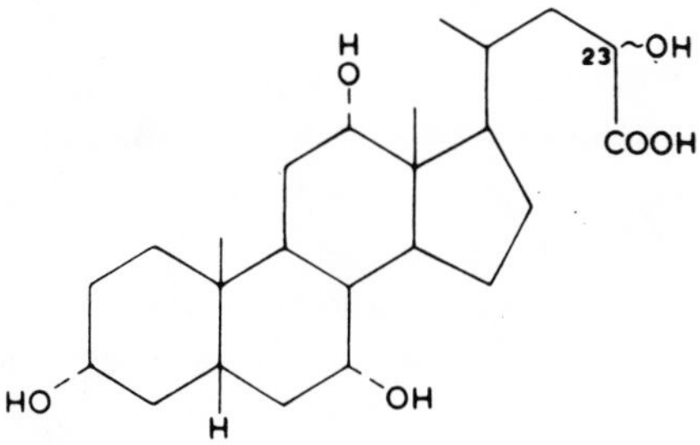

XIII. 3α,7α,12α,23ξ-Tetrahydroxy-5β-cholan-24-oic acid.

References p. 117

Among osteichthyeans, *Latimeria chalumae*, the coelacanth, has the bile alcohol latimerol, the 3β,5α enantiomer of VI and the Dipnoi, other primitive osteichthyeans have C_{27} bile alcohols. The elasmobranchs have scymnol sulfate (XII) as a principal bile alcohol, whereas the majority of teleosts have taurocholate, the most advanced type.

The chief bile alcohol of *Chimaera monstrosa* is 5β-chimaerol (VII, 5β). It is less advanced than scymnol.

The primitive teleosts have C_{27} bile alcohols while the most advanced have C_{24} bile acids. The bile salts of amphibians and fishes are more primitive than those of higher vertebrates which have no bile alcohols.

The Chelonia are separated by paleontological evidence, from other reptiles. They have a C_{27} acid of a very specific type. In reptiles Archosaures and Varanids have C_{27} acids while advanced lizards and snakes have cholic or allocholic acid.

In birds and mammals no more than traces of other than C_{24} bile acids are found (for detailed literature on comparative biochemistry of bile salts, see Haslewood[161,162]).

When we consider the pathway 6c→8, from cholesterol for which only a single biosynthetic pathway is known it appears that the biosynthesis of steroid hormones by the adrenal cortex, the ovary or the placenta depends on a previous synthesis of cholesterol followed by the implication of a number of enzymes present in the above cited organs of vertebrates: 3β-dihydrogenase, 17-hydroxylase, 21-hydroxylase, 11β-hydroxylase, 6β-hydroxylase, etc.

31. Mechanisms in diachronic molecular epigenesis

(a) Point mutations and commutations

What we have designated as diachronic molecular epigenesis we recognize, in the context of biosemiotics, as a change of configuration, *i.e.* of significant. When the change is also accompanied by a radical change of signified, we recognize the process of a commutation (see Section 22, p. 54).

The flow of point mutations through diachronic molecular epigenesis may be considered as the fringe of the conservative replication process and it may be accompanied by little or no change of signified (orthologous molecules) at least at the molecular integrative level, though polymorphic proteins present differences at the kinetic level.

On the contrary, in commutations, both significant and signified of the bioseme are modified. It was once tempting to consider the slow process of amino acid point mutations as in time, gradually producing the change of signified observed in the cases of commutation.

Nevertheless it clearly appears now that the commutations of lactose synthetase or of ACTH which are gross changes, are concordant with recent divergences in phylogeny, and that they imply a greater introduction of information than can be the case in point mutations.

(b) Gene duplication

Gene duplication and independent evolution of the resulting genes (review in Ohno[129]) may lead to marked changes in significant and signified, resulting in modifications in catenary pathways. Partial gene duplication has been considered in Section 23 (p. 57).

Horowitz[163] has proposed that the evolution of pathways may have developed by retrogressive evolution of one related enzyme to another. In Procaryotes, the diachronic epigenesis of biosynthetic catenary biosyntagms must have resulted from the origination of new genes. The genes forming an operon are presumed to have originated from tandem duplication followed by functional differentiation (Lewis[164]).

Dixon[106] proposes a model of diachronic epigenesis at the level of an operon, by gene duplication (Fig. 29).

"It is imagined that an organism possessed a single gene for a particular enzyme, *1*, which could convert a metabolite, C, abundantly present in the original environment, into one of its vital cell constituents. A gene duplication occurs with production of a new gene which can mutate and perhaps eventually produce as a minor component an altered enzyme protein.

2. Eventually the environment becomes deficient in metabolite C but there is a great deal of a structurally related metabolite B; enzyme *1* possesses a site for binding B but cannot convert it to C. Enzyme *2*, however, can both bind B, and as a result of its mutationally altered structure, convert it to C. This is of no selective advantage while C is abundant, but as C becomes deficient, mutants in which increased synthesis of enzyme *2* can occur (*e.g.* those derepressed for this enzyme) are selected and the organism can now live on the hitherto untapped supplies of B. Similarly, as B becomes exhausted, another minor constituent, enzyme *3*, derived by duplication from enzyme *2*, can convert A_1+A_2 to B. The original binding site for C in enzyme *1* might still be retained in enzyme *3* and could provide a regulatory site for the allosteric control of *3*, the enzyme catalyzing the initial step in the pathway, by the final product C"[106] (Fig. 29)

On the other hand, in the Eucaryotes, evidence exists in favour of the diachronic epigenesis of catenary pathways as a result of a commutation of a protein or by the introduction of enzymes in existing or in new

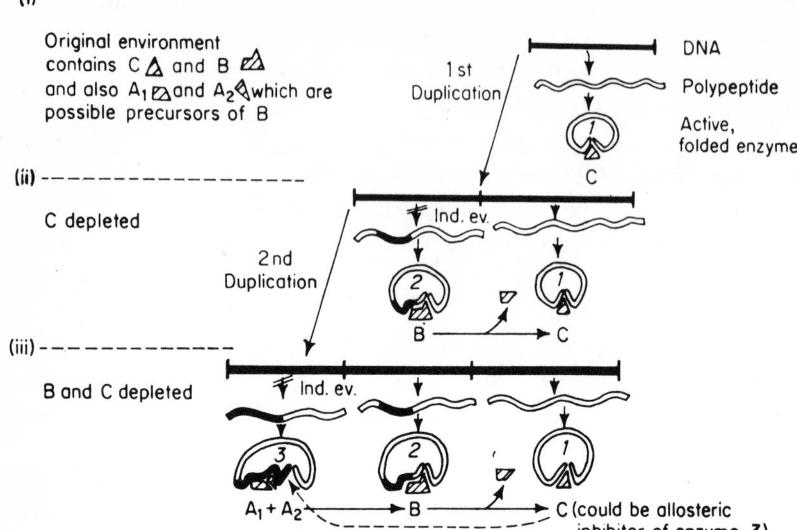

Fig. 29. Cistron duplication and the evolution of a metabolic pathway. (Ind. ev.: independent evolution). (Dixon[106])

catenary biosyntagms. The linearity which is observed in the flux of point mutations as observed for instance in cytochrome *c*, does not apply to proteins during the period in which a commutation, implying a radical change of signified, emerges.

It is recognized, for instance, (see Chapter 4) that α-lactalbumin (one of the components of the catenary biosyntagm of lactose synthesis) is homologous with lysozymes and was derived in phylogeny from a precursor of lysozyme when, approximately 100–150 million years ago, milk-producing animals appeared.

As stressed by Dickerson[145] the UEP (Unit Evolutionary Period: time in millions of years for one percent change in amino acid sequence to show up between two divergent lines of evolution) is, for lysozyme, of the order of 5.3 million years. Assuming a linearity between lysozyme and α-lactalbumin evolutions, Dickerson calculated that α-lactalbumin would have diverged from lysozyme 390 million years ago, *i.e.* in early Devonian, which is absurd.

To quote Dickerson:

"A more reasonable explanation is that the *change* in pressures exerted on α-lactalbumin when a polysaccharide synthesizing role led to an increase in the rate at which mutational changes were retained in the sequence rather than being weeded out as harmful to the *status quo*."[145]

The linear approximation is valid only if the flux of point mutations concerns proteins which have kept their biological activity (in other words proteins which have not radically evolved from the point of view of the signified) and can not be applied to those whose significant and signified were commutated, as in the case of lysozyme and α-lactalbumin.

When considering the aspects of the signified of heteropolymer quaternary biosyntagms, we have examined the case of the tryptophan synthetase of *E. coli*. This represents a case in which the multichain synthetase we described in the case of *E. coli* can be considered as an efficient multichain which has resulted from diachronic epigenesis from two low efficiency enzymes catalyzing sequential steps in the biosynthesis of tryptophan (see Dixon[106]).

The replacement of polar residues by non-polar ones in a mutating polypeptide may enhance a formation of an aggregate enzyme with a different signified leading to the replacement of the separate enzyme molecules by the enzyme aggregate or by a number of different aggregates such as may be observed in the case of isoenzymes.

(c) Hybrid genes

Although the acceptance in evolution that this category of mutation is rare in appearance, an example can be found in Dayhoff's *Atlas*[124]. The example given is the case of herring clupeins YI and YII giving rise to the gene-producing clupein Z by a hybridisation resulting from misalignment of homologous chromosomes during synapsis, and accompanied by a crossing-over with a non-allele gene.

(d) Deletions and insertions

The necessity of inserting one or more "gaps" in aligning homologous proteins reveals the existence (in the genetic history) not only of insertion or deletion of triplets in DNA, but also the acceptance of the shortened or lengthened protein.

"Gap events" appear less frequently than replacements (point mutations). Deletions and insertions appear to be more easily accepted if the gene duplicated and the resulting genes are evolving in the direction of a

References p. 117

commutation. In the evolutionary tree of lactalbumin and lysozyme, for instance (Fig. D-17 in ref. 124), five of the seven "gap events" occurred on the line between the lysozyme and lactalbumin ancestors, while, on the 236 estimated amino acid replacements, only 63 occurred on that line.

The available information (Table 5-3 in ref. 124) reveals that terminal residues are more liable to deletion or insertion than internal ones.

(e) Heterotypic expressions

Even without change of signified a bioseme may become inserted in a new catenary biosyntagm.

As stated above, in Eucaryotes, some producer genes are active in different tissues, and in a general fashion the same biosemes may be involved in different catenary biosyntagms in several species of organisms. An example of such a call on a diversity of producer genes depending on other integrators in other varieties of cells is provided by the case of the phenolase complex, as presented by Mason[165] as a case of what he calls "heterotypic expressions."

The phenolase complex may act as one of the terminal oxidases in plant tissues. The synthesis of flavonoids and among them the hormones controlling the ability of male and female gametes of some green algae to copulate, must also be classified among the characters due to phenolase, as the hydroxylation of tyrosine and the formation of the *o*-dihydroxybenzenoid pattern in flavonoid is to be ascribed to the presence of this enzyme. Lignins are present as framework substances in all plants to the extent of 15–30 per cent of their dry weight. They contain phenylpropanoid nuclei in several degrees of hydroxylation, methylation and glycosidation. It is now clear that the products of phenol-*o*-hydroxylation in the presence of phenolase form a part of the phenolic pool from which the precursors of the flavonoids, the tannins, the alkaloids and the lignins are drawn.

The browning of plant tissues, *i.e.* during the formation of seed coats (in apple, grape, chestnut, etc.), in bark, in mushrooms, etc. is due to the enzymic oxidation of phenols.

In arthropods, the process of sclerotization or tanning of cuticular proteins is due to quinones derived from the polyphenol layer (see Chapter 4). The phenolase complex is the specific catalyst for the formation of melanins, responsible for the black, brown, buff and Tyndall blues which colour the teguments, feathers, hair and eyes of chordates: fish, salamanders, toads and frogs, turtles, lizards, snakes, crocodiles, birds and mammals.

Another case of heterotypic expression is provided by alcohol-dehydrogenase, a universally distributed enzyme which is found in a special quality in the rhodopsin system where it acts as retinene reductase (Fig. 30).

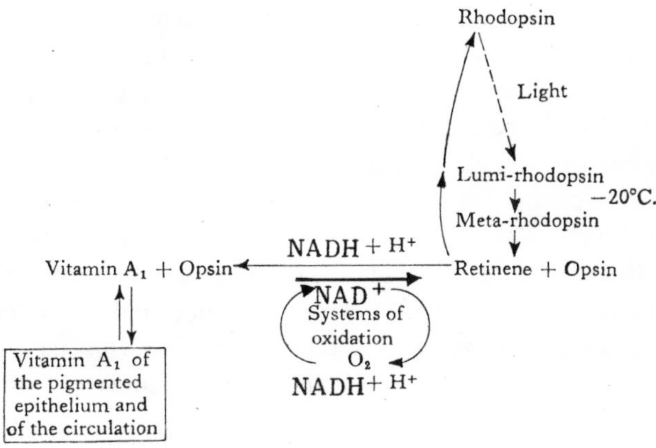

Fig. 30. The rhodopsin system. (after Wald[166])

The secretion of milk, due to a biochemical differentiation in one type of mammalian cell is released and controlled by prolactin (see Chapter 4) resulting from a biochemical differentiation of another type of cell, in the adenohypophysis of fishes, amphibians and reptiles. Its insertion in the secretion of milk in mammals is thus an insertion into another biochemical system situated at the molecular integrative level. Another example is the action of pitocin on the uterus. The hormone is found in mammalian vertebrates and acts in the control of water metabolism. Aldolase is one of the components of the catenary biosyntagm of glycolysis and is also involved in the catenary biosyntagm of CO_2 reduction in photosynthesis. Several enzymes belong to the latter as well as to the metabolic biosyntagm of pentoses.

Heterotypic expressions correspond (in the context of the model of Britten and Davidson) to the occurrence of identical producer gene redundancies in several batteries of genes. This requires receptor genes corresponding to each battery, another aspect of redundancy of DNA as mentioned in Section 17 (p. 38) and a reason for considering the facts of heterotypic expression as aspects of the evolution of regulation at the genetic level.

32. Deletion and reshuffling

Another regulatory aspect is afforded by the case of catenary biosyntagm reshuffling, in the case of a deletion, in descendance, of significant–signified units. Such a neointegration due to a change of regulation at the genome level is observed in the parasite of the pigs intestine *Ascaris lumbricoides*. In this species, the adult stage, in contrast with free nematodes, obtains its energy through the anaerobic pathways of catabolism. Terminal oxidation is similar to that of the red blood cell of mammals, operated through a flavin system. *Ascaris* is insensitive to cyanide. Its muscles or female reproductive system contain no cytochrome *c* nor cytochrome-oxidase. In the presence of air it forms peroxide and, as its catalase is deficient, hydrogen peroxide kills the animal. It forms the same end-products in atmospheres either of air or of mixtures of nitrogen and carbon dioxide.

The end products of the anaerobic metabolism of *Ascaris* is not lactic acid which is produced in traces, but a mixture of volatile fatty acids among which the most important quantitatively are α-methylbutyric, α-methyl-valeric and succinic acids. The pathway proposed by Saz and Weil for the formation of these acids in adult *Ascaris* muscle strips is shown in Fig. 31. From glucose to pyruvate, the Embden–Meyerhof scheme is operative, but the lactic dehydrogenase, which reduces pyruvate with NADH to form lactate is at low concentration and for the amount present, almost inactive. In the familiar form of muscle glycolysis of mammalian muscles, lactic dehydrogenase has an important role. In the course of glycolysis the first oxido-reduction (oxidation of 3-phosphoglyceraldehyde to 3-phosphoglyceric acid) requires the reduction of NAD. As the coenzyme exists in very small amounts, the NADH must be reoxidised to serve again. This is the function of lactic dehydrogenase. In *Ascaris*, other reactions are called upon to accomplish this function. Saz and Vidrine[167] showed that succinate is formed *via* the fixation of CO_2 into pyruvate and in conformity with the proposed pathway, malonate inhibits the incorporation of CO_2 in succinate. Each of the enzymes involved in the scheme have been shown to exist in adult *Ascaris* muscle. It appears that CO_2 is fixed in phosphoenolpyruvate

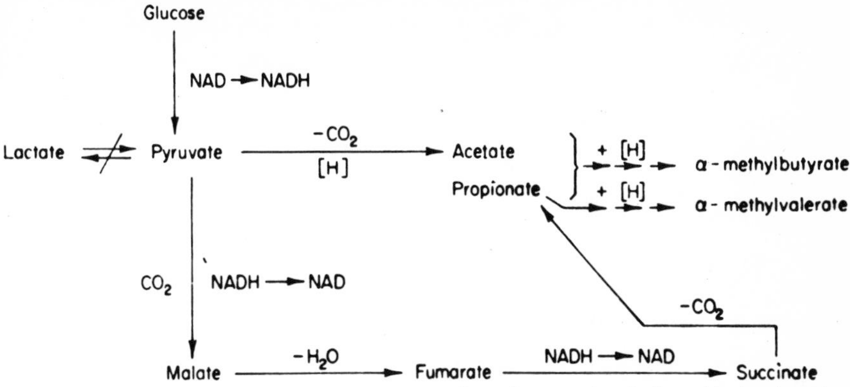

Fig. 31. Proposed pathway for the formation of succinate and volatile acids in *Ascaris* muscle (after Saz and Weil[168]).

to form oxaloacetate which is, in turn, reduced to malate in the extra-mitochondrial regions of the cell. Malate is used in mitochondria where a dismutation takes place, one molecule generating mitochondrial NADH by being oxidatively decarboxylated to pyruvate by the NADH-linked malic acid enzyme, another molecule being reduced to succinate with the generation of ATP. The high malic dehydrogenase activity would have to pull the phosphoenolpyruvate carboxykinase reaction in the direction of CO_2 fixation. There is no lactic acid production because the phosphoenolpyruvate is converted into succinate before reaching the stage of pyruvate.

How are the other products of *Ascaris* so-called fermentation produced in muscle strips? Succinate is decarboxylated to propionate and the condensation of the propionate and of one acetate gives α-methylbutyrate (Fig. 32). The acetate involved results from the decarboxylation of pyruvate. Two molecules of propionate condense into an α-methylvalerate (Fig. 33).

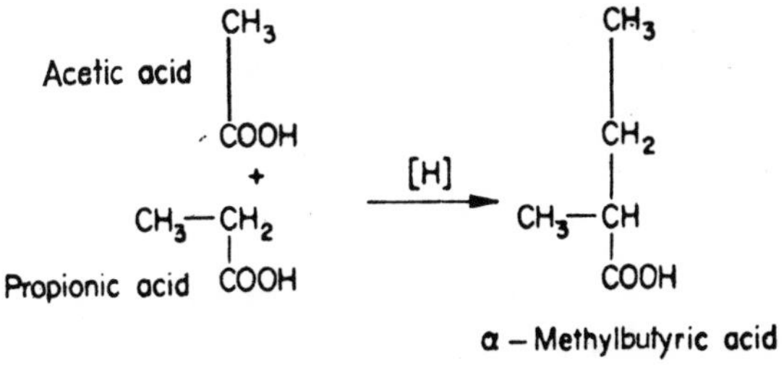

Acetic acid CH_3–COOH + Propionic acid CH_3–CH_2–COOH $\xrightarrow{[H]}$ CH_3–CH_2–CH(CH_3)–COOH

α – Methylbutyric acid

Fig. 32. Condensation of propionate and acetate into α-methylbutyrate.

References p. 117

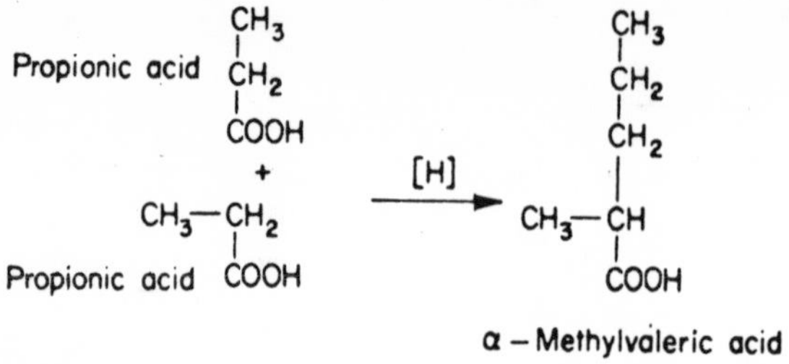

Propionic acid CH_3-CH_2-COOH + Propionic acid CH_3-CH_2-COOH $\xrightarrow{[H]}$ $CH_3-CH_2-CH_2-CH(CH_3)-COOH$

α – Methylvaleric acid

Fig. 33. Condensation of two molecules of propionate into α-methylvalerate.

The succinic dehydrogenase which has been described in *Ascaris* in fact functions as a fumaric reductase. Fumarate reduction goes to completion and the sequence of electron transport is shown in Fig. 34.

Anaerobically, electrons from NADH are accepted by a flavoprotein and directed to succinic dehydrogenase which, in turn, reduces the terminal electron acceptor, fumarate, to succinate. If oxygen is present, succinate or NADH can be oxidized with formation of H_2O_2. The regressive epigenesis of lactic dehydrogenase in adult *Ascaris* is therefore to be seen as being compensated by a whole rearrangement of enzyme systems providing an advantage since in the stage between NADH and flavoprotein, a supplementary high-energy phosphate bond is produced.

There is no system of phosphorylating respiratory chain in *Ascaris*, which in the adult stage has no cytochrome oxidase. The suggestion of the

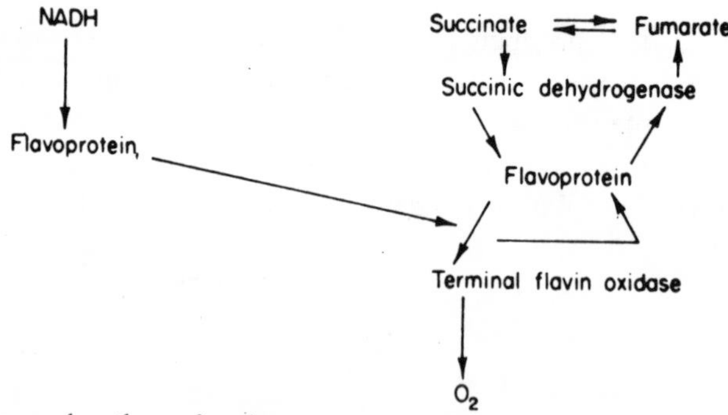

Fig. 34. Proposed pathway for electron transport in *Ascaris* muscle (after Kmetec and Bueding[169] and Bueding[170]).

presence of cytochromes *b* and *c* was erroneous (see Saz[171]). In adult *Ascaris*, the tricarboxylic acid cycle has also regressed through the low activity of aconitase, condensing enzyme and pyruvate dehydrogenase, as well as by the lack of the terminal oxidase system. Regressed as it is the tricarboxylic acid cycle can still be used as a source of precursors for biosynthesis[171]. In contrast with the adult *Ascaris*, its eggs have an obligatory requirement for oxygen and are endowed with a cytochrome-linked terminal respiratory pathway. This need is satisfied in spite of another regressive aspect of *Ascaris* biochemistry, the inability to synthesize the porphyrin nucleus, as the hemoglobin of the enteric fluid is produced in the cells of the gut, from porphyrin derivatives, including those of the chlorophyll in the gut contents of the pig. This hemoglobin is the source of all the hemoproteins of the organism, including the cytochrome of the egg[171].

When considering at the molecular integrative level the aspects of the diachronic molecular epigenesis, the cases collected in this review allow recognition of a number of ways through which the diachronic changes do occur in phylogeny.

Some of these aspects concern the mutations, due to a number of forms of modification of the protein controlling DNAs, and other aspects, also at the level of the genome, involve a redundancy of regulator DNAs which leads to the establishment of new integrations. At the phenotypic level changes of the molecular system generally involve ligand-induced configuration changes as the basis of integration and regulation.

At the level of the catenary metabolic biosyntagms, diachronic epigenesis takes the form of commutations, lateral or terminal extensions or deletions, heterotypic expressions and reshufflings.

There is no chance for the outcome of diachronic molecular epigenesis to be accepted by natural selection *i.e.* for the new biosyntagm to spread into the species without a corresponding epigenesis of a counterpart, for instance as substrate in the case of an enzyme and as target in the case of a hormone. The biosemiotic viewpoint adopted in this review emphasizes the necessity for a sign to be associated with a receptor.

It is commonly accepted that, at least in bacteria, one of the two strands of DNA is selected by an unknown mechanism for transcription, at least when highly purified RNA polymerase of *E. coli* is used, while with a crude preparation of the enzyme, both strands of the DNA are used in transcription. As, *in vivo*, RNA polymerase is not the enzyme involved, it cannot be absolutely excluded that complementary molecular epigenesis is taken advantage of in shaping parts of the amino acid sequence of the significant of a sign at the signal or receptor site of the receptor.

References p. 117

No invention of configuration will reach the status of a bioseme except by entering into a couple of *relata* and it is only in this case that it will reach the threshold of the impact of natural selection at the level of individual organisms.

33. Biosyntagm and biosystem

The biosyntagm is a metonymical aspect, while the biosystem (p. 57) is a metaphoric one, not dealing with relations of contiguity and solidarity as in biosyntagms, but with relations of substitutive association and diachronic epigenesis.

The distinction between the syntagmatic and systematic domains may be illustrated by a comparison borrowed from architecture by F. de Saussure[42]. A unit of signification may be compared to a column in an antique temple, in relation of continuity with other parts of the edifice, the architrave for instance (syntagmatic relation in the syntagm of a temple).

On the other hand a comparison of columns in different temples leads to classifying them in the doric order, for instance, or to differentiate the doric and the corinthian orders.

The basic concept of the systematic aspect of biosemiotics is homology, as defined in previous sections.

In chemical (prebiological) evolution (see Chapter 5) we are concerned with an extraction, from the *inorganic continuum*, of the biosyntagms resulting in the emergence of dissipative structures maintaining the steady state of the system far from equilibrium conditions (Prigogine[24]; Eigen[99]), while in biochemical evolution, these dissipative structures are maintained by heredity, and in the course of time, a diversification resulting from progressive extraction of biosyntagms *from the biosystem* is accomplished. Considered, as was the case in the preceding pages, at the molecular integrative level, molecular diachronic epigenesis in phylogeny is a matter of diversification of metabolic catenary biosyntagms, a diversification which originates in variations of the informative catenary biosyntagms governing the flux of information from DNA to proteins.

IV. INSTANCES OF MOLECULAR BIOSEMIOTIC ASPECTS AT HIGHER INTEGRATIVE LEVELS

34. Self-assembly in supramolecular structures

In this review, we have been concerned so far with couples of *relata* established between biomolecules at the integrative molecular level; *i.e.* essentially at the level of metabolic pathways implementing the flow of matter and energy through an open system, and with their integration, regulation and diachronic epigenesis.

The meshwork of catenary metabolic biosyntagms concurs in the obtainment of free energy in the form of ATP and in the production of biomolecules exerting a feedback on these biosyntagms. It also produces biomolecules contributing by the association of their configurations to the establishment of the different levels of structural units (cell organelles, cells, organisms), or by their biological activities, to the integration of the functions of cells, of organisms and of ecosystems. The self-association of biomolecules in higher configurations is an expression of the coupling of biosemes carried by these biomolecules. The first clear demonstration of a self-assembly of biomolecules was the one of collagen[172]. Such forms of biosemes are involved in the aggregation of macromolecules in organelles (Asakura *et al.*[173]; Tilney and Porter[174]) as well as in the aggregation of cells in multicellular structures. Membranes can assemble themselves from their components. Phosphoglycerides and cholesterol associate to form a lipid bilayer between two water compartments. When supplemented with certain proteins or polypeptides, such a bilayer mimics the electrical behaviour of nerve cell membranes[175]. The membrane of *Mycoplasma laidlawii* can be dispersed under the action of detergents. When these are removed, lipids and proteins reassociate to form a membrane sheet (references in Lehninger[176]). Ribosomes can also be self-assembled from their constituent biomolecules[177] and such is also the case with the constituents of the tobacco mosaic virus[178] and with the T_4 bacteriophage of *E. coli*[179] (for recent literature on the subject of the assembly of intracellular structures, see ref. 179a).

35. Physiological level

By function, the physiologists mean a collection of molecular activities, recognizable, at a higher integrative level, in the phenomenology of their integrated aspect.

If we consider, for instance, the electroplax membrane of the electric eel *Electrophorus electricus*, we recognize at the membrane level the existence of a receptor for acetylcholine, ligand-inducing configuration changes (changes of signified) at the level of the membrane fabric, the result of which is a change of membrane potential (literature, see Podelski and Changeux[180]).

If we consider, again at the level of cell membranes, the effectors of the adenylate cyclase system we recognize the adenylcyclase as the primary enzyme system through which hormonal peptides alter the metabolism, the structure and the function at the level of target cells. In the target cell, information is carried by the hormone to a receptor and it is received, decoded and transduced into the expression of the endocrine action at the molecular integrative level. In such a system the receptor is coupled with a response mechanism, while in the case of the hormones active at the genomic level, the receptor is selected among multiple receptors.

To quote Rodbell *et al.*[181]:

> "In informational terms, this system selectively receives, translates and amplifies the "message" contained with the structure of the hormone to give a new messenger."

This kind of regulatory catenary biosyntagm is divided by Rodbell into several biosemes: hormone-binding sites (discriminators), coupling processes (transducers) and catalytic component (amplifiers).

As underlined above, the orthologous proteins show little or no variation of signified in the course of the evolution of organisms. This is the case with cytochromes *c*, all yielding to whatever organism they belong, the same first-order rate constant for their reactions with bovine cytochrome *c* oxidase over a wide range of cytochrome *c* concentrations (Smith[182]). But there are large variations, among the cytochromes *c* of different species, of the binding of various inorganic anions by the ferric form of the protein (Barlow and Margoliash[183]).

Margoliash *et al.*[184] have suggested the existence of a correlation between the ion-binding properties of cytochrome *c* and the transport properties of mitochondria and proposed a model for the ion-carrier function of cytochrome *c* at the level of the mitochondrial inner membrane (see the *Harvey Lecture* by Margoliash[185]).

Considering the integrative level of the "functions" at the higher level of physiological integration we may recognize what the present author has called "physiological radiations"[13,36,103,105,117,120,186–188] or "physiological epigenesis"[105]. In such "radiations" (a term used in a meaning akin to the familiar biological concept of adaptive radiations) while the configuration of a biomolecule is not modified, advantage is often taken in the evolution of organisms at the physiological level, of different properties and activities of the biomolecule, *i.e.* of other aspects of an already present signified.

Carotenoids, for instance, absorb light. In the case of the phototropism of plants, they transfer the electromagnetic energy to an enzyme, while in photosynthesis, they transfer it to chlorophyll. In animals, the absorption of light by the carotenoids, obtained from plants and more or less modified, are used in many ethological services: attraction of insects and birds towards flowers and fruits, sexual dichronism in insects, feather colours in birds resulting from different states of oxidation of carotenoids of alimentary origin. On the other hand carotenoids may undergo *cis–trans* isomerization which is utilized in vision as well as in the perception of odours[189]. In fish, carotenoids are involved in photo-response[190]. In the trout, for instance, xanthophores are coloured by lutein and erythrophores by astaxanthin, two carotenoids involved in the nature of skin colour[191]. But in certain species of fish, as for example in *Salmo iridaeus*, another property of the same molecular unit is taken advantage of, the molecule of astaxanthin being also used as a fertilization hormone[192]. Many such physiological radiations, active at the physiological level, can be identified.

Thyroid hormones exert their main signification as growth stimulators but they are involved in metamorphosis, in the regulation of basal metabolism, in protein metabolism, in osmoregulation, etc.

Dopamine appears in the biosynthetic syntagm as a precursor of norepinephrine (Kirshner[193]). On the other hand it seems to have assumed a neurohumoral role at the level of the brain of mammals (Bertler and Rosengren[194]) and of *Helix* (Kerkut and Walker[195]). It also plays a regulatory function at the level of the heart of *Venus mercenaria* (Greenberg[196]). Finally the fact that dopamine is the only catecholamine which can be identified in the ganglia of different mollusks (lamellibranchiates or gastropods) (Sweeney[197]) is in accordance with the notion that it plays in some organisms the same role as epinephrine or norepinephrine in others.

References p. 117

Enzyme systems and their associated components (substrates, coenzymes, cellular regulators, hormonal regulators) frequently contribute to changes at the higher physiological level by the introduction of one or several of their components into a more specialized system. The visual function of the vitamins A is the only one it has been possible to demonstrate in animals other than mammals and birds. In the latter two, without showing commutation, the carotenoid configuration obtained a new signified in controlling the formation of the mRNA commanding the keratin synthesis in epithelial cells. In the visual cycle in which vitamins A are involved, retinene is produced by the reduction of vitamin A_1. The enzyme involved is the common alcohol dehydrogenase. This universally distributed enzyme is inserted here in a new system which is extremely specialized.

Peroxidase, in *Balanoglossus* is used as a luciferase (Dure and Cormier[198]).

We propose to define physiological radiations as the insertions of the same molecular significant (with possible uses of different aspects of the signified) in several biosemiotic aspects at levels of integration higher than the molecular level. This concept differs from the concept of commutation in which the molecular signified is radically changed in phylogeny, and from the concept of heterotypic expression (Mason) in which a biomolecule is involved by its unmodified biosemes in different catenary biosyntagms at the molecular integrative level.

Physiological radiation (a concept which obtains at the physiological integrative level) must also be distinguished from the concept of different utilizations of a same biomolecule as receptor in the formation of couples of *relata* at the molecular level of integration. We may designate such biomolecules as polygamic. Besides their role as endobiosemes in protein macromolecules, amino acids show a high degree of polygamy. As underlined by Krebs[199].

"... Glycine takes part in the synthesis of purine bases, porphyrins, creatine, glutathione, bile acids, and hippuric acid. Serine is a precursor of ethanolamine and choline, both of which are constituents of phospholipids. Methionine plays a role in the synthesis of creatine and choline, aspartic acid is a precursor of pyrimidines, and tyrosine is a precursor of adrenaline, noradrenaline and thyroxine."

36. Ecological level

A number of end-products of the biosynthetic catenary biosyntagms are neither significant at the molecular integrative level nor at the structural or physiological levels of the organism, but are significant as ecomones (see Section 8, p. 10).

At the level of integration of ecosystems, the biosemiotic structure of ecomones and of the structures with which they contract their coupling are problems for the future. As in the case of messenger hormones, the contact of a coactone with a receptor is at the starting point of a catenary informative syntagm.

Wilson and Bossert[200] and Wilson[201,202] have described ways in which coactone actions may have been modified in the evolution of organisms. One of these ways is the adjustment of fading time. For instance, by a reduction of the emission rate or a raising of the threshold concentration, the lapse of time separating the emission of a pheromone and the disappearance of the active space can be shortened. As Wilson[203] remarks this has been "a chief design feature in the evolution of alarm and trail systems" affording an increase in information per signal and an "opportunity for the transmission of more discrete signals".

As the duration of action of acetylcholine is shortened by its inactivation by cholinesterase, a pheromone can also be deactivated. This is the case, for instance, among bees; the pheromone 9-ketodecenoid acid, the queen substance ingested by worker bees is enzymatically deactivated (Johnston *et al.*[204]). Instead of being restricted, the active space of a coactone may be expanded. This is the case for the insect sex pheromones, pinpointing very small targets within very large spaces.

Another evolutionary acquisition is the multiplication of coactone-producing exocrine glands in the same species. In another aspect which we may call ecological radiation of coactones, their signified is used in different ways in different ecological contexts. *trans*-α-Keto-2-decenoic acid in the beehive, acts as a caste-inhibitory pheromone while in the nuptial flight it acts as the primary female sex-attractant. Combinations may also introduce new signifieds. When confined with queens for hours, beeworkers, combining scents acquired from the queen with their own, are attacked by nest mates (Morse and Gary[205]).

37. Pluribiomolecular nature of biochemical adaptations

The concept of adaptation at the molecular level (biochemical adaptation) has been introduced by the present author[6] who showed how complex ecological responses can be explained, in part at least, on biochemical grounds (respiratory function of blood, regulation of glycemia, digestion, osmoregulation). In several cases, the adaptive aspect was identified at

References p. 117

the level of a molecular species. If for example, we consider[6] the oxygen absorption curves of the few species for which we have a knowledge of the respiratory cycle, we see that in the different cases, even if the external medium is the same, very different values of the gradient of oxygen pressures obtain between the environment and the internal medium. The partial pressure of oxygen corresponding to "arterial" conditions is obviously related to the rate of exchange through the respiratory epithelium as well as to that of the circulation of the oxygen carrier. In spite of this, the degree of saturation of the arterial blood corresponds in each case to the higher region of the dissociation curve. As a matter of consequence, any lowering of P_{O_2} in the medium results in the delivery of oxygen from the carrier. Such a biochemical adaptation, while accounting in part for the relationship of a species with its surroundings, is only one among the many features that account for this relationship.

Specific adaptive aspect of a molecule is not common and this is one of the reasons for the occasional use of expressions, such as "non Darwinian evolution" or "neutral changes". We should keep in mind that adaptation at the molecular level is not necessarily obvious at the level of a single molecular species. Rather, such adaptations more commonly result from changes at the level of molecular species involved in a complex polygenic adaptive mechanism. This concept has been formulated by the present author[36]. An example of the polygenic nature of biochemical adaptations is provided by the cocoon of the silkworm (literature in Florkin[36]). The physical properties of fibroin insuring the protective function of the cocoon, depend on the peculiarities of protein synthesis in the distal part of the silkgland. The structure of fibroin, the main cocoon protein, may be explained by the specific nature of a category of mRNA liberated therein. But a detailed analysis shows that if the explanation of the *composition* and *structure* of fibroin is situated at the level of a given gene, the very *existence* of the cocoon depends on other factors as well: the nature of the specific food of *Bombyx mori*, the mulberry leaves; its attraction by these leaves (endocoactones); the behaviour leading to masticate and swallow the leaves (endocoactones); the enormous appetite of the silkworm during the second part of the fifth instar; the characteristics of the removal of the amino acids from the hemolymph, the regulation of the nymphal weight, etc. All these factors contribute to the very existence of a cocoon.

The cocoon formation in the silkworm appears originally as a secretion of a protective sheet for the prenymphal molt and the nymphal life in the conditions met in nature by the wild ancestor of the domestic form. The protein fibroin, the function of which is related to its physical properties, largely derives from a great excess of specific food rich in glutamic acid, glutamine, aspartic acid and asparagine. This very specific food is ingested in great bulk during the period of activity of the silk gland and thus provides it with a series of amino acids selectively absorbed. The peculiar structure of fibroin, with its recurring sections made up of small amino acids, depends on the nature of the messenger RNA specific of the distal part of the gland.

The period of activity of the silk gland is accompanied by a derivation of the pyruvate-producing constituents of the food towards the posterior part of the gland in which pyruvate is used to synthesize alanine, glycine and serine, according to pathways corresponding to those found in other forms of animal life. Some glycine and serine can also be included directly in the fabric of fibroin, and this is also the case for tyrosine, taken from the hemolymph. The aspartic and glutamic acids and their amides, forming an important proportion of the proteins of mulberry leaves, are among the main pyruvate producers in the silk glands. The final section of the silk thread is of endogenous origin. Its production is not only part of the accomplishment of an adaptation subserving ecological functions, but it also subserves the physiological function of regulating the normal size of the nymph by a derivation of a part of the substance of the larva towards the production of silk.

At the time of hatching, another adaptation at the cocoon level sets in. It depends on the chemical structure of sericin, a protein synthesized in the distal segment of the silk gland and coating the fibroin fibers. Sericin is the specific substrate of an enzyme secreted by the moth and by which sericin is specifically dissolved, permitting the separation of fibroin threads and the passage of the imago (Kafatos and Williams[206]).

38. Coevolutionary diachronic epigenesis

Species evolve in relation to aspects of the environment but they do not evolve in isolation.

The term "coevolution" was coined by Ehrlich and Raven[207] to designate *evolutionary reciprocal interactions* between different kinds of organisms among which there is *no exchange of genetic information.* Molecular aspects of biochemical coevolution depend on the formation of couples

References p. 117

of *relata* and can be identified among groups of organisms living in a close ecological relationship.

The clearest demonstration of a coevolutionary diachronic molecular epigenesis has been accomplished by Ehrlich and Raven in their paper mentioned above[207]. An extensive survey of patterns of plant utilization and information on factors affecting food plant choices shows that the emergence of the biosynthesis of secondary plant substances has been critical in the evolution of angiosperm subgroups and perhaps of the angiosperms themselves.

Reciprocal selective responses appear "as a factor in the origination of organic diversity".

Angiosperms have, in diachronic molecular epigenesis, developed catenary biosyntagms leading to the biosynthesis of a variety of "secondary substances" (see Chapter 2 of this Volume). These compounds are not inimical to the plant's maintenance and growth. But some reduce the palatability of the plant for animals. This protection would, so to speak, introduce the plant into a new adaptive zone, with a possible evolutionary radiation and a characterization of new families. Another new adaptive zone would be entered when a population of insects, for instance, would become enabled to feed on some previously protected plant group. Here, in the absence of competition from other phytophages the population of insects considered would be free to diversify freely.

Ehrlich and Raven[207] underline that in situations in which the supply of the "preferred" plant is limited enough to become a factor of survival of a larva, the change of food plant would be favoured. After the restriction of certain groups of insects to a narrow range of food plants the formerly repellent substances of these plants would become attractants.

In this field the diachronic aspect must be kept in view, as a molecular innovation may have had a selective advantage in the context of a remote period and that the nature of the adaptive zone may have been modified by the organisms that entered it.

We refer the reader to the paper of Ehrlich and Raven for a wealth of arguments in favour of the thesis that

> "the fantastic diversification of modern insects has developed in large measure as the result of a stepwise pattern of coevolutionary stages superimposed on the changing pattern of angiosperm variation."[207]

Before the molecular coevolutionary aspects of food plant selection in butterflies as documented by Ehrlich and Raven were described, the present author (Duchâteau *et al.*[208]; Florkin[36,117]) had described what we now call biochemical coevolution in the case of the "specialized" type of pattern of inorganic constituents of the hemolymph in Lepidoptera, as well as in the larval stages of Hymenoptera which they considered as an evolutionary aspect parallel to the evolution of angiosperms (see Chapter 4, Vol. 29B).

Interesting cases of biochemical coevolution consist in the predator–prey relationship and the parasite–host relationship, the plants and pollinators relationship or the relationship of partners in symbiosis. Allelopathy (a term coined by Molish[209] to designate the effects on the growth of a plant of products of other plants) is active in the distribution patterns of plants (Muller[210]). All these aspects inserted in the biochemical continuum provide examples of biochemical coevolution, characterized by reciprocal evolutionary interactions and confirming that, even at the molecular integrative level, species do not evolve in isolation.

39. Biochemical taxonomy

It is to Aristotle that we owe the invention of the first practical system for the classification of living beings. More than 2000 years ago this naturalist of genius proposed to classify organisms according to the *degree of similarity* of their morphological and anatomical characteristics. Despite the appearance of other systems claimed to be more natural, but today forgotten, the system of Aristotle has survived and is today referred to as "the natural system". A certain number of species are collected together in a new group, defined by the possession of common characteristics, and several of these groups together form a more general category, etc. By the accumulation of taxonomic data it was possible to formulate a principle which was the subject of much admiration and satisfaction among naturalists; this was, that classification, based upon a few diagnostic characteristics, of a species and the placing of it in a given class enables one to forecast the existence of a whole range of morphological and biological traits in individuals of the same species. This idea, to which we are accustomed and which appears self-evident and banal, must have stirred those few who glimpsed the possibilities which at that time were still unsuspected by the majority. When finally the merit of this system was admitted, it was recognized that an essential criterion was the existence of a fundamental similarity, and not merely a superficial one. It also led

References p. 117

to the recognition of a common basic plan in the chief natural groups. During the first half of the 19th century, this idea appeared in numerous guises depending on the personal preferences of its protagonists. For the most theoretically inclined, like Goethe or Oken, "these ideal patterns which the creative principle set before itself were, so to say, Platonic ideas in the mind of the creative spirit" (Sherrington[211]). The more objective, as Julian Huxley[212] said of Thomas Huxley, "simply *assumed* that structural homology (or common archetypal plan) was the right key to unlock classificatory secrets". We might as well say that during the first half of the 19th century, the idea of a plan was purely descriptive and had no explanatory basis, and that Schiller was right when he is said to have replied to the exposition of Goethe: "Das ist keine Erfahrung, das ist eine Idee."

The publication of Charles Darwin's great book changed all this. Since the advent and generalization of the Evolution Theory, the basic criterion for natural classification, which up till then was the *degree of similarity*, has become the *degree of phylogenetic kinship*. The aim of present-day taxonomy is to build up a classification consisting of classes based upon phylogenetic relations. This new point of view has not led to any fundamental modification of classification. Here is an important point, the consideration of which leads to an increased confidence in the system established by taxonomists.

When discussing the significance of various characteristics of plants useful in their taxonomy at the supraspecific level, Erdtman[213] wrote:

> "Much attention is paid by taxonomists to minor characteristics such as the form, structure and arrangement of epidermal cells because they are essentially indifferent to external factors and consequently conservative.
>
> Such properties are therefore handed down from generation to generation and from species without suffering much change. Comparatively recent specializations possess little taxonomic interest."[213]

We find here what we have, at the molecular integrative level, called the *system*. From the biochemical aspect, as with other aspects, if it is true that the characteristics of the species and its sub-divisions are naturally inserted into the higher integrative level of the physiology of the organism and in its relation with its ecological niche, it is to be expected that the biochemical traits which are transmitted from organism to organism and bear witness to a common line of ancestors will not necessarily be those which show marked changes with respect to physiological or ecological aspects. The utility of supraspecific categories is above all to enable us to express our views on the probable nature of phylogeny.

As Calman[214] says,

"the characters most important in taxonomy are those which maintain themselves unchanged through the greatest range of variations."

Finding his knowledge of the descendance of organisms in phylogeny, as repeatedly claimed by the present author, does not prevent the comparative biochemists of exploring it by entirely different ways when in the case of orthologous proteins the opportunity is afforded to him to explore by entirely different methods the phylogeny of such a molecule as cytochrome *c*, as has been brilliantly accomplished by Margoliash and by Dayhoff. This is in complete intellectual agreement with the philosophy of taxonomy as illustrated by the quotations cited above.

But let us return to biochemical taxonomy. The Austrian chemist Rochleder[215], as early as 1854, clearly stated the taxonomic importance of the biochemical characteristics of plants, when the more general classes were concerned, thus:

"Die Familieähnligkeit der Pflanzen is bedingt durch das gleichzeitige Vorhandensein mehrerer Stoffreihen". Much information on the relations between the biochemistry and the taxonomy of plants was already compiled by Molish in a book which appeared in 1933[216]. The field has tremendously increased since then and is covered by an extensive literature (see Chapter 2 of this volume).

When we come to the higher categories in animal taxonomy, the idea of biochemical character is by no means a new one. For a long time, the taxonomists have considered the siliceous skeleton of Radiolaria as a characteristic of that order, and within this order the presence of strontium sulphate in the skeleton as characteristic of the sub-order Acantharia. In the phylum Porifera the species belonging to the class Calcispongiae have long been distinguished from other sponges, by the calcareous nature of their spicules. The phylum Brachiopoda has been divided into the Ecardines, characterized by a calcareous shell, and the Testicardines. These concepts were based on inorganic constituents. The consideration of biomolecules in the field of animal taxonomy was exemplified by the extensive enquiry of Reichert and Brown[217] on the crystalline forms of hemoglobins. The present author[6] has, in 1947 collected a large number of

References p. 117

data on the biochemical systematic of animals and has defined in biochemical terms such taxonomic categories as vertebrates, mammals, tunicates, cyclostomes, elasmobranchs, sipunculids, arthropods, insects and inquired into the hierarchy of biochemical characters, identifying for instance among the biochemical aspects of the water beetle *Hydrophilus piceus* such characters as belong to arthropods, insects, endopterygotes, Coleoptera or the genus *Hydrophilus*. Other aspects of the biochemical taxonomy of animals were collected in other publications of the same author[13,218]. The field of the biochemical taxonomy of animals has grown as much since as to permitting the publication of the extensive series of collective volumes of *Chemical Zoology*[219] and the general field of chemical taxonomy, to become the subject matter of a new periodical, *Chemical Taxonomy*.

Considering the ever growing mass of data it may be pointed out that the presence in organisms of a kind of biomolecule should not be taken as a taxonomic character without proper criticism. In the perspective developed in this Chapter, biomolecules of indirect homology which are often referred to in biochemical taxonomy, particularly in plant taxonomy, should not be considered as homologous without the demonstration of their production by a common biosynthetic catenary biosyntagm situated in the field of direct protein homology. What changes in phylogeny is not the chemical composition, which is not inherited, but the information governing the production of a specific array of metabolic syntagms characteristics of the species.

Among the reproaches currently formulated against chemical taxonomy, it is sometimes heard that it is mere platitude, confirming classical taxonomy. It is difficult to imagine the confusion which would occur if a different taxonomy arose from biochemical data. This would mean a collapse of the whole fabric of biology. Another reproach is that biochemical taxonomy tends to impose its dictates on systematicians. As said above, it does not often happen that contradictions arise between taxonomy and biochemical taxonomy. Sometimes, biochemists, in their comparative studies, have noted gross discrepancies between organismic and molecular data. In certain cases, at least, these discrepancies have been taken into account by competent taxonomists in the process of revision of their phyletic sequences of organisms and it has happened that this revision has brought a change in the tree of phylogeny, not by a dictate of comparative biochemistry but by an indication of possible error of taxonomists who, like all men, occasionally make mistakes.

V. SUMMARY

It is imperative to distinguish generalized evolution from restricted evolution. Besides the concept of energy conservation and other concepts of thermodynamics, the concept of generalized evolution is of an extensive nature, uniting the whole reality of the material universe. Generalized evolution at the biomolecular level appears as inevitable, and due to "the presence of certain matter with specified autocatalytic properties and under the maintenance of the finite (free) energy flow necessary to compensate for the steady state production of energy (Eigen[39]).

Restricted evolution is a concept of intensive nature, concerned with specific cases of the phylogeny of organisms.

In parts I to III of the present review, concerned with molecular events identified in the course of phylogeny of organisms, the accent has been placed on biosemiotic aspects, *i.e.* the intensive aspects of information involved in the impact of natural selection acting at the level of the organism. These aspects identified at the integrated molecular level are designated as diachronic molecular epigenesis, a denomination avoiding confusion with the evolution of populations and giving precisely a reference to the molecular level.

Biochemical evolution, an aspect of molecular evolution, has been essentially considered at the level of metabolic catenary biosyntagms, on the basis of the diachronic molecular epigenesis of their effectors. These are sequenced biosyntagms and their diachronic molecular epigenesis results from the diachronic molecular epigenesis of their sequencing biosyntagms, transmitted in descendance. The widely-confirmed colinearity of the phylogeny of biomolecules and of their biosynthesis leads to suggest that the essential aspect of "biochemical evolution" resides in the diversification of a restricted number of catenary metabolic biosyntagms, by terminal extension or deletions, or lateral extensions of deletions. This diachronic epigenesis of catenary metabolic biosyntagms and of their regulation appears as the result of point mutations and of resulting polymorphism due to the feedback, of selection on the genetic structure of population, of the more radical changes of signified designated as commutations and often deriving from gene duplications, of novel batteries of producer genes, of deletions and insertions, of heterotypic expressions, etc. These diversifications can

References p. 117

be grouped in the notion of the extraction of biosyntagms from the molecular biosystem, the basic concept of which is homology.

That the signified of biomolecules may be recognized as being involved at levels of integration higher than the molecular one, for instance at the level of the self-assembly in supramolecular structures, at the physiological or ecological levels is reviewed in part IV.

While the concept of the evolution of organisms as an expression of biochemical evolution remains a subject of study, it has received a great deal of support from the recognition of the concordance of the phylogenetic trees of sequencing or sequenced biomolecules and of the organisms concerned. This supports the concept according to which the structure and nature of organisms are expressions of the biomolecular order implemented by the heterocatalytic command of the specification of amino acid sequences. The pressure of natural selection should not be considered as being exerted on specific biomolecules, but on organisms, *i.e.* on the biomolecular order expressed in the structure and nature of the whole collection of biomolecules of which the organism is made. This molecular order is subject to changes in descent and to selection pressure. In such a perspective the evolution of organisms and the evolution of the biomolecular order (biochemical evolution) are recognized as interrelated facets of the same natural reality.

REFERENCES

1 S. Prakash, R. C. Lewontin and J. L. Hubby, *Genetics*, 61 (1969) 841.
2 G. Teissier, *Scientia*, 52 (1962) 1.
3 O. von Fürth, *Vergleichende Chemische Physiologie der niederen Tiere*, Fischer, Iena. 1903.
4 E. Baldwin, *An Introduction to Comparative Biochemistry*, Cambridge University Press, London, 1937.
5 A. J. Kluyver and H. J. L. Donker, *Chemie Zelle Gewebe*, 13 (1926) 134.
6 M. Florkin, *L'Évolution Biochimique*, Paris, Masson, 1944.
7 M. Florkin, *Biochemical Evolution* (translation by S. Morgulis), Academic Press, New York, 1949.
8 H. C. Dessauer, in *Systematic Biology*, Natl. Acad. Sci. (U.S.), 1969, p. 325.
9 G. G. Simpson, *Science*, 146 (1964) 1535.
10 H. F. Blum, *Amer. Scientist*, 49 (1961) 474.
11 E. Mayr, *Federation Proc.*, 23 (1964) 1231.
12 P. Weiss, in A. Koestler and J. R. Smythies (Eds.), *New Perspectives in the Life Sciences*, Hutchinson, London, 1969.
13 M. Florkin, *Unity and Diversity in Biochemistry*, (translation by T. Wood), Pergamon, London, 1960.
14 T. Dobzhansky, *Genetics and the Evolutionary Process*, Columbia Press, New York, 1970.
15 E. Schrödinger, *What is Life?*, Cambridge University Press, London, 1948.
16 L. Brillouin, *Amer. Scientist*, 37 (1949) 554.
16a L. Brillouin, *Vie, Matière et Observation*, A. Michel, Paris, 1959.
17 C. W. F. McClare, *J. Theoret. Biol.*, 30 (1971) 1.
18 L. J. Henderson, *The Fitness of the Environment*, MacMillan, New York, 1927.
19 C. Brinton, in E. W. Forbes and J. H. Finley (Eds.), *The Saturday Club: A century completed: 1920–1956*, Houghton Mifflin, Boston, 1958.
20 J. Parascandola, *J. Hist. Biol.*, 4 (1971) 63.
21 J. C. Speakman, *Molecules*, McGraw Hill, New York, 1966.
22 A. J. Lotka, *Elements of Physical Biology*, Williams and Wilkins, Baltimore, 1925.
23 H. F. Blum, *Time's Arrow and Evolution*, 1st edn. 1951, 3rd edn. 1968, Princeton University Press, 1968.
24 I. Prigogine, in M. Marois (Ed.), *De la Physique Théorique à la Biologie*, C.N.R.S., Paris, 1971.
25 P. Glandsdorff and I. Prigogine, *Physica*, 30 (1964) 351.
26 I. Prigogine, *Introduction to Thermodynamics of Irreversible Processes*, 3rd edn.. Wiley, New York, 1971, *Nature*, 246 (1973) 67.
27 R. Buvet, E. Etaix, F. Godin, P. Leduc and L. le Port, in R. Buvet and C. Ponnamperuma (Eds.), *Chemical Evolution and the Origin of Life*, North-Holland, Amsterdam, 1971.
28 L. Szilard, *Z. Physik*, 53 (1929) 840.
29 C. Shannon and W. Weaver, *The Mathematical Theory of Communication*, University of Illinois Press, Urbana, 1949.
30 L. Brillouin, *Science and Information Theory*, Academic Press, New York, 1956, 2nd edn. 1962.
31 H. Quastler (Ed.), *Information Theory in Biology*, University of Illinois Press, Urbana, 1953.
32 C. E. Shannon, *Bell Syst. Techn. J.*, 27 (1948) 379.
33 H. J. Morowitz, *Bull. Math. Biophys.*, 17 (1955) 81.
34 H. A. Johnson, *Science*, 168 (1970) 1545.
35 M. Florkin, *Acad. Roy. Belg., Bull. Classe Sci.*, 5th ser. 51 (1965) 239.

36 M. Florkin, *Aspects Moléculaires de l'Adaptation et de la Phylogénie*, Masson, Paris, 1966. (The section of this book dealing with adaptation has been translated and brought up to date in: M. Florkin and E. Schoffeniels, *Molecular Approaches to Ecology*, Academic Press, New York, 1969).
37 E. Sondheimer and J. B. Simeone (Eds.), *Chemical Ecology*, Academic Press, New York, 1970.
38 M. Florkin, *Acad. Roy. Belg., Bull. Classe Sci.*, 5th ser., 55 (1969) 257.
39 M. Eigen, *Naturwissenschaften*, 58 (1971) 465.
40 P. Karlson and M. Lüscher, *Nature*, 183 (1959) 55.
41 W. L. Brown and T. Eisner, in W. L. Brown, *Amer. Naturalist*, 102 (1968) 188.
42 F. de Saussure, *Cours de Linguistique Générale*, 3rd edn., Payot, Paris, 1971.
43 A. Richards, *The Meaning of Meaning* (with R. G. Ogden), Routledge and Kegan Paul, London, 1924.
44 U. Eco, *La Structure Absente*, Mercure de France, Paris, 1972.
45 R. Barthes, *Eléments de Sémiologie*, Du Seuil, Paris, 1964.
46 W. Büchel, *Nature*, 213 (1967) 319.
47 J. P. F. J. Ryan, *J. Theoret. Biol.*, 36 (1972) 139.
48 J. A. Wilson, *Nature*, 219 (1968) 534.
49 I. Harris, *Brit. Med. Bull.*, 16 (1960) 183.
50 A. C. J. Burgers, *Endocrinology*, 68 (1961) 698.
51 O. Hechter and Th. Braun, in M. Margoulies and F. C. Greenwald (Eds.), *Structure–Activity Relationships of Proteins and Polypeptide Hormones*, Excerpta Medica, Amsterdam, 1971.
52 R. Schwyzer, *J. Mond. Pharm. (The Hague)*, 11 (1968) 254.
53 R. E. Dickerson and I. Geis, *The Structure and Action of Proteins*, Harper and Brown, New York, 1969.
54 R. E. Dickerson, in H. Neurath (Ed.), *The Proteins*, Vol. 2, Academic Press, New York, 1964.
55 J. Monod, J. Wyman and J. P. Changeux, *J. Mol. Biol.*, 12 (1965) 88.
56 E. R. Stadtman, *Harvey Lectures*, 65 (1971) 97.
57 A. R. Williamson, *Essays Biochem.*, 5 (1969) 139.
58 I. P. Crawford and C. Yanofsky, *Proc. Natl. Acad. Sci. (U.S.)*, 44 (1958) 1161.
59 J. C. Gerhard and H. K. Schachman, *Biochemistry*, 4 (1965) 1054.
59a N. O. Kaplan, *Ann. N. Y. Acad. Sci.*, 151 (1968) 382.
59b J. M. Trujillo, B. Walden, P. O'Neil and H. B. Anstall, *Nature*, 213 (1967) 88.
59c J. V. Kilmartin and J. B. Clegg, *Nature*, 213 (1967) 269.
60 D. Prescott and P. Kuempel, *Proc. Natl. Acad. Sci. (U.S.)*, 69 (1972) 2842.
61 A. Kornberg, *Science*, 163 (1969) 1410.
62 G. H. Stent, *Advan. Virus Res.*, 5 (1968) 95.
63 B. M. Lewin, *The Molecular Basis of Gene Expression*, Wiley, London, 1970.
64 E. Fuchs, R. L. Millette, W. Zillig and G. Walter, *Europ. J. Biochem.*, 3 (1967) 183.
65 B. C. Baguely and M. Staehelin, *Biochemistry*, 7 (1968) 45.
66 R. S. Hayward, G. L. Elicieri and S. B. Weiss, *Cold Spring Harbor Symp. Quant. Biol.*, 31 (1966) 459.
67 W. Fuller and A. Hodgson, *Nature*, 215 (1967) 217.
68 S. H. Kim, G. J. Quigley, F. L. Suddath, A. McPherson, D. Sneden, J. J. Kim, J. Weinzierl and Alexander Rich, *Science*, 179 (1973) 285.
69 F. H. Bergmann, P. Berg and M. Dieckmann, *J. Biol. Chem.*, 236 (1961) 1735.
70 F. Lipmann, in V. V. Koningsberger and L. Bosch (Eds.), *Regulation of Nucleic Acid and Protein Biosynthesis, [BBA Library, Vol. 10]*, Amsterdam, Elsevier, 1967, p. 177.
71 J. Lucas-Lenard and A. L. Haenni, *Proc. Natl. Acad. Sci. (U.S.)*, 59 (1968) 554.

72 K. A. Marcker, B. F. C. Clark and J. S. Anderson, *Cold Spring Harbor Symp. Quant. Biol.*, 31 (1966) 279.
73 B. F. C. Clark and K. A. Marcker, *J. Mol. Biol.*, 17 (1966) 394.
74 F. Chapeville, F. Lipmann, G. von Ehrenstein, B. Weisblum, W. J. Ray and S. Benzer, *Proc. Natl. Acad. Sci. (U.S.)*, 48 (1962) 1086.
75 M. S. Bretscher, *Nature*, 220 (1968) 1088.
76 M. S. Bretscher, *J. Mol. Biol.*, 42 (1969) 595.
77 J. S. Anderson, M. S. Bretscher and B. F. C. Clark, *Nature*, 215 (1967) 490.
78 J. C. Brown and P. Doty, *Biochem. Biophys. Res. Commun.*, 30 (1968) 284.
79 A. G. Cairns-Smith, *The Life Puzzle*, Oliver and Boyd, Edinburgh, 1971.
80 H. Y. K. Chuang, A. G. Atherly and F. E. Bell, *Biochem. Biophys. Res. Commun.*, 28 (1967) 1013.
81 T. Ohta, I. Shimada and K. Imahori, *J. Mol. Biol.*, 26 (1967) 519.
82 M. Yarus and P. Berg, *J. Mol. Biol.*, 42 (1969) 171.
83 A. N. Baldwin and P. Berg, *J. Biol. Chem.*, 241 (1966) 839.
84 F. H. Crick, *Nature*, 234 (1971) 25.
85 F. Jacob and J. Monod, *J. Mol. Biol.*, 3 (1961) 318.
86 J. Monod and G. Cohen-Bazire, *Compt. Rend.*, 236 (1953) 530.
87 R. A. Yates and A. B. Pardee, *J. Biol. Chem.*, 227 (1957) 677.
88 E. J. Britten and E. H. Davidson, *Science*, 165 (1969) 349.
89 B. Pullman and A. Pullman, *Nature*, 196 (1962) 1137.
90 B. Pullman, in E. Schoffeniels (Ed.), *Biochemical Evolution and the Origin of Life*, North-Holland, Amsterdam, 1971.
91 H. L. Kornberg, *Essays Biochem.*, 2 (1966) 1.
92 E. R. Stadtman, *Advan. Enzymol.*, 28 (1966) 41.
93 D. E. Koshland, Jr., in A. Åkeson and A. Ehrenberg (Eds.), *Structure and Function of Oxido-Reduction Enzymes*, Pergamon, Oxford, 1972.
94 R. G. Shulman, E. Ogawa, K. Wüthrich, Y. Yamane, J. Peisach and W. E. Blumberg, *Science*, 175 (1969) 251.
95 K. P. Chakraborty and R. B. Hurlbert, *Biochem. Biophys. Acta*, 47 (1961) 607.
96 C. W. Long and A. B. Pardy, *J. Biol. Chem.*, 242 (1967) 4715.
97 D. E. Koshland, Jr, *Harvey Lectures*, 65 (1969–1970) 33.
98 G. A. Robison, R. W. Butcher and E. W. Sutherland, *Cyclic AMP*, Academic Press, New York, 1971.
99 M. Eigen and L. de Maeyer, *Naturwissenschaften*, 53 (1966) 50.
100 J. D. Watson and F. H. C Crick, *Nature*, 177 (1953) 964.
101 J. D. Watson, *Molecular Biology of the Gene*, 2nd edn., Benjamin, New York, 1970.
102 G. S. Stent, *Molecular Genetics*, Freeman, San Francisco, 1971.
103 M. Florkin, *Actes Soc. Helv. Sci. Nat.*, (1957) 35.
104 T. Dobhzansky, *Evolution, Genetic and Man*, Wiley, New York, 1955.
105 M. Florkin, in E. Schoffeniels (Ed.), *Biochemical Evolution and the Origin of Life*, North-Holland, Amsterdam, 1971.
106 G. H. Dixon, *Essays Biochem.*, 2 (1966) 147.
107 T. H. Jukes, *Molecules and Evolution*, Columbia Univ. Press, New York, 1966.
108 C. Manwell and C. M. A. Baker, *Molecular Biology and the Origin of Species: Heterosis, Protein Polymorphism and Animal Breeding*, Univ. of Washington Press, Seattle, 1970.
109 S. Granick, in V. Bryson and H. J. Vogel (Eds.), *Evolving Genes and Proteins*, Academic Press, New York, 1965.
110 A. Lehninger, *Biochemistry, The Molecular Basis of Cell Structure and Function*, Worth, New York, 1970.

111 S. Cohen, *Science*, 139 (1963) 1017.
112 E. A. Zeller, *Advan. Enzymol.*, 8 (1948) 459.
113 G. Wald, in A. I. Oparin (Ed.), *Evolutionary Biochemistry*, Pergamon, Oxford, 1963.
114 A. Redfield, *Am. Naturalist*, 70 (1946) 110.
115 H. M. Fox, *Proc. Roy. Soc. (London). Ser. B*, 136 (1949) 378.
116 M. Florkin. *Experientia*, 4 (1948) 176.
117 M. Florkin, *A Molecular Approach to Phylogeny*, Elsevier, Amsterdam, 1966.
118 F. H. Crick, *Symp. Soc. Exptl. Biol.*, 12 (1958) 138.
119 C. B. Anfinsen, *The Molecular Basis of Evolution*, Wiley, New York, 1959.
120 M. Florkin, *Acad. Roy. Belg. Classe Sci*, 48 (1962) 819.
121 I. Ingram, *The Hemoglobins in Genetics and Evolution*, Columbia Univ. Press, New York, 1963.
122 E. Margoliash and W. M. Fitch, *Ann. N. Y. Acad. Sci.*, 21 (1968) 217.
123 W. M. Fitch and E. Margoliash, *Evolutionary Biol.*, 4 (1970) 67.
124 M. O. Dayhoff (Ed.), *Atlas of Protein Sequence and Structure 1972*, Vol. 5, Nat. Biomed. Res. Found., Georgetown, 1972; Vol. 5, Suppl. I, 1973.
125 E. Margoliash and W. M. Fitch, in M. Marois (Ed.), *De la Physique Théorique à la Biologie*, C.N.R.S., Paris, 1971.
126 H. Neurath, K. A. Walsh and W. P. Winter, *Science*, 158 (1967) 1638.
127 C. S. Wright, R. A. Alden and J. A. Kraut, *Nature*, 221 (1969) 235.
128 J. Monod, *Le Hasard et la Nécessité. Essai sur la Philosophie Naturelle de la Biologie Moderne*, Du Seuil, Paris, 1970.
129 S. Ohno, *Evolution by Gene Duplication*, Springer, Berlin, 1970.
130 J. D. Watson and F. H. C. Crick, *Cold Spring Harbor Symp. Quant. Biol.*, 18 (1953) 193.
131 E. Freese, *J. Mol. Biol.*, 1 (1959) 87.
132 E. Freese, in J. H. Taylor (Ed.), *Molecular Genetics*, Part I, Academic Press, New York, 1963.
133 R. L. Watts and D. C. Watts, *J. Theoret. Biol.*, 20 (1968) 227.
134 R. J. Britten and D. E. Kohne, *Carnegie Inst. Wash., Year Book*, 66 (1968) 83.
135 R. J. Britten and E. H. Davidson, *Quart. Rev. Biol.*, 46 (1971) 111.
136 R. L. Watts, in E. Schoffeniels (Ed.), *Biochemical Evolution and the Origin of Life*, North-Holland, Amsterdam, 1971.
137 T. Ohta and M. Kinamura, *Nature*, 233 (1971) 118.
138 T. Dobhzansky, *Science*, 175 (1972) 49.
139 B. J. McCarthy, *Progr. Nucleic Acid Res.*, 4 (1965) 129.
140 T. H. Jukes, *Biochem. Biophys. Res. Commun.*, 24 (1966) 744.
141 R. L. Hill and J. Buettner-Janusch, *Fed. Proc.*, 23 (II) (1964) 1236.
142 E. L. Smith and E. Margoliash, *Fed. Proc.*, 23 (II) (1964) 1243.
143 E. Margoliash and A. Schjejter, *Advan. Protein Chem.*, 21 (1966) 113.
144 E. Margoliash, W. M. Fitch and R. E. Dickerson, in E. Schoffeniels (Ed.), *Biochemical Evolution and the Origin of Life*, North-Holland, Amsterdam, 1971.
145 R. E. Dickerson, *J. Mol. Evolution*, 1 (1971) 26.
146 M. Kimura, *Nature*, 217 (1968) 624.
147 J. L. King and T. H. Jukes, *Science*, 164 (1969) 788.
148 E. Zuckerkandl and L. Pauling, in *Problems of Evolutionary and Technical Biochemistry*, Science Press, Acad. Sci. U.S.S.R. (1964) in Russian. The original English version has been published in *J. Theoret. Biol.*, 8 (1965) 357.
149 W. D. McElroy and H. H. Seliger, in A. I. Oparin (Ed.), *Evolutionary Biochemistry*, Pergamon, Oxford, 1963.
150 W. J. Rutter, *Fed. Proc.*, 23 (1964) 1248.
150a C. Y. Lai and B. L. Horecker, *Essays Biochem.*, 8 (1972) 149.

151 E. Broda, *Progr. Biophys. Mol. Biol.*, 21 (1970) 145.
152 E. Broda, in E. Schoffeniels (Ed.), *Biochemical Evolution and the Origin of Life*, North-Holland, Amsterdam, 1971.
153 M. Florkin, *Introduction à la Biochimie Générale*, Masson, Paris, 1941.
154 S. Granick, *Harvey Lectures*, 44 (1950) 220.
155 L. Bogorad, *Biogenesis of Natural Products*, Pergamon, Oxford, 1963.
156 T. Robinson, *The Organic Constituents of Higher Plants*, Burgess, Minneapolis, 1963.
157 J. H. Richards and J. B. Hendrickson (Eds.), *The Biosynthesis of Steroids, Terpenes and Acetogenins*, Benjamin, New York, 1964.
158 K. Bloch, in C. A. Leone (Ed.), *Taxonomic Biochemistry and Serology*, Ronald, New York, 1964.
159 G. A. D. Haslewood, in E. Schoffeniels (Ed.), *Biochemical Evolution and the Origin of Life*, North-Holland, Amsterdam, 1971.
160 K. Amimoto, *Hiroshima J. Med. Sci.*, 15 (1966) 213.
161 G. A. D. Haslewood, *Biol. Rev. Cambridge Phil. Soc.*, 39 (1964) 537.
162 G. A. D. Haslewood, *Bile Salts*, Methuen, London, 1967.
163 N. H. Horowitz, *Proc. Natl. Acad. Sci (U.S.)*, 31 (1945) 153.
164 E. B. Lewis, *Cold Spring Harbor Symp. Quant. Biol.*, 16 (1951) 159.
165 H. Mason, *Advan. Enzymol.*, 16 (1955) 105.
166 G. Wald, *Science*, 113 (1951) 287.
167 H. J. Saz and A. Vidrine, Jr., *J. Biol. Chem.*, 234 (1959) 2001.
168 H. J. Saz and A. Weil, *J. Biol. Chem.*, 237 (1962) 2053.
169 E. Kmetec and E. Bueding, *J. Biol. Chem.*, 236 (1961) 584.
170 E. Bueding, *Fed. Proc.*, 21 (1962) 1039.
171 H. J. Saz, in M. Florkin and B. Scheer, (Eds.), *Chemical Zoology*, Vol. 3, Academic Press, New York, 1969.
172 F. O. Schmitt, *Proc. Am. Phil. Soc.*, 100 (1956) 476.
173 S. Asakura, G. Eguchi and I. Ino, *J. Mol. Biol.*, 10 (1964) 42.
174 L. G. Tilney and K. R. Porter, *J. Cell. Biol.*, 34 (1967) 327.
175 P. Mueller and D. O. Rudin, in D. R. Sanadi (Ed.), *Current Topics of Bioenergetics*, Vol. 3, Academic Press, New York, 1969, p. 175.
176 A. Lehninger, *Naturwissenschaften*, 53 (1966) 57.
177 P. Traub and M. Nomura, *Proc. Natl. Acad. Sci. (U.S.)*, 59 (1968) 777.
178 H. Fraenkel-Conrat and B. Singer, *Biochim. Biophys. Acta*, 24 (1957) 540.
179 W. B. Wood, R. S. Edgar, J. King, I. Lielausis and M. Henninger, *Fed. Proc.*, 27 (1968) 1160.
179a Assembly of intracellular structures, *Fed. Proc.*, 31 (1972) 10.
180 T. R. Podelski and J. P. Changeux, in J. F. Danielli, J. F. Moran and D. J. Triggle (Eds.), *Fundamental Concepts in Drug-Receptor Interactions*, Academic Press, New York, 1970.
181 M. Rodbell, L. Birnbauer, S. L. Pohl and H. J. M. Kraus, *Acta Diab. Lat.*, 7 (1970), Suppl. I, p. 9.
182 L. Smith, cited by E. Margoliash *et al.*[184].
183 G. H. Barlow and E. Margoliash, *J. Biol. Chem.*, 241 (1961) 1473.
184 E. Margoliash, G. H. Barlow and V. Beyers, *Nature*, 228 (1970) 723.
185 E. Margoliash, *Harvey Lectures*, 66 (1972) 177.
186 M. Florkin, in F. Clark and R. L. M. Synge, (Eds.), *The Origin of Life on the Earth*, Pergamon, Oxford, 1959.
187 M. Florkin, in A. I. Oparin (Ed.), *Evolutionary Biochemistry*, Pergamon, Oxford, 1963.
188 M. Florkin, *Année Biol.*, 6 (1967) 499.
189 M. H. Briggs and R. B. Duncan, *Nature*, 191 (1961) 1310.

190 T. W. Goodwin, *Biochem. Soc. Symp.*, 6 (1951) 63.
191 D. M. Steven, *J. Exptl. Biol.*, 25 (1948) 369.
192 M. Hartmann, F. G. Medem, R. Kuhn and H. J. Bielig, *Z. Naturforsch.*, 3 (1947) 330.
193 N. Kirshner, *Pharmacol. Rev.*, 11 (1959) 350.
194 A. Bertler and E. Rosengren, *Acta Physiol. Scand.*, 47 (1959) 350.
195 G. Kerkut and R. J. Walker, *Comp. Biochem. Physiol.*, 3 (1961) 143.
196 M. J. Greenberg, *Brit. J. Pharmacol.*, 15 (1960) 365.
197 D. Sweeney, *Science*, 139 (1963) 1051.
198 M. S. Dure and M. J. Cormier, *J. Biol. Chem.*, 238 (1963) 790.
199 H. A. Krebs, in N. O. Kaplan and E. P. Kennedy (Eds.), *Current Aspects of Biochemical Energetics*, Academic Press, New York, 1966.
200 E. O. Wilson and W. H. Bossert, *Recent Progr. Hormone Res.*, 19 (1963) 673.
201 E. O. Wilson, *Science*, 149 (1965) 1064.
202 E. O. Wilson, in T. Sebeok (Ed.), *Animal Communication*, Indiana Univ. Press, Bloomington, Md., 1968.
203 E. O. Wilson, in E. Sondheimer and J. B. Simeone, *Chemical Ecology*, Academic Press, New York, 1970.
204 N. C. Johnston, J. H. Law and N. Weaver, *Biochemistry* 4 (1965) 1615.
205 R. A. Morse and N. E. Gary, *Bee World*, 42 (1961) 197.
206 F. C. Kafatos and C. M. Williams, *Science*, 146 (1964) 538.
207 N. R. Ehrlich and P. H. Raven, *Evolution*, 18 (1964) 586.
208 Gh. Duchâteau, M. Florkin and J. Leclerq. *Arch. Intern. Physiol. Biochim.*, 61 (1953) 518.
209 H. Molish, *Der Einfluss einer Pflanze auf die andere. Allelopathie*, Fischer, Jena, 1937.
210 C. H. Muller, *Bull. Torry Botan. Club*, 93 (1966) 332.
211 Sir Ch. Sherrington, *Goethe on Nature and Science*, Cambridge, Univ. Press, 1949, p. 24.
212 J. Huxley, *Evolution, The Modern Synthesis*, Allen and Unwin, London, 1942.
213 H. Erdtman, *Progr. Org. Chem.*, 1 (1952) 22.
214 W. T. Calman, *The Classification of Animals*, Methuen, London, 1949.
215 F. Rochleder, *Phytochemie*, 1854.
216 H. Molish, *Pflanzenchemie und Pflanzenversandschaft*, Springer, Vienna, 1933.
217 E. T. Reichert and A. P. Brown, *The Crystallography of Hemoglobins*, Carnegie Institution, Washington, 1909.
218 M. Florkin, in *First International Congress of Biochemistry, 19–25 August 1949*, Congress Lecture, 23 Aug. 1949.
219 M. Florkin and B. T. Scheer (Eds.) *Chemical Zoology*, Vol. I–VII, Academic Press, New York, 1967–1972.
220 K. Danna and D. Nathans, *Proc. Natl. Acad. Sci. (U.S.)*, 69 (1972) 3097.
221 E. Gyurastis and R. Wake, *J. Mol. Biol.*, 73 (1973) 55.
222 H. G. Callan, *Proc. Roy. Soc. (London), Ser. B*, 181 (1972) 19.

GLOSSARY

Alleles. Alternative forms of a gene occurring at a definite place on a chromosome.

Allelopathy. Effects of products of a plant on the growth of another plant.

Allomones. Coactones (see below) transmitting chemical signals to individuals of other species.

Analogy (biochemical). Biomolecules having similar functions in different groups of organisms are designated as analogous.

Anticodon. A sequence (on RNA) complementary of a codon (see below).

Battery of producer genes (Eucaryotes). Unlinked structural (producer) genes coordinated by integrator genes.

Bioseme. Minimal configuration aspect, carrier (significant) of molecular signification (signified), either sequential, structural, functional, protective, connective, motive, signaling, catalytic, processing, regulating, priming, repressing, releasing, etc.

Biosyntagm. An associative configuration of biosemes composed of significant units in a relation of reciprocal solidarity.

Biosynthetic primary pathways. Biosynthetic catenary pathways of general occurrence.

Biosystem. The expression of the diachronic (evolutionary, hereditary) relations of biomolecules.

Catenary metabolic biosyntagms. Biosyntagms composed of chains of enzymes involved in a catabolic or an anabolic metabolic pathway.

Central metabolic biosyntagm. A catenary biosyntagm found in the majority of cells and allowing for the acquisition of free energy in the form of ATP high-energy bonds, and for the provision of the starting points of biosynthetic pathways.

Cistron. Portion (biosyntagm) of the nucleotide sequence of DNA which provides the sequencing information for a single polypeptide chain.

Coactones. Ecomones (see below) active in the process of relationship of an active and directing organism and a passive and receiving organism.

Codon. A bioseme consisting of three adjacent nucleotides and coding for an amino acid.

Coevolution. Evolutionary reciprocal interactions between different kinds of organisms among which there is no exchange of genetic information.

Commutation. Radical change of the significant and of the signified of a biomolecule in diachronic epigenesis.

Continuum (biochemical). All molecular aspects of organisms and their molecular extensions.

Diachronic (molecular) epigenesis. Diachronic (in the phylogeny of organisms) epigenesis of the synchronic (limited to life span) epigenesis. Results in biomolecular changes along the branches of phylogeny.

Ecomones. Non-trophic molecules contributing to insure, in an ecosystem, a flux of information between organisms.

Endocoactones. Coactones which are not liberated by the coactor in the environment.

Evolution. Divergence in the gene pool of populations of organisms, resulting in differential selection, by differential reproduction, of genetic variations in the populations concerned and governed by interaction of population size, migration rate and selection intensity.

Evolution (biochemical). Trivial expression designating the pattern of changes accomplished through diachronic molecular epigenesis along the phylogeny of organisms.

Evolution (molecular). Trivial expression including chemical (prebiological) evolution and biochemical evolution (see above).

Exocoactones. Coactones liberated in the medium by a coactor and reaching the coactee.

Heteropolymer. Quaternary biosyntagm resulting from the association of different tertiary biosyntagms.

Heterotypic expression. Insertion of a bioseme in a new catenary biosyntagm.

Homology (biochemical). Common ancestry at the biomolecular level.

Homopolymer quaternary biosyntagm. An association of identical protomers, the signified of which differs from the signified of each protomer, as a result of subunit interaction.

Isology (biochemical). Chemical kinship among biomolecules.

Modulation (of a signified). Quantitative changes introduced in a signified.

Molecular biosemiotics. Intensive aspect of biomolecular systems of signification, considered besides the thermodynamic viewpoint and besides the quantifying (extensive) viewpoint of the information theory.

Non-allelic genes. Contrary to the allelic ones (see above), they no longer occur at definite places on the chromosomes, and do not pair at synapsis.

Operon (Procaryotes). A biosyntagm composed of linked cistrons.

Orthologous (genes or amino acid sequences). Sequences whose genes have a direct lineage corresponding to the lineage of species (antonym: paralogous, see below).

Paralogous (genes or amino acid sequences). Following a gene duplication, have evolved different forms of the signified (antonym: orthologous, see above).

Pheromones. Exocoactones produced by a coactor and acting on a coactee, both coactor and coactee belonging to the same species.

Phylogenesis. The process of the advent of new species and new branches of the phylogenetic tree and by extension the process of the production of new biomolecules or modified biomolecules, along these branches.

Phylogeny. Theory expressed by the phylogenetic (or phylogenic) tree in which is embodied our knowledge of the succession or organisms in degree of specialization.

Physiological radiation. Insertions of the same biomolecular significant in several biosemiotic aspects of levels of integration higher than the molecular one.

Point mutation. Result of random (see below) change in the course of recombinations, crossing over, etc. of DNA sequences, and consequently of polypeptide sequences.

Polygamy (biomolecular). Different forms of utilization of a same biomolecule as receptor in the formation of couples of *relata* at the molecular integrated level.

Primary biosyntagm. A biosyntagm in which biosemes are directly related to an amino acid sequence.

Quaternary biosyntagm. Established through spatial relationships between the biosemes (subunit interaction) carried by the polypeptidic chains of a multichain protein.

Random changes. Depending on statistical causality resulting from molecular and supramolecular structures, and not on a functional or teleological organization.

*Regulatory enzymes (*or *allosteric enzymes).* Enzymes, the signified of which may be modulated.

Saltatory replication. Production of families of repeated DNAs in evolutionary events which are initiated and terminated, rather than continuing through the course of the evolution of organisms.

Secondary biosynthetic pathways. Limited, in opposition to primary pathways (see above), to special categories of organisms.

Sequencing biosyntagms. Biosyntagms composed of nucleotides or as codons as biosemes, and the signified of which is of sequencing nature, governing the sequence of nucleotides or of amino acids in their polymers.

Semiotics. The general science of signification. Biosemiotics is one of its sections.

Significant and signified. (see Bioseme, above).

Synchronic epigenesis. Epigenetic development and maintenance of an organism, resulting from the determination, by the genes, of the patterns of developmental processes.

Tertiary biosyntagm. In these biosyntagms, the biosemes are established by folding, bringing different biosemes together.

Chapter II

Biochemical Evolution in Plants

T. SWAIN

Royal Botanic Gardens, Kew (Great Britain)

1. Introduction

Plants are no more ancient than animals and certainly less so than bacteria[1–16], but, from the biochemical point of view they are much more sophisticated than either. Well over four-fifths of all presently known organic compounds of natural origin have been isolated from plants[17–27]. Moreover, this fraction is likely to increase as modern methods of separation and structural determination[28] are applied to the ever increasing number of newly investigated plant species[29]. The preponderance in the numbers of compounds of plant origin is reflected in the variety and complexity of their structures[30–38], and hence in the heterogeneity of the underlying biochemistry[37–41]. This is not to suggest that there are only a few novel compounds found in the Animal Kingdom[42,43], nor that animal biochemistry is monotonously repetitive[44], but the major classes of animals, such as Mammalia, show only subtle variation, in their chemistry and biochemistry[45,46]. This is in contrast to the extreme diversity and intricacy of their controlled behaviour patterns[47,48], a characteristic which plants do not possess. Indeed, we might sum up the major differences between plants and animals in the aphorism "animals act, plants produce".

The effect of such a dichotomy on the biochemical evolution of animals and plants can be seen from the matter presented in the various chapters of this volume. We must remember, however, that the evolution of any group of organisms is dependent on the evolution of all. Species do not evolve in isolation[1,2,5,49–58]. We must also take into account the vast changes in geological environments which have occurred since the earth

References p. 287

began[11,59,60]. Without changes of some sort, evolution could not have taken place: without evolution, life would have undoubtedly ceased to exist almost as soon as it began. For plants, the most important features of their evolution have been contingent on their ability to change and diversify their biochemistry as the environment changed around them.

We do not know why biochemical changes took place in the past, nor in many cases, even whether they have occurred. Nor do we know what importance to attach to a single change; furthermore we have little idea about the exact time at which any given biochemical transformation was introduced. We cannot, therefore, present a continuous spectrum of biochemical evolution, including the order and time of the individual events. All that can be done is to pinpoint certain areas where the course seems clear. The rest must await an increase in our knowledge and understanding.

This Chapter first outlines present ideas about the evolution of the Plant Kingdom based wholly on our knowledge of the palaeontological record[1,3,7,49,50,55–62] and on the assumptions underlying the classification of the existing flora[63–72]. Against these proposals, the biochemical evolution of several groups of compounds, biosynthetic pathways and cellular structures will be examined.

2. The Plant Kingdom

(a) Introduction

The concept of dividing living organisms into two main groups, animal and plant, is extremely ancient, undoubtedly older than language itself[1–3]. However, during the last hundred years or so, the limitations have been much criticised[15,73–82] as microscopists uncovered the abundance of unicellular organisms which did not readily fit into either of these divisions[16,82]. Nevertheless the two Kingdom system has remained the basis of many comprehensive biological classifications, the formal outlines of most being based on that in *Systema Naturae* which was published by Linnaeus[83] in 1735.

The main characters which are normally used by biologists to differentiate the two Kingdoms are[1,64]: (*i*) plants contain chlorophylls and are photosynthetic (autotrophic) or are derived from photosynthetic ancestors, whereas animals do not contain chlorophylls and obtain their food by ingesting other organisms (heterotrophic); (*ii*) in contrast to animals, plants are non-motile and are usually firmly fixed in one position all their adult life, have no nerves or muscles, and lack a system for excreting any solid organic waste products of metabolism; (*iii*) individual plant cells, unlike those of animals, have a rigid wall, and usually contain a large watery vesicle called a vacuole; and (*iv*) growth in many plants is often not limited, and the number of individual organs, leaves, flowers, roots and so on, is not absolutely fixed, extra elements being added from time to time, whereas animals have a fixed form and individuals of a given species show little variation in size at any given age, regardless of habitat.

Using such criteria, it is generally quite easy to assign any one of the larger living organisms to one or the other of the two Kingdoms, even when it lacks one or more of the typical features outlined for an animal or a plant respectively. For example, the higher fungi, such as the common mushroom (*Psalliota campestris*), are non-photosynthetic and obtain nutrients from dead organic matter by first breaking it down with extracellular enzymes and then ingesting the products (saprophytic). Yet they are usually never considered as anything but members of the Plant Kingdom because they are non-motile, have rigid cell walls and lack nerves and the ability to excrete waste products[64,70].

However, attempts to classify unicellular organisms on the bases listed above quickly gives rise to absurd difficulties[16,75,82,84]. Some are motile and digest food (*e.g. Amoeba proteus*) and are thus clearly animal like, while others are non-motile and photosynthetic (*e.g. Spirogyra longata*) and are likewise clearly plant like. But there is a vast number of unicellular organisms, including most bacteria, which have combinations of features which make it more or less impossible to assign them to one Kingdom or the other. For example, *Euglena gracilis*, although photosynthetic, is motile and has no rigid cell wall. Furthermore, when deprived of light, this unicellular organism adapts a saprophytic habit. As more organisms of this indeterminate type were discovered in the nineteenth century, several proposals were advanced to overcome the limitations of the two Kingdom system[73,74]. The most widely accepted of these was the scheme put forward by Haeckel[74] in 1886 who added a third Kingdom, the Protista, to include all organisms lacking somatic tissue differentiations, including those which formed colonies (*e.g. Volvox aureus*). More recently, it has been suggested that the Protista be widened to include all animals and plants more primitive than sponges and mosses respectively[75,85].

The bacteria and blue-green algae are readily distinguished from all

References p. 287

organisms by the fact that they lack a nucleus and other organelles such as mitochondria, endoplasmic reticulum, chloroplasts, Golgi apparatus, internal membranes and vacuoles[1,16,77–82,84]. DNA is present in these procaryotes as one single fibril uncomplexed by histones[16,82,84], whereas in the nucleated procaryotes it is divided among several usually unequally sized chromosomes, all of which contain histones and other proteins[1,86–90]. There is also a wide difference in the structure of the flagella in motile forms of pro- and eu-caryotic cells; those of the former being much simpler and consisting of a single fibril, while the eucaryote flagellum contains 11 internal protein strands arranged in a '9+2' pattern[78–81]. These differences, but chiefly the lack of a nucleus, led several systematists to segregate the bacteria and blue-green algae from other organisms. Haeckel placed them in a separate sub-Kingdom, Monera of his Protista[74]. More recently, Copeland[15,75] elevated-them to Kingdom rank, having a four Kingdom system: Monera (procaryotes), Protoctista (protozoa, red and brown algae, fungi), Plantae (green algae and all land plants) and Animalia (multicellular animals). This system has been modified by later workers, notably in the transfer of the green algae to the Protoctista, thus restricting the Plantae (usually named Meta- or Embryo-phyta) to photosynthetic land plants[15,76].

In all these extended systems there has been some disagreement about the place of the fungi. Most authorities have placed them with the protists[73,75]. Grant[91] and Whittaker[15,76], on the other hand, arguing that the presence of multinucleate cells in the mycelium, the absorptive mode of nutrition and the diversity of their chemical constituents show great difference from other organisms, have raised them to Kingdom rank.

No system can be wholly satisfactory. In Whittaker's five Kingdom system[15], for example, the Protista is restricted to unicellular eucaryotic organisms, thus excluding *inter alia* the Rhodophyta which are classed along with the higher plants, although most authorities regard them as being quite primitive[81]. It should be noted that raising the fungi to Kingdom rank[15,85] above the Protista, makes their unicellular members as advanced as the Animalia and Plantae which is plainly not the case. Furthermore, some algae, like the Chrysophyta, are grouped with typical, though primitive, animals like the Ciliophora. The utility of such systems, therefore, is somewhat limited for phylogenetic purposes.

In this volume a modified three Kingdom division has been followed: micro-organisms and fungi, animals and plants. In discussing the evolution of and within any of these groups, of course, developments in the other two must be taken into account. The Plant Kingdom, which is dealt with in this Chapter is taken to include all autotrophic (chlorophyll-containing) eucaryotic organisms and those clearly related to them. Chlorophylls are of course absent from all obligate heterotrophes and saprophytes including the whole of the Animal Kingdom and the fungi[81]. Chlorophyll *a* is present in the blue-green algae and related compounds, the bacteriochlorophylls, occur in the photosynthetic bacteria[16,81,84]. But in neither of these groups of procaryotes is the pigment confined to a clearly differentiated photosynthetic organelle (chloroplast or chromatophore) which is separated from the rest of the cell by a double-layered unit membrane[92–94]. Although these features exclude the procaryotic photosynthesisers from the Plant Kingdom as defined above, a discussion of their biochemistry must, of course, enter into any consideration of the biochemical evolution of photosynthesis in plants. Similarly, although the fungi are excluded from the Plantae on the basis of their mode of nutrition, certain aspects of their cell-wall structure[95,96] and other features of their biochemistry must be included in any well rounded account of the plant evolution.

(b) Classification of plants

In the context of this Chapter, any useful scheme for the classification of the Plant Kingdom must be based on an orderly sequence of evolutionary relationships between the various groups of plants. However simple this statement might appear, it assumes both that the plants can be arranged together in an unambiguous fashion and that evolutionary relationships can be fairly easily discerned. Neither of these statements is strictly true. In pre-Darwinian times, classification of living things was largely based on the application of Aristotelian logic[57]. A number of selected characters was chosen, such as the shape of the leaves, and each was then arranged in order of presumed importance according to the ideas of the taxonomist concerned. The application of these principles led to the production of purely artificial systems of classification, like that of Linnaeus himself[57,64,83]. Such systems are extremely useful and have their modern counterparts in the dichotomous keys found in many modern floras. These allow for the relatively rapid identification of any plant but have the disadvantage that certain closely related plants, perhaps differing only in flower colour, may be widely separated.

References p. 287

TABLE I

Classification of Lower Plants[a,b]

Division[c,d,e] *(number of species)*	*Class*	*Common name*	*Homonyms*	*Age of oldest known fossil*[f] *(years)*
Schizophyta (1200)	Schizomycetes	Bacteria		3.2×10^9
Cyanophyta (1200)	Cyanophyceae	Blue-green algae		3.2×10^9
Chlorophyta (6500)	Chlorophyceae	Green algae		1.0×10^9
	Charophyceae	Stoneworts	Charales Charophyta	450 million
Chrysophyta (5700)	Xanthophyceae (400)	Yellow-green algae	Heterokontae	140 million
	Chrysophyceae (300)	Golden-brown algae	Coccoliths	100 million
	Bacillariophyceae (5000)	Diatoms		130 million
Pyrrophyta (1000)	Desmophyceae (100)			?
	Dinophyceae (900)	Dinoflagellates		450 million (0.9×10^9)
Cryptophyta (100)	Cryptophyceae	Cryptomonads		400 million
Euglenophyta (400)	Euglenophyceae	Euglenids		140 million
Phaeophyta (1500)	Phaeophyceae	Brown algae		400 million (600 million)
Rhodophyta (3000)	Bangiophyceae Florideophyceae	Red algae		600 million (0.9×10^9)
Fungi (3200)	Myxomycetes[g] (500)	Slime moulds	Mycophyta	460 million (2.0×10^9)
	Phycomycetes (1300)			
	Ascomycetes (1500)			
	Basidiomycetes (1500)			

By the end of the eighteenth century the drawbacks of the artificial systems had already become recognised. It was realised that in order to gain a better understanding of the relationships between different plants, they had to be grouped on the basis of having in common a relatively large number of different characters, none of which was assumed *a priori* to be more important than any other. The application of these ideas led to the development of the so-called natural classifications and, thanks to botanists like Jussieu and De Çandolle, by the middle of the nineteenth century most of the major divisions of higher plants recognised today had been grouped together in more or less the same way as found in modern taxonomic schemes[57,63–72,97–100].

The original system of Linnaeus, however, gave a logical framework of nomenclature essential for the elaboration of future classifications. Linnaeus assigned each individual organism to a named *species*. A species can be defined as a group of organisms having closely similar external and internal characters, including chemical and biochemical ones, existing in one or more natural populations, individual members of which reproduce their kind in sufficient numbers to maintain the population in a relatively stable state and, conversely, are genetically isolated from individuals of other species. Previous definitions of the term[57,101–103] have implied that the biological species only occur in sexually reproducing forms, but asexual species are equally valid and here genetic isolation presumably results from the inability of the individual genome to recombine with foreign DNA[86–90].

Species which possess in common a number of characters not shared by closely related plants are grouped together into a *genus*. The combination

---

Notes to Table I

[a] Also called Thallophyta (thallus, undifferentiated).

[b] The ending '*-phyta*' is given to all groups of Divisional rank; the ending '*-phyceae*' for Classes of Algae; '*-mycetes*' for classes of fungi: for higher plants the ending '*-opsida*' has been used for sub-divisions, although strictly it should be restricted to Classes (see Table II).

[c] Numbers in brackets refer to the approximate number of species in each Division or Class: usually these are greatly underestimated.

[d] Organisms in the Divisions Schizophyta and Cyanophyta are procaryotes (no nucleus or organelles): all other organisms here and among higher plants (see Table II) are eucaryotes.

[e] The divisions from Cyanophyta to Rhodophyta (excluding Euglenophyta) are usually grouped together as Algae.

[f] Ages shown are the earliest agreed dates when fossil of the Division in question were laid down, but earlier forms probably existed. Ages in brackets are based on micro-fossils in pre-Cambrian sediments.

[g] Myxomycetes is often given either Divisional rank or classed with the Bacteria.

TABLE II

Classification of Higher Plants[a,b]

Division (number of species)	*Sub-division*	*Class*	*Common name*	*Age of oldest fossil (years)*
Bryophyta[c] (23 000)		Hepaticae (9000)	Liverwort	380 million
		Anthocerotae (30)	Hornwort	?
		Musci (14000)	Moss	280 million
Tracheophyta[d] (297 000)	Psilopsida (3)		Psilophytes	400 million
	Lycopsida (1300)		Lycopods	380 million
	Sphenopsida (25)		Horsetails	380 million
	Pteropsida (295 600)	Filicinae (9300)	Ferns	375 million
		Gymnospermae (640)	Gymnosperms	370 million
		Angiospermae (286 000)	Angiosperms	135 million

[a] Grouped together as Metaphyta, Embryophyta or Cormophyta.
[b] See Notes to Table I.
[c] Non-vascular cryptogams (non-seed or spore bearing plants).
[d] Often divided in Pteridophyta, fern-like plants or vascular cryptogams which includes all of the first three sub-divisions and the Filicinae (Pteropsida), and Spermatophyta, seed-bearing plants including Gymnospermae and Angiospermae.

of the names of the genus and the species is then the epithet applied to the organism itself. Genera are grouped into Families, Families into Orders, Orders into Classes and Classes into Divisions of the Plant Kingdom (see Tables I and II). There is no limit to the size of any hierarchical category or to the number or nature of the characters used to delineate it[57,101]; this depends only on the authority dealing with the problem. Nor is there any general agreement about the specific rank to be given to any of the higher taxa. Thus the Angiospermae are variously regarded as a Class of the Sub-division Pteropsida of the Division Tracheophyta of the Kingdom Plantae[63], or as a Division of the sub-Kingdom Embryophyta[64].

Regardless of the varying interpretations as to the relative importance of these higher categories of plants, they are more or less unambiguously separated from one another by differences in morphological organisation and their mode of reproduction. All Divisions of the Plant Kingdom must, of course, have had a common origin in some remote ancestral form and the distinctions between Divisions are interpreted as major evolutionary events[57,101]. The Divisions and Classes are usually arranged in two groups on the basis of organisation in a similar way to the division of the protists from multicellular forms in the four Kingdom systems. The Thallophyta, or undifferentiated plants are shown in Table I, and the Metaphyta, or higher plants, all of which grow on land, are shown in Table II. It will be noted that the fungi are here shown as a Division of the Thallophyta although, as discussed above (p. 128), they might be better elevated to Kingdom rank. Also included in Table I are the procaryotic bacteria (Schizomycetes) and blue-green algae (Cyanophyta) which, as previously mentioned, are best regarded as a separate Kingdom (Monera) and the Euglenophyta which are often assigned to the Protozoa.

The arrangement of the Divisions shown in Tables I and II is approximately according to increases in the complexity of organisation and implies a rough evolutionary hierarchy. The earlier entries in each Table are presumed to be more ancient, or at least to represent organisms which show less modification from ancient stocks, than do later entries. This assumption is based on the generally accepted proposition that the more primitive forms of organisation have preceded the more complex and that evolution is irreversible[1,2,49–58]. It must not be assumed directly, however, that any group of organisms shown is necessarily truly ancestral, that is patristically related, to any other. Indeed, as will be seen in the following Sections, some relatively primitive groups have evolved only quite recently.

(c) The changing Earth

The course of evolution must be seen against the enormous changes which have taken place in the land, oceans and atmosphere of the earth over the aeons.

It is generally accepted that the earth originated over four and a half billion years ago (4.5×10^9)[11,104] but there is no agreement as to how it was formed[105]. The most widely held theory is that the planets of our solar system were formed by the condensation of matter contained in rings of a rotating gaseous nebula, the center of which condensed to form the sun. Alternatively, the planets originated from matter pulled

References p. 287

out from the sun by the catastrophic approach of a second dense star. In either event, the earth must have originally had a composition similar to that of the sun with a hydrogen–helium atmosphere[4,105–109]. There is reasonable evidence from the state of oxidation of most primitive rocks that the early atmosphere of the earth was reducing, but it has been suggested that the first helium-rich atmosphere was lost by some calamitous event, such as the capture of the moon, and replaced by a secondary atmosphere produced by outgassing volatile products trapped in the earth's mantle. It has been postulated that this secondary atmosphere contained ammonia and either methane or carbon dioxide together with water and perhaps, some hydrogen[105–109]. The various constituents of such an atmosphere could condense under the influence of the then more intense u.v. radiation, through radioactivity, or by shock waves from micro-meteorites to form amino acids, purines, pyrimidines and other organic molecules which gave rise to primitive pre-biotic proteins, nucleic acids and other polymers necessary as a prelude to life[4,78,107–109]. Other scientists hold that these primary organic molecules were formed from carbon-rich deposits in volcanic rocks[108]. In either case it should be noted that a reducing atmosphere must have been present and it seems highly likely that there was little free oxygen present even after the first primitive procaryotes had originated 3.5 billion years ago[78]. Some oxygen must have occurred in a combined form if the energy present in the reduced pre-biotic molecules was to be used for the living processes of growth and replication[108]. It seems highly likely that this would arise by the photolysis of water by the u.v. light from the sun. This process would be self-limiting, since the water vapour would be protected from solar radiation when the amount of oxygen[106] in the atmosphere rose to about 0.02%. All further increases in oxygen in our air have come from the activities of photosynthetic autotrophs. It seems probable that the most primitive organisms were like today's iron bacteria and did not produce free oxygen, but oxidised ferrous iron to ferric[4,10,78]. This would account for the haematitic iron formations which are found in all old rocks between 3.2 and 1.8 billion years old. Free oxygen presumably started to increase from 1.8 b. years and this corresponds to the earliest primitive multicellular organisms. Oxygen would be converted by the sunlight into ozone and molecular oxygen (O) which would extensively oxidise the surface rocks, and it is noteworthy that many thick oxidised iron beds are dated about 1.8 to 2.0 b. years old. When the level of oxygen reached about 0.2%, sufficient ozone would be formed in the upper atmosphere to produce a screen which would exclude much of the DNA-inactivating radiation[106,107] around 260 nm. This would allow organisms to exist near the surface of the seas and thus allow a more efficient photosynthesis. The increase in oxygen presumably also led to a vast increase in the utilisation of other more efficient energy-yielding biological systems and gave the opportunity for the widespread radiation of organisms which happened 580–600 million years ago[1–10]. At the onset of the Silurian period, 450 million years ago, the oxygen level reached over 1% and colonisation of the land could begin.

So much for the atmosphere. What about the land masses of the earth, since it is here that the prime events in the evolution of our present-day fauna and flora have taken place? Have there been any changes in the general disposition of the land masses? And what has caused the process of mountain building and when did it occur? Fortunately, during the last few years, a new unifying theory of the structure of the earth's crust has been adumbrated and gives a satisfying explanation of all the changes which have taken place since life conquered the land 400 million years ago[110,111].

At the beginning of the 17th century, when the outlines of the continents were becoming known, several scientists pointed out the complementary shapes of the coastlines on each side of the Atlantic. It was not until eighty years ago, however, that the Austrian geologist, Suess, suggested that Africa, South America, Australia and India were once joined together to form a super-continent which he called Gondwanaland. Nearly twenty years later it was suggested that about the start of the Mesozoic, Gondwanaland was joined to Europe and North America to form one land mass, Panagea. These proposals, based on similarities in geological structure and fossil distribution, did not gain acceptance until the last decade or so[110,111]. It is now generally agreed that the surface of the earth is comprised of a series of six major and six minor "plates", which are relatively thin (about 100–150 km thick) and are constantly in motion with respect to each other. These plates may be regarded like pieces of a cracked eggshell and their motion is due to convection currents in the earth's mantle below the crust. These movements cause the plates to move apart at certain points which are always in mid-ocean and cause the formation of ridges due to the creation of new crust. At other points, one of the two plates slips under the other and this occurs at deep ocean trenches. Lateral movement of plates may also occur giving rise to fault lines which are the centers of volcanic and earthquake regions. Finally, collision of one plate with another is held to account for the formation of mountain ranges.

This theory of plate tectonics is supported by evidence from the comparative ages of orogenic belts in present-day continents that are postulated to have been joined, ice erosion patterns of the Permo-Carboniferous glaciation, palaeomagnetic measurements of ocean depths, seismatic observations, and evidence from the distribution of fossils and present-day flora[112].

The results of plate movements on the land mass and ocean beds of the world have been dramatic. Evidence points to the fact that Gondwanaland was formed at the start of the Ordovician, 500 m. years ago, and existed at least until the beginning of the Jurassic, 190 m. years ago. The northern continents, North-America (including Greenland), Europe and Asia, were joined from the mid-Devonian (370 m. years ago) to the late Cretaceous (70 m. years ago) in a super-continent known as Laurasia, and it seems highly likely that this was joined in the Panagean land mass for 50 m. years or so between 280 m. years ago and the breakup of the southern continents[110].

This association of continents means, of course, that the Atlantic Ocean is relatively new by palaeontological time standards, and this is borne out by the fact that fossil remains in the core taken from the sea bed are never older than 200 m. years.

Besides moving relative to each other, the whole plate system has drifted compared to the earth's a is. The British Isles, for example, have moved from 40° South 550 m. years ago through 5° S. from 425 to 300 m. years ago to 55° N. today[113]. The movement of plates against each other has been sporadic as judged by mountain building activities. Thus there have only been three main orogenic developments in Europe since the early Devonian. These are the Caledonian (400 m. years ago) which was responsible for the ranges of Norway, Eastern Greenland, Scotland and Nova Scotia; the Hercyanian in the late Carboniferous (270 m. years ago) which accounted for ranges in Spain, South-Wales, North-Africa and the Appalachians in the U.S.A.; and the Himalayan–Alpine about 50 m. years ago, which caused the upthrust of most of today's high mountain ranges throughout the world[60].

(d) The origin of plants

At the end of the nineteenth century, the apparent presence of filamentous microfossils observable under the microscope in thin slivers of rock was reported from certain Pre-Cambrian cherts, which are hard early sedimentary rocks[3,4,7,8]. About this time, it was also suggested that the calcareous laminated stromatolite formations found in some cherts pointed to the existence of organisms similar to present-day lithographic blue-green algae. Nevertheless, palaeontologists appeared reluctant to accept such findings as definite evidence for Pre-Cambrian life[3,8]. It was not until 1954, when Tyler and Barghoorn[114] described the occurrence of structurally preserved microorganisms in the Pre-Cambrian stromatolites cherts of the Gunflint Iron formation of the Canadian shield (Table III), dated at 1.9 billion years old, that the scepticism of most palaeontologists was overcome[7,8].

The earliest rock formation in which microfossils have been reported is the Onverwacht carbon-rich fine ground silaceous chert from South Africa tentatively dated at 3.2 billion years old. It is possible however that these incompletely preserved forms may be non-biological artefacts like those found in some ancient basaltic lava[4,7,8,78]. The latter, being igneous in origin, could not have contained any living organisms. The somewhat later Fig Tree deposits, which overlie the Onverwacht series, and are dated at 3.1 billion years, have proved to be a richer source of more varied and complete microfossils which were unicellular and presumably procaryotic, judging from their appearance in the light and electron microscopes. Such microorganisms were probably embedded in a thin silica sheath which preserved them from deformation.

Support for the presence of living Pre-Cambrian forms comes from the detection of accompanying chemical substances which are of presumed biological origin[114a]. The success of earlier studies on the compounds associated with Phanerozoic fossils, coupled with the development of ultrasensitive analytical methods, led to the chemical examination of Pre-Cambrian sedimentary rocks, especially those associated with microfossils. The compounds which were sought are those known to be chemically stable such as alkanes, isoprenoid-reduction products, steranes and porphyrins[4,7,8]. In all cases, it has been important to demonstrate that the compounds are not artefacts produced either through later contamination or by being formed abiogenetically[114a]. Contamination from more recent organisms is likely to be localised and can usually be eliminated by

References p. 287

TABLE III

Biological Activity in the Precambrian[3,b]

Age (million years)	Era	Sediment examined	Organisms present			Geochemistry			
			Procaryotes simple	Procaryotes filamentous	Eucaryotes	Alkanes	Isoprenoids	Porphyrins	12C/13C
0–600	Phanerozoic								
600	Late Precambrian								
		Sinian, China	+	+					
		Bitter Springs, Australia	+	+	+	+	+	+	+
1000		Nonesuch Shale	+	+		+	+	+	+
		Belt Series, Montana	+	+	+ ?				+
		Munos Shale	+			+	+		+
1500		Huronian	+	+					
		Me Minn Formation				+	+		+
	Middle Precambrian								
		[a]Gunflint Iron	+	+	+ ?	+	+		+
2000		[a]Transvaal Super	+	+		+		+	+
		[a]Witwatersrand Group	+	+		+		+	+
		Cobalt Series	+						
2500	Early Precambrian								
		Soudan Iron	+			+ ?	+ ?		+
		[a]Bulawayan	+			+	+		+
3000		[a]Fig Tree Series	+			+	+	+	+
		[a]Onverwacht Series	+ ?			+	+	+	
3500		Oldest known sediments							
4500–5000		Formation of the Earth							

[a] Reported to contain bound amino acids
[b] Modified from Echlin[7] and Schopf[8].

examination of several replicates of deep-core samples from the rock formation at issue. There are, of course, difficulties when the amount of organic material is extremely small and large rock samples have to be taken for analysis. Abiogenetically formed compounds are less easily distinguished except in cases where the molecules detected have either chiral groupings (with different stereo forms) as do the amino acids, are likely to be formed biosynthetically in varied amounts as with even- and odd-carbon number alkanes, or have other defined features arising from their biosynthesis such as the regularity of methyl side-chains in polyisoprenoids[4]. All such features would be unexpected from random abiogenetic reactions, although it is possible that if the reactions took place on certain types of regular silicate surfaces, some *enrichment* of apparently biogenetic products might occur. In addition, since the fixation of carbon dioxide gives an increase in the proportion of the ^{12}C isotope[7] a measure of the $^{12}C/^{13}C$ isotopic ratio in carbon-rich rocks by mass spectrometry can provide evidence for biological activity. Such an enrichment, again, might occur abiogenetically but appears to be unlikely[7,108].

One must be careful, however, not to interpret the evidence obtained too optimistically and draw conclusions about the evolution of cellular metabolism and organisation which are not wholly warranted. For example, the presence of the isoprenoid compounds pristane (I) and phytane (II) in the organic residues of Onverwacht chert has been taken to mean that photosynthesis developed over 3.0 billion years ago on the grounds that the

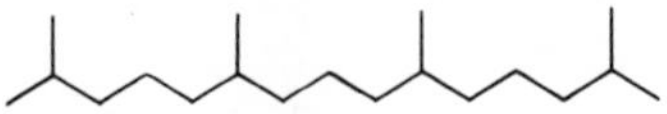

(I) Pristane

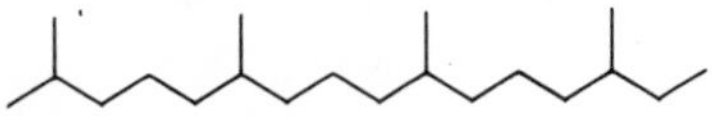

(II) Phytane

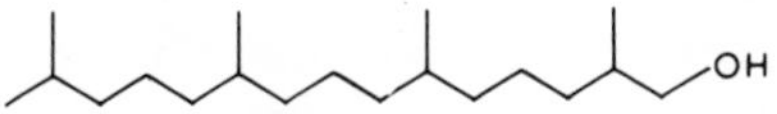

(III) Phytol

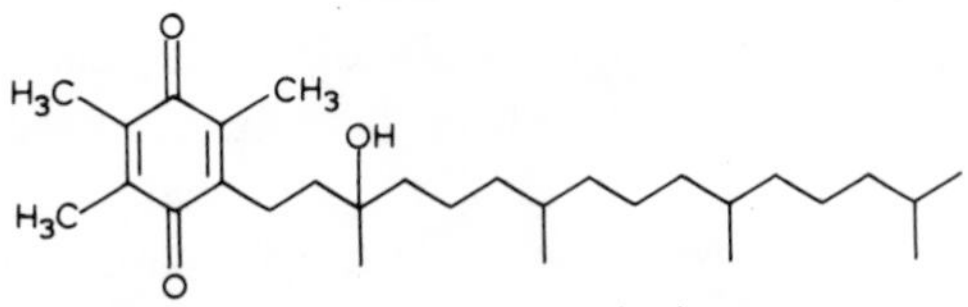

(IV) α-Tocopherolquinone

residues arose from the phytol side-chain of chlorophyll (III). However, it is equally possible that the two isoprenoids came from the side-chains of primitive prenylquinones, such as α-tocopherolquinone (IV), or from a polyprene alcohol, like that known to be important as a carbohydrate carrier in cell wall synthesis in present day bacteria (see p. 197). Other evidence, such as the low oxygen tension at that time, as deduced from the greater degree of reduction of iron in the rock formations (see p. 134), indicates that photolysis of water was probably not involved in any photosynthetic process 3.0 billion years ago, although the stromatolite formation found in the Bulawayan group (2.8 billion years old) indicates that organisms similar to present day oxygen-producing blue-green algae developed quite early. It should also be remembered that $^{12}C/^{13}C$ ratios give evidence only of carbon dioxide fixation[7], which need not necessarily be associated with light-requiring reactions (see p. 194).

Taking the morphological and geochemical evidence together (Table III) it seems highly probable that living unicellular procaryotes existed 3.2 billion years ago or even earlier. Even the most conservative estimate would agree that living forms including blue-green algae were present at the time the cherts of the Transvaal Supergroup and the Gunflint Iron formation were laid down, *i.e.* about 2 billion years ago.

The largest advance in the development of living organisms since their inception was the establishment of the eucaryotic cell. The discontinuity between procaryotes and eucaryotes is undoubtedly the most important in the history of life, for, although some complex procaryote colonial forms are known, no anuclear organism has ever developed multicellular organisation[77–79,94]. In order to produce multicellular forms, individual cells in the organism have to differentiate in a regular and controlled manner and this presumably requires a more complex nuclear organisation such as that found only in eucaryotes.

The time at which eucaryotes arose is as problematical as the way in which their characteristic organisation of organelles developed. The Bitter Springs fossil material, which is about 1.0 billion years old, contains evidence of relatively advanced eucaryotes somewhat like modern green algae in having nuclei, double cell walls, organelles and apparently exhibiting mitosis and meiosis. The earlier microbiota of the Gunflint Iron formation show no trace of such eucaryote characteristics[114–118]. It seems probable that the development of the eucaryotic cells, all of which are almost entirely aerobic, had to await the increase in oxygen tension which took place in the later Pre-Cambrian era (see p. 134). Some fossils suggesting eucaryotic affinities, however, have been reported in the Beck Spring Dolomite of 1.3 billion years of age, and possibly in the deposits of the Belcher Group which are 1.7 billion years old[7].

The mode of origin of the eucaryotes cannot, of course, be deduced from the fossil record. There are no intermediate forms between them and the procaryotes, containing for example, nuclei but not mitochondria. This tells against the idea that the change between the two was gradual, although it should be noted there are no large biochemical discontinuities between the two groups[78]. It has been suggested that the cytoplasmic organelles arose from an area in the procaryotic cell like the bacterial mesosome[94]. Even if that were so, the theory fails to account for the most important feature of the eucaryotic cell, the organisation of the genetic material in the nucleus. Another theory, which has recently received support, is that the mitochondria, chloroplasts and basal bodies of the eucaryotic flagella arose from free-living specialised procaryotes by endosymbiosis[77–79,119]. The theory implies, however, that these proto-organelles have lost many of the functions of free-living organisms, and lost them to exactly the same extent, in all present-day eucaryotes. It also denotes that either the symbionts acquired all the organelles in a single step or that intermediate forms were too inefficient to have survived. Again, the theory does not adequately explain the development of the nucleus[78,119]. It seems possible, indeed, that this single evolutionary step might have been sufficient to allow adequate *control* of compartmentalized cell functions with concomitant economy of energy which then permitted all future development of multicellular forms.

The evolution of the multicellular organism led to the most dramatic expansion in living forms. As shown by the fossil record, this happened in a relatively short length of time at the start of the Cambrian period 600 million years ago. The earliest multicellular forms are the soft-bodied complex coelentera and annelids in the South Australian Edicara sandstone[3]. This suggests that more primitive multicellular forms must have developed much earlier. There is, however, no unambiguous record of multicellular plant life until the mid-Ordovician, where remains of several marine Chlorophyta and Rhodophyta have been found in sediments of that age (Table IV). Strangely, fossils of the rather primitive silaceous and calcareous Chrysophyta (diatoms and coccolithophores) do not occur until the Jurassic, 180 million years ago.

References p. 287

The first vascular land plants appeared at the Silurian–Devonian boundary, 400 million years ago. Earlier land plant fossils, found in the Ordovician and Cambrian, have been discounted on the ground that the remains do not possess sufficient essential details to allow them to be unambiguously identified[120,121]. Nevertheless, as with animals, it is unlikely that complex forms, like the tree-like lycopod, *Baragwanathia longifolia*, could have arisen 400 million years ago from the earlier multicellular algal organisms without many intermediate forms developing over several aeons.

(e) Plant evolution

The main evidence for evolutionary processes in both plants and animals comes from a study of the fossil record[3,5,14,54,61,62,108,120–122]. To be sure, a study of the differences between present-day organisms throws much light on probable hierarchical relationships. But the interpretation of such variations relies to a great extent on speculations which are unlikely to be confirmed. In any event, the actual time at which biological discontinuities occurred can only be discerned from palaeontological studies.

Except for relatively recent fossil remains, where actual undegraded organic matter of the organism may still be present, plant fossils are either the coalified or mineralised impressions of the plant part or petrified structures[61,120,121]. In the former case, the part must have been compressed and buried in sediments before complete decay of the outer wall and cuticle took place, while in the latter case, the original substances in the cells must have been replaced by deposition of a mineral, like silica, from solution with preservation of either part or all of the outline of the plant fragment. Unlike higher animals, where the bony skeletal structures are extremely resistant to decay, the entire fabric of most plant parts are susceptible to rapid bacterial and fungal attack. Only under special circumstances, such as extremes of heat and cold, submergence under water, deposition in highly acidic or basic soils, are microbial activities sufficiently inhibited to allow products of ancient plants to be preserved intact, albeit perhaps contaminated with remains of microorganisms[120,121]. Thus latex residues and chlorophyll breakdown products have been obtained from 50 million-year old Geiseltal lignite in Germany, and complex polysaccharides have been shown to be present in a number of Palaeozoic plant fossils over 230 million years old[121]. Some parts of plants are much more resistant to decay than others, especially those having lignified, or mineralised cell walls, or those which contain tannins or have a tough integument like pollen (see p. 252).

It should be noted also that even the less resistant plant parts may take longer to decay than do soft parts of animals, as shown by the finding of the ancient polysaccharides mentioned above. This fact is important in the short-term decay process, and undoubtedly accounts for the fact that plant remains often show greater detail of external and internal structure than do animal fossils. Because of this, palaeobotanists have been able to pinpoint the emergence in time of certain crucial evolutionary events such as the development of the vascular system, of secondary xylem elements or of the seed habit.

It must be clearly recognised, however, that the proportion of the world's flora fossilised at any one time was always extremely small and likely to have been highly unrepresentative. The chances of montane plants, for example, being preserved by fossilisation is remote and probably accounts for the paucity of the record of certain groups. To be fossilised, a plant, which did not itself contain mineral elements must have grown in, or been transported to, an environment such as a lake, delta or marsh, where mineralisation and burial were sufficiently rapid in relation to microbial decay to ensure its preservation. Even where fossilisation does occur, the sedimentary rocks containing them have to be exposed before any palaeontologist can get to work. Probably less than one per cent of sedimentary deposits have been subjected to proper examination. One further difficulty concerns dating of the deposits[122]. Precise radioactive dating can be used for most igneous rocks and, in certain instances, be applied to sedimentary environments[4,7,104]. But in most cases, the age of plant fossils has been determined mainly from associated animal remains whose chronology is much better known[1,3,104], together with supporting evidence from the relationship of strata and the general palaeobotanical record. Since this record reflects the more or less dominant flora at any one era, the time of emergence of any new diversified group of plants is unlikely to be discovered with certainty. The fragmentary and biased nature of the plant fossil record needs to be borne in mind in considering the evolutionary outline which is given in Table IV.

(i) The evolution of non-vascular plants

As outlined earlier, it is generally agreed that the procaryote blue-green algae (Cyanophyta) were derived in the early Pre-Cambrian from primitive

TABLE IV

Evolution of the Plant Kingdom

Era	*Period*	*Epoch*	*Beginning of age in millions of years*	*Plant developments*	*Animal and geologic developments*
Cenozoic	Quaternary	Recent	0.01	Cultivation of plants	Neolithic to modern times
		Pleistocene	2.5	Extinction and re-invasion of glaciated areas. Speciation of herbs	Succesive glaciations: Origin of man
	Tertiary	Pliocene	7	Extinction of many early species due to climate change	Continental uplift in Americas
		Miocene	26	Rise of herbaceous angiosperms	Elevation of European Alps: Origin of Hominidae
		Oligocene	38	Polar spread of relict conifers and angiosperms	Elevation of Pyrenees: Origin of monkeys
		Eocene	58	Extensive afforest-ation of near polar areas	Lignites deposited: Cat and horse ancestors
		Paleocene	63	Notable expansion of grasses	Earliest primates
Mesozoic	Cretaceous		135	Woody angio-sperms: Rise of grasses and pines: Diatoms appear	Last of giant reptiles Rise of mammals
	Jurassic		180	Origin of angio-sperms: ginkgoales and conifers abundant: Seed ferns die out	Rise of birds and higher insects
	Triassic		225	Rise of cycads and spread of conifers	Rise of dinosaurs
Palaeozoic	Permian		280	Rise of conifers	Expansion of reptiles

(continued)

TABLE IV (continued)

Era	*Period*	*Epoch*	*Beginning of age in millions of years*	*Plant developments*	*Animal and geologic developments*
	Pennsylvanian		325	Widespread coal swamps of club-mosses, horsetails and seed ferns. Ferns abundant	Rise of reptiles. Insects abundant
	Mississippian		345	Seed ferns appear with gymnosperms and giant lycopods. First seed plants	Rise of amphibia
	Devonian		400	Rise of vascular plants including ferns	Decline of trilobites; Fish widespread
	Silurian		430	First traces of land plants. Triradiate spores	Jawless fish abundant
	Ordovician		500	Multicellular algae prevalent	Rise of fish
	Cambrian		600	Marine algae forming calcareous reefs	Trilobites abundant. Rise of most marine invertebrates
Precambrian			3500?	Bacteria, blue-green algae	
Prebiotic			4500		Age of oldest rocks

photosynthetic bacteria with which they have several features in common. There are marked differences in pigment composition, however, and the absence of ubiquinones (p. 237) from contemporary Cyanophyta indicates that the connection may be more remote that some authorities hold. However, as stressed previously, the relationship between both these groups of organisms, and the earliest eucaryote is obscure, although it is unlikely to be polyphyletic[63,64,78,81,123]. Otherwise one would expect intermediate forms of organisation between the two, and some quite discrete variation in their biochemistry.

In the Plant Kingdom, the most primitive eucaryotes are believed to be unicellular members of the green algae (Chlorophyta)[3, 14, 16, 61–64, 123], although on biochemical grounds, the red algae (Rhodophyta) with carbohydrate reserves and phycobilin pigments like the blue-green algae (p. 190), and having no flagella, are apparently much less advanced[70, 81]. Indeed all species of contemporary Chlorophyta are biochemically quite similar in several respects to higher plants, especially in their carotenoid pigmentation, possession of chlorophylls *a* and *b* and, in having starch as the reserve carbohydrate and a cellulosic cell wall[63, 64, 70, 81].

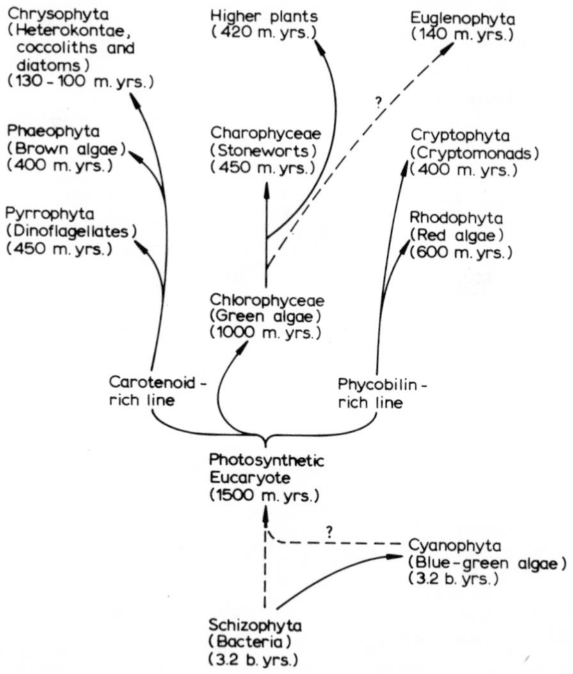

Fig. 1. Possible evolution of lower plants. Figures in brackets give age of oldest fossil record.

The fossil record is obscure on the nature of the earliest algal forms. Most authorities ascribe them to the Chlorophyta but some believe that the Rhodophyta are almost equally ancient (Table I). Nevertheless, the two Divisions must be assumed to have arisen along different lines; the Chlorophyta giving rise to the later Charophyceae (Stoneworts) and eventually to land plants (Fig. 1), whereas the Rhodophycean line perhaps gave rise to the Cryptophyta (cryptomonads) which also have phycobilins. The latter, however, do have flagella arranged laterally in pairs like the brown algae (Phaeophyta), although a direct connection between these two Divisions is unlikely[63,64,81].

The Phaeophyta are morphologically the most highly developed of the algae, with branching leaf-like filaments and some members which occupy salt-marsh habitats. The fossil record, however, suggests an origin a little later than the Rhodophyta, about 400 million years ago. Biochemical evidence links the Phaeophyta with the dinoflagellates (Pyrrophyta), which are probably older, and with the more recent Chrysophyta (diatoms), since several species in all three Divisions have similar carbohydrate reserves, chlorophyll c_2 and the unique carotenoid pigments fucoxanthin (V) and the acetylenic, diatoxanthin (VI) and diadinoxanthin (VII)[70,124] (see p. 248).

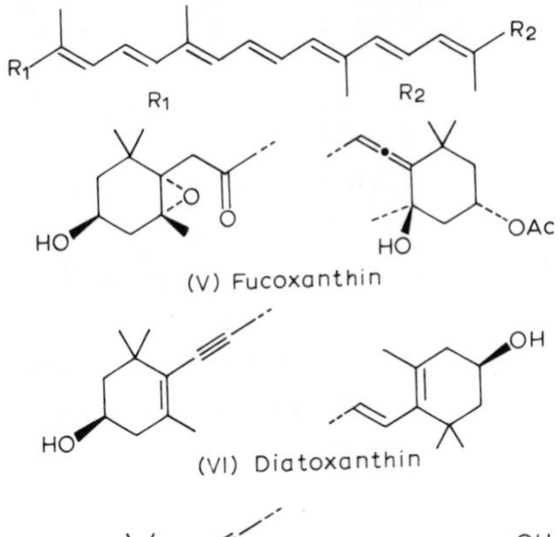

(V) Fucoxanthin

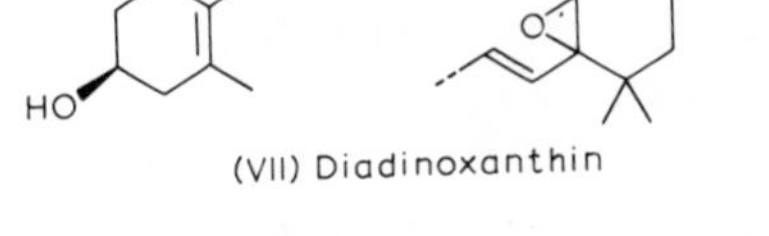

(VI) Diatoxanthin

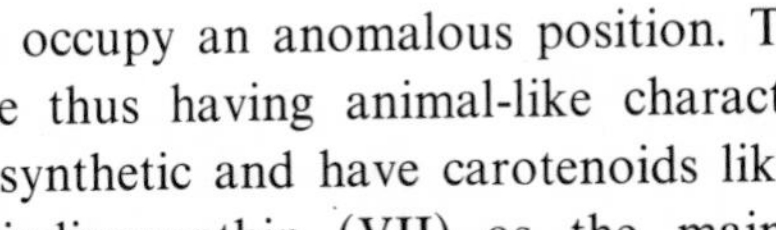

(VII) Diadinoxanthin

The Euglenophyta occupy an anomalous position. They have no cell walls and are motile thus having animal-like characteristics[63,64]. The organisms are photosynthetic and have carotenoids like the green algae but also contain diadinoxanthin (VII) as the main xanthophyll[124]. On the other hand, the carbohydrate reserve material, paramylum, is more like that in brown algae being a β-1,3-linked glucan[70,81].

Finally, mention must be made of the fungi. The fossil record is again ambiguous, fungal-like microfossils being found in the Gunflint-Iron sediments, 1.9 billion years old[8]. Most of the undisputed fungal remains, however, date from 400 million years ago. Several investigators have suggested that the fungi were derived from photosynthetic algal ancestors by the loss of chlorophyll: often the Rhodophyta are pinpointed as the possible ancestral line[70]. Many mycologists believe, however, that the fungi originated from the protozoa and therefore may not be related to the green plants at all. The wide differences between the cell walls of most fungi and those of algae and higher plants[95,96] coupled with the wide variety of unique secondary compounds synthesised by fungi preclude any close link[125–127]. It seems possible that the fungal line is a vestigial remnant of wall-synthesising eucaryote, in which photosynthetic capability, whether by symbiosis or other means, had never been gained. At least this view supports the hypothesis mentioned earlier (p. 128) that the fungi should be treated as a separate eucaryote kingdom on par with plants and animals.

(ii) The evolution of vascular plants

The earliest traces of vascular plants (Tracheophyta) in the fossil record were found, as mentioned previously, in several sediments which were deposited roughly 400 million years ago. These early plants are considered to be related to the present-day most primitive sub-division of the tracheophytes, the Psilopsida (Table II), of which only two genera exist today[61–64,123]. These plants are very primitive, but the fossils show dichotomous branching, stomata, tracheids, non-vascularised leaf scales and terminal sporangia containing cutinised spores typical of living *Psilotum nudum.* The Psilopsida show no relationship to any known algae or to any of the Bryophyta. Members of the latter Division probably represented the first land plants, although fossils are rare until about 350 million years ago. Present-day Bryophytes share many features with the Chlorophyta, but differ in having a differentiated cutinised thallus, heteromorphic alternation of generations with the gametophyte as the more conspicuous, and enclosed sex organs producing cutinised aerial-borne spores. All these features represent necessary evolutionary advancement for colonisation of the land, but it is possible that some of them evolved later when the epiphytic habit of present-day Bryophytes became common.

By the end of the Devonian period, 345 million years ago, representatives of the three other main sub-divisions of vascular plants had developed, the lycopods (Lycopsida), the horsetails (Sphenopsida) and the ferns (Filicinae). These soon dominated the land surface of the earth and gave rise to the massive coal deposits of Carboniferous times. It is clear that all three were derived separately from more primitive Psilopsida and that none are directly ancestral to higher taxa. The most primitive of the three, the Lycopsida, show advancement over the earlier vascular plants in having phloem, vascular connections to the leaves, and heterospores confined to sporangia arranged in cone-like stobili protected by leaf-like structures, the sporophylls. There are only five living genera. Only one present-day genus of the Sphenopsida exists, *Equisetum*. From this, and the rich fossil record, this sub-division differs from the lycopods in having the microphyllous leaves borne in whorls and having the sporangia arranged in strobili without sporophylls[61–64,123]. The ferns, the least advanced of the three classes of the Pteropsida, which includes the gymnosperms and angiosperms, differ from the other two groups in having large branched leaves which bear the sporangia, and generally lacking heterospory. The possession of large leaves is paralleled by anatomical changes in the conducting system and so-called "leaf gaps" appear in the vascular system of the central cylinder of the stem. In all but one genus of contemporary ferns, *Botrychium*, secondary growth is absent. This is paralleled by observations of the fossil record, and even the tree ferns appear to consist entirely of primary apically derived tissue[61–64,123].

The next major taxon of the Pteropsida, the Gymnospermae, are a diverse class, with seven Orders, three of which are now extinct. They are grouped together on the basis of producing seeds containing the embryonic sporophyte, developed from the ovule where fertilization takes place. The heterosporous gametophyte generation thus shows no independent existence, as with the plant groups dealt with above. Most gymnosperms are trees and show much secondary thickening. Representative fossil protogymnosperms showing secondary thickening and other characters have been found in sediments 370 million years old. Such fossils, however, have typical fern-like leaves and are usually referred to as tree-ferns or pteridosperms. The earliest pteridosperms to show a typical ovule-like structure are not found in rocks more than 330 million years old. It appears likely that these protogymnosperms arose from a similar group of plants as did the ferns, or even from the early ferns themselves where heterosporous fossils are known. The present-day gymnosperms are typically divided into the Orders Cycadales, Coniferales, Ginkgoales and

Gnetales. The cycads closely resemble later pteridosperms and probably had their origin in such stock. The Coniferales, the most predominant of present-day gymnosperms, are well represented in the fossil record of Carboniferous times. These early forms show a close relationship with a now extinct order, the Cordaitales, whose fossils are not dissimilar to the leaves of the contemporary *Araucaria*. The Cordaitales appear to have evolved in the same way as, but independently from the pteridosperms from axial heterosporous forms. The Ginkgoales also may have had an independent origin or have evolved from a Cordaitalean line. The Gnetales, on the other hand, are quite different from other gymnosperms and appear to be intermediate between the latter and the angiosperms. The three genera which are contained in the Order are also distinct from each other and are often raised to higher rank. Their affinities are not clear, some authorities suggesting origins from the Cordaitales and others from the Pteridosperms. The differences are many, however, and probably the result of prolonged evolutionary change (Fig. 2).

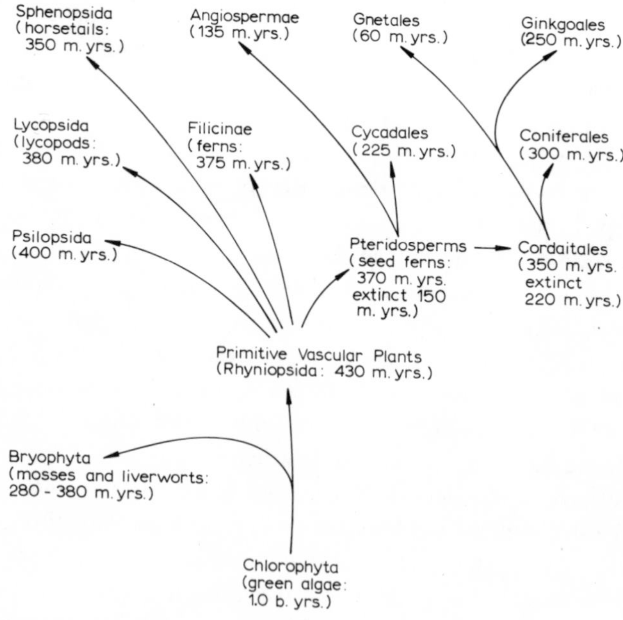

Fig. 2. Possible evolution of higher plants. Figures in brackets give the age of the oldest fossil record.

The angiosperms, or flowering plants, are by far the largest class in the Plant Kingdom. The fossil pollen record shows that they probably arose 135 million years ago[56,58,128], although claims have been made for a much earlier evolution. By 60 million years ago, the angiosperms had become the dominant group in the world's vegetation. They possess the advantage over the gymnosperms of a more specialized and more rapid means of sexual reproduction whose evolution was apparently correlated with that of pollinating animal species, birds and insects[56,58,128]. The vascular system is more highly developed and the ability to adopt a deciduous or annual habit allows the development of leaves which have greater efficiency in photosynthesis during the summer growth period. The ancestors of the angiosperms are thought to be some primitive protogymnosperms which showed a tendency for the ovules to be enclosed in carpet-like structures. Other authorities, however, believe that the angiosperms developed from some pteridosperm ancestor closely related to the Cycadales, the other gymnospermous Orders arising via the Cordaitales from the psilophyte

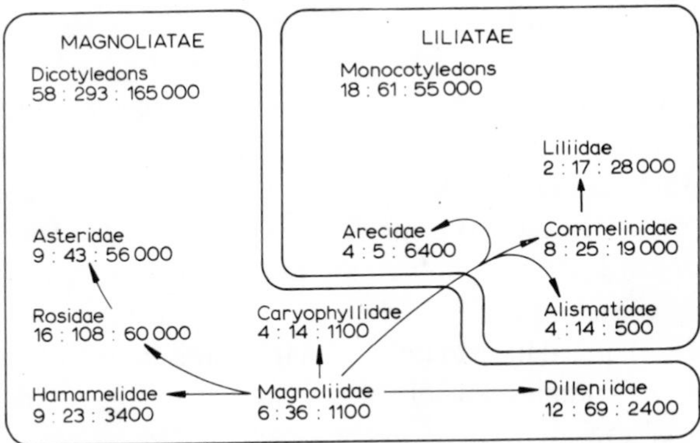

Fig. 3. Evolution of angiosperms after Cronquist[56]. The numbers under each of the headings indicate the orders, families and species in each. The Magnoliidae conforms to what has often been called the Ranalian complex but includes the Aristolochiales and Papaverales. The Hamamelidae includes a number of families traditionally placed at the beginning of the dicots in the Amentiferae because of their reduced wind-pollinated flowers. Caryophyllidae consists of the families of the Englerian Centrospermae plus the Cactaceae, the Polygonales, Plumbaginales, and Batales. The Rosidae includes several Orders accepted in the Rosalian complex, and the Asteridae has all the sympetalous Orders excluded from the more primitive Dilleniidae. The Alismatidae can be equated to the aquatic Helobiae of Engler and Prantl. The Commelinidae includes the Cyperales, with the Graminae, and related Orders. The Arecidae contains the Palmae (Arecales) and the Liliidae, the Liliaceae, Orchidaceae. Takhtajan's arrangement is similar[58] except he splits the Magnoliidae into two separate sub-classes (Magnoliidae and Ranunculidae).

stock. The possible evolution of the higher plants is shown in Fig. 2[56,58,64,123].

From the fossil record it is not easy to determine the phylogenetic relationships within the angiosperms themselves. It is usually postulated that the most primitive members are represented today by the woody members of the Magnoliales and that the predominantly herbaceous monocotyledons were derived from aquatic stocks, such as the present-day Nymphales. One arrangement of the angiosperms is shown in Fig. 3 based on the proposals of Cronquist[56] and Takhtajan[58]. Other authorities have proposed a polyphyletic origin of the angiosperms[129], or that taxa other than the Magnoliales are the most primitive[56–58,130] (see p. 265).

3. Biochemical evolution

The evolution and classification of the Plant Kingdom has been described in some detail in order to set the scene for the rest of this Chapter. An important word of warning is needed, however, before proceeding. No present-day organism can be regarded as any more ancient in the phylogenetic sense than any other. While it is undoubtedly true that all contemporary species are ultimately patristically related to ancient bacteria-like procaryotes, this does not mean we should seek, or expect, any such relationship between modern bacteria and plants in other divisions. The only conclusion we are entitled to draw is that the present-day bacteria appear to have changed little morphologically and anatomically from their Pre-Cambrian forebears. We may perhaps assume with some conviction that they have also changed little in intracellular organisation and biochemistry. In addition, the combination of the evidence from the fossil record and from the form and function of modern-day representatives of all classes of higher plants permits us to construct an approximate hierarchical series such as those shown in Figs. 1, 2 and 3, in which the higher members show an advancement in overall organisation to that of the lower members. But, as will be seen, this is not necessarily translated directly or indirectly into biochemical terms, *except* that the size of the genome progressively increases in the more advanced groups[46,130–137a] (see also Chapter I).

If we assume that the biochemistry of an organism underlies not only its metabolic activity but also its forms, function and behaviour patterns, then the biochemical abilities determine to a large extent, if not wholly, the organism's fitness to exist in any given environment. In any given taxa, of course, including species or even populations, there is a good deal of genetic plasticity which allows a range of biochemical expression and a reserve of unexpressed characters which may be called into play through sexual selection (or recombination) following subtle changes in the environment. For example, climatic changes have been immense over the whole history of living organisms (see p. 133) and even during the last one million years there have been four ice ages of great severity. It thus seems likely that only those taxa possessing or acquiring a wide range of biochemical capabilities would have been able to survive such vast changes.

It must be realised that some biochemical features, such as a desirable temperature optimum for certain enzymes, may only be achieved in a limited number of ways. Where this is so, these aspects of the biochemistry of the organisms would appear to be conservative. Equally, other biochemical characters might be produced in several quite different ways, any one of which might be used giving a, perhaps, spurious sense of evolutionary advance. We must assume, also, that because of certain constraints, some aspects of biochemistry cannot be carried out by any combination of genomes in a taxa. Providing the lack of such features is not lethal, this might well be exemplified in contemporary descendants. The bacteria, for example, perhaps lack sufficient DNA in the genome to allow for sufficient to be used for control purposes, rather than the synthesis of structural proteins. This probably is the reason, along with the lack of nuclear organisation, which has prevented the procaryotes from producing multicellular forms.

Finally we must remember that as with other features[49], any new biochemical element will only be retained in the organism if it confers some selective advantage. In most instances, the selective advantages which operated at the time the element appeared may now have passed, and the element may be regarded as a relic like man's appendix.

In any event, the evolution of a new biochemical feature is unlikely to entail any single drastic change, but rather several slight modifications of existing pathways. Even so, its incorporation into the genome may affect that of other biochemical characters, especially if there is competition for limited intermediates in biosynthesis, and some existing characters might be displaced completely. In other cases where no interaction occurs, different biochemical features will probably evolve at different rates.

All such effects must be borne in mind as we examine the variation in plant biochemistry outlined in the rest of this Chapter.

4. Nucleic acids

(a) Introduction

It is now accepted that the genetic information which any given organism possesses is entirely carried by the DNA, or in the case of some viruses, the RNA it contains[49,50,86–90]. In all normal cells, the whole of the DNA is replicated by relatively well understood processes before cell division and this enables each of the daughter cells to receive the full complement of information which determines their activity and development[89,90]. In procaryotes, this process leads to the formation of very uniform colonies of cells having exactly the same genotype. In sexual eucaryotes, on the other hand, where overall control rests almost entirely with nuclear, chromosomal DNA, the situation is more complex. Extranuclear DNA in the mitochondria and chloroplasts and other cytoplasmic factors may play a part in determining the biochemistry of the individual cell but their inheritance during cell division is ill-understood[89]. Furthermore, the independent assortment of the genes which determines individual characters takes place during meiosis, which means that each parent's genetic information is not necessarily present to the same extent in each haploid gametophyte. Fusion of gametophytes can thus give rise to mixtures of genotypes among the progeny. Such genotypic variation is usually reflected in the phenotype and accounts for most of the overall heterogeneity within populations of a given species of sexual eucaryotes[49].

It is generally recognised that differences between species, and hence the main underlying processes of evolution, rest in the dissimilarity of their respective gene pools; that is, ultimately, the variation in the sequences of nucleotides in their chromosomal DNA[49]. Such dissimilarities are due to mutations, caused either by mismatching during replication or by the destructive effect of natural radiation, which are incorporated into the genome[138]. Mutations at the molecular level undoubtedly occur extremely frequently[138–143], but normally do not contribute variability to the gene pool because they are not fixed by selection[49].

Mutations are most likely to be incorporated into the genome in populations which are geographically isolated reproductively from other populations of the same species. In such cases certain mutations may lead to partial or complete reproductive isolation of the population even when the geographical barrier subsequently ceases to exist, and thus eventually leads to speciation[49,50,56,58,88,102].

(b) Deoxyribonucleic acid

It is not surprising from the above considerations, that most scientists believe that evolution will only become properly understood when the sequence of bases of the deoxyribonucleic acid of the entire genome of a number of selected taxa is fully known. But such data are unlikely to be obtained with present-day technology, since even if one had a machine which could sequence 10 bases per second along the DNA chain, it would take 10 years' continuous operation to obtain the results for one average haploid genome of a higher plant or animal (*ca.* $3.2 \cdot 10^9$ nucleotide pairs[46,144]). It should be noted that some plants and amphibia may even have twenty times this amount[137a]. Of course it is possible to contemplate the examination of the sequence of bases in the viral DNAs which are about one million times smaller (*ca.* 3000 base pairs[46]), but even this task is extremely difficult. In order to obtain information about the base sequence and the degree of homology between the DNAs of different organisms, therefore, different approaches have had to be adopted. The three main methods are: (*a*) to determine the degree of hybridisation (association) between single-stranded DNAs or of defined RNAs or mixtures of the two prepared from different organisms; (b) to determine the base sequence of specified RNAs such as ribosomal or transfer RNA which are transcribed from the DNA; and (*c*) to determine the amino acid sequence of specified proteins, which thus reflect the sequence of bases, albeit imperfectly because of degeneracy, of the DNA.

All these methods have their advantages and disadvantages which will be outlined shortly. Before doing so, however, it might be as well to reiterate the facts outlined in the introduction. That is, in the main, higher animals differ from simple microorganisms not so much in their biochemistry as in their multicellular organisation and behavioural patterns, while higher plants produce a vast array of compounds over and above those involved in primary metabolism. This means that plants must also possess a large number of extra enzymes and control mechanisms necessary for the synthesis and degradation of these secondary substances. Now, the average haploid genome of higher plants and animals contains, as mentioned above, about $3.2 \cdot 10^9$ nucleotide pairs, while that of the average bacteria has only $1.3 \cdot 10^6$ base pairs[46,144]. One must assume, however, that the bacterial genome contains sufficient information to produce all the enzymes necessary for its existence, together with the requisite control mechanisms for the various metabolic sequences, and any unique information which distinguishes one bacterial taxa from another. Most bacteria can also synthesise all the amino acids, reserve polysaccharides, lipids and so on of common metabolism, and possess all the facets of protein and nucleic acid synthesis, besides producing certain structures unique to the bacterial cell, such as the wall (see p. 198).

It would seem, therefore, that since the average bacterial genome codes for less than 1000 proteins of average size containing 500 amino acids (each protein being equivalent to 1500 base pairs), and that probably one third at least of this information is used for control mechanisms or the synthesis of unique compounds, that a genome of *ca.* 10^6 base pairs is sufficient to maintain primary metabolism in the simple bacterial cell. Higher organisms on average thus contain over 3000 *times* more DNA in their genome than is necessary for primary biochemical sequences. Some of this extra DNA is obviously concerned with the more sophisticated control mechanisms required by the eucaryotic cell[16,87]; another part relates to the complex intercellular interaction, differentiation and ontogeny required in a multicellular organism[86]; and it also seems probable on current evidence that as much as 30–50% of the DNA consists of sequences which are repeated between 10^3 and 10^6 times, and thus merely reinforce the same information[145–150].

There is a trend for the more advanced species of families and genera of higher organisms to have less DNA than the more primitive members, and it has been suggested that this represents an actual loss of 'redundant' DNA[137a]. Although DNA which is not expressed in any way can probably by dispensed with by an *organism* without much loss of viability (indeed there may be a gain because of the possible ensuing increases in the rate of nuclear and cell division which the reduced DNA complement brings[150a]) there must be a reduction in the overall evolutionary potential since 'redundant' DNA must be part of the capital from which future variation may be drawn[146]. It seems possible, therefore, that the major evolutionary changes might arise from the more primitive groups in a taxon rather than from the most advanced. It could also be concluded that if all the primitive members of a given taxon have died out, that taxon itself will eventually succumb because the advanced members which remain have insufficient genomic plasticity to overcome any large changes in their environment. This may account for several extinctions of both animals (*e.g.* trilobites and dinosaurs) and plants (*e.g.* seed ferns).

If it is assumed that about 10 times more information than is present in the bacteria is required by higher organisms for running primary metabolism plus an equivalent amount for eucaryotic cell control[151], and a further 1000 times more DNA is needed for multicellular organisation, this would still only account for roughly half the total non-repetitious DNA present. It is suggested that in animals, most of the excess non-redundant DNA functions as an information store which determines innate and learned behaviour patterns. In plants, where the amount of repetitious DNA may be higher[145], part of the residual genome at least is undoubtedly responsible for the biosynthesis, organisation and metabolism of secondary compounds. The possible function of repetition DNA is discussed below.

The easiest method for comparing the degree of similarity between DNAs isolated from different species is by hybridisation. There are several procedures which may be used[145,148,152,153]. In one, a mixture of the two DNAs, one labelled and in low concentration and usually subjected to a preliminary mechanical shearing, is treated under conditions which induce both to dissociate into single strands. The mixture is then incubated under reassociating conditions and the rapidly reassociated DNA separated from the still single-stranded DNA by filtration through a specific adsorbent[145]. The proportion of radioactivity in the reassociated fraction is a measure of the relatedness of the DNAs from the two species. Using a related technique, it was shown that a plot of the log percent relatedness of a number of animal species, using the mouse as standard, against time since divergence, gave a straight line[145]. However, as indicated above, up to 50% of higher animal or plant DNA consists of repetitious sequences and these reassociate more rapidly together than does any unique single copy part of the nucleotide chain[145,146]. Thus, the degrees of similarity found between animals or plants in the early DNA hybridisation experiments were subject to error because of the content of repetitous DNA fragments which may be of little importance in determining the phenotype[150].

It is necessary to point out that there are certain conclusions which arise from the data on the repeated copies of DNA[154]. For example, mouse genome is reported to contain 10% of its DNA repeated 10^6 times[145,146]. Now 10% of the mouse haploid genome is approximately $300 \cdot 10^6$ nucleotide pairs, so that the length of each copy is about 300 nucleotide pairs long, certainly much smaller than the 1500 or so bases which code for the average protein. Some satellite DNAs, as these repeating units are called, may be even shorter[148]. A further 15% of mouse DNA ($450 \cdot 10^6$ nucleotide pairs) is repeated 10^3 to 10^5 times, and here the average number of pairs for each copy would be 45000. It is possible that the short copies of the repeated DNA code for transfer RNA (see p. 162) or might, in animals, represent memory units and as such be connected with behavioural patterns based on learning. The larger sized copies may be concerned with ribosomal RNA, or part of differentiation programming[135], ageing[154], or as spacer genes[148,149]. If so, it could mean that a good deal of the DNA would not code for protein and might well have a nonsense sequence (in the protein sense). There is evidence that the smaller repeating units have evolved in the Rodentia fairly recently[155,155a] and yet none of the present closely related members of the order have satellites in common. The same appears to be true for insects[156] and probably for plants also[157,158].

In any case, the presence of multiple copies of certain parts of the DNA of higher organisms makes it difficult to interpret hybridisation experiments. It is presumably for this reason that few have been carried out on higher animals or plants since 1970. Indeed, only one short paper on the subject dealt with plants[153], and the information obtained could not be regarded as throwing any new light on the relationships of the species examined[159]. However, hybridisation studies have been of great value in microbiological taxonomy (see Chapter III), although here the large variation in the actual base ratios between the various taxa perhaps determines the large divergences in hybridisation of DNA which have been so useful[160–163].

All in all, it is obvious that the present data on DNA homologies obtained by the hybridisation technique do not allow many definite phylogenetic speculations to be made on higher organisms. Even where hybridisation occurs between homologous sites of the non-repetitious part of the genome, it is possible that this may be due to perfect matching over a relatively short portion of the strands, perhaps as little as 50 bases in a sheared fragment of 500 base pairs[147]. Such a situation might be obtained in 95% of cases with 25 overall base substitutions. Of course, due to degeneracy, this might or might not affect the sequence of amino acids which are coded for, and one could have then cases where the section of DNAs of two organisms showed homology but the coded protein did not and *vice versa*. It is known, for example, that the messenger RNAs for histone IV from various organisms are quite different as judged by hybridisation experiments, yet the histones do not differ from each other by more than 3 or 4 amino acids[164]. It seems also likely that similar structures could well apply to that part of the DNA which does not code for protein (*e.g.* spacer DNA[149] or even ribosomal RNA[164–169]) and certainly until we know more about the way in which genetically determined behaviour patterns in animals or secondary product metabolism in plants is controlled, it would be wise to concentrate on other approaches than hybridisation to DNA homologies in the higher eucaryotes.

In conclusion, it might be noted that many higher plants and animals have more or less the same order of quantities of DNA in the diploid nucleus[144] and this indicates that they may have an approximately equal degree of advancement in terms of potential genetic information. However, the more common occurrence of polyploidy and higher chromosome numbers in plants[49] may represent important differences which need to be examined if a better insight is to be obtained about the variation in the biochemical evolution of DNA in the two kingdoms.

(c) Ribonucleic acid

Ribonucleic acids (RNAs) are directly concerned with the translation of the genetic code carried by DNA into the sequence of amino acids found in proteins[170]. The code is directly transcribed onto a special form of RNA, messenger RNA, by base matching of mRNA to DNA in a similar way to the replication of DNA itself. Each protein thus has its own mRNA, each molecule of which, depending on the circumstances, can act as the template for several thousand exact copies of the protein[170]. The individual amino acids are transported to the site of synthesis and arranged along the template of mRNA in the correct sequence by attachment to a second form of ribonucleic acid, transfer RNA (tRNA), to which the amino acid is joined by a covalent bond[171–174]. Each organism, so far examined, has been shown to contain multiple iso-accepting tRNAs specific for each single amino acid and these do not show a uniform correlation with the known degeneracy of the code[171]. That is, tRNAs for those amino acids with four or more triplet codes, do not appear to be more frequent than the other tRNAs.

Proteins are synthesized at special sites within the cell, the ribosomes, which contain from 40–65% of RNA (rRNA), the rest being comprised of protein, amines and membrane components[164–170]. The exact function

References p. 287

of rRNA is not really known, although part must be concerned with binding the mRNA[175] and, perhaps, the tRNAs in a suitable manner for protein synthetases to act efficiently (see Chapter I).

The three types of RNA are of quite different molecular size and exist in unequal amounts in the cell. Several ribosomal RNAs have been isolated with molecular weights of between $4.0 \cdot 10^4$ (5S) to $1.75 \cdot 10^6$ (28S)[46,165,169,176]; overall, rRNAs comprise 75–80% of all ribonucleic acid in the cell. The molecular weight of mRNAs depends on the size of the protein for which it is a template. For an average protein of 500 amino acids (60000 MW) mRNA has an MW of $4.8 \cdot 10^5$ (14 S, 1500 bases) but the whole fraction varies from about $1.0 \cdot 10^5$ to $2.0 \cdot 10^6$ in molecular weight and accounts for 4–5% of the total RNA[177]. Transfer RNAs make up the bulk of the remaining RNA in the cell (15%) and all have a molecular weight of $2.5 \cdot 10^4$ (that is, 75 or so bases). All three types are presumably synthesised on DNA since hybridisation experiments have shown that they are complementary for stretches of the DNA backbone[177].

Because of its overall heterogeneity, the mRNA fraction has not been used to compare ribonucleic acids (and hence DNA) from different organisms with a view to determining their relationships. There appear to be similarities, however, in the base sequences of the polypeptide initiating sites[177]. Ribosomal RNA, although not homogeneous within the cell can be readily separated into discrete molecular size fragments with good precision by electrophoresis on polyacrylamide gels and comparison of these has been used to adduce relationships between taxa[165]. The individual tRNAs, in spite of their similarity, can be separated and the actual base sequence of about 40 of these has been determined, enabling direct comparisons of the genetic code differences between organisms to be made[171–174].

The ribosomes of living organisms can be divided into three main types depending on their size: procaryotes and higher plant chloroplasts contain ribosomes with sedimentation coefficients of 70S; cytoplasmic ribosomes, both plant and animal, have sedimentation coefficients of 80S and animal mitochondria ribosomes have values of 50–60S. They all contain two sub-units, one roughly twice the size of the other[167,176,178]. The smaller sub-unit of the ribosomes of all eucaryotes contains rRNA of molecular weight $0.7 \cdot 10^6$ (18 S) and thus shows a strong evolutionary conservation of size within the two kingdoms (Table V)[165,169]. In the

TABLE V

Molecular Weights of Ribosomal rRNAs and Transcription Units[165,169,178]
Mol.wt. × 10^{-6} (average taken from sources quoted)

Phyla	*Transcription unit*	*rRNA*		*Number of rRNA genes*
		Large	*Small*	
Mammalia	4.27	1.72	0.67	400–1250
Birds	3.92	1.61	0.63	700–1300
Reptiles	2.74	1.51	0.62	
Amphibians	2.76	1.58	0.61	1000–2500
Fish	2.70	1.55	0.65	
Insects	2.85	1.40	0.65	260
Echinoderms		1.40	0.68	
C[illegible]ophora		1.31	0.69	
Angiosperms	2.5	1.30	0.70	1600–1330
Ferns		1.34	0.72	
Green algae		1.29	0.69	
Fungi		1.30	0.71	200–240
Blue-green algae		1.07	0.56	
Bacteria		1.09	0.57	1–45
Chloroplasts		1.09	0.56	
Mitochondria				
fungal		1.26	0.72	
animal		0.80	0.43	

ribosomes of procaryotes, on the other hand, the smaller sub-unit has rRNA with a lower MW, $0.56 \cdot 10^6$. The procaryotes are also uniform in the size of the rRNA in the larger sub-unit (Table V) and again this is lower in molecular weight than the rRNA in the larger sub-units of eucaryotic cells. In all the plants examined, the large sub-unit rRNA has MW of about $1.30 \cdot 10^6$, but in the Animal Kingdom it appears as if the molecular weight of this rRNA fraction increases in the more recently evolved species[165]. It may also be noted that in all eucaryotes examined, the rRNAs of both sub-units are synthesised as one initial transcription unit in the nucleolar organiser region of the DNA[169,179]. The transcription unit is larger than the two rRNAs formed from it, and again it appears that the amount of non-ribosomal sequences is higher in the more recently evolved animals than it is in plants (Table V: compare the size of the transcription unit, column 2, with the sum of rRNA's, column 3). Bacteria

References p. 287

on the other hand, apparently have no very high MW precursor of the two rRNAs, although they are coded on the same cistron[165]. One other evolutionary feature of rRNA transcription is the large increase in the numbers of copies of the genes responsible in higher organisms (Table V). Whereas *Escherichia coli* has only one gene for rRNA, the onion (*Allium cepa*) has 13300 per telophase nucleus; that is about 1% of total DNA in this plant is occupied by multiple copies of a single gene[169]. This could certainly account for a part of the repetitious DNA (see p. 156).

There is no uniformity, however, in the amount of RNA in cells of polyploid series[169]. In *Tradescantia ohioensis*, the diploid and tetraploid have the same amount of RNA, although the DNA is expectably doubled. In barley, however, both RNA and DNA approximately double from the diploid to the tetraploid. The increase in DNA with ploidy in hyacinth cultivars is paralleled by increases in the number of rRNA genes and nuclear-organising regions per nucleus[169]. All this points to the need for many more comparative studies.

It is interesting to note that the size of the blue-green algal ribosomes and rRNAs is exactly the same as those in the chloroplasts of higher plants[165]. There are several other parallels between the two: neither the ribosomes from Cyanophyta nor from higher plant chloroplasts yield a fragment of RNA MW $5.0 \cdot 10^4$ (5.85 S) on heating to 60°C, unlike the cytoplasmic ribosomes of all eucaryotic plants including green algae[176]. The ribosomal RNA from several blue-green algae has been shown to hybridise with the DNA from *Euglena gracilis* indicating a significant genetic homology between the two[181]. Furthermore, the protein-synthesising activity of the 70 S bacterial chloroplast ribosomes is inhibited by chloramphenicol, lincomycin and spectomycin, but not by cycloheximide[180]. These data strongly support the suggestion that chloroplasts originated in eucaryotes by symbiosis from a procaryotic ancestor[181,182]. However, the same is not true for the mitochondria. Both animal and plant mitochondrial ribosomes are different in several respects from these in procaryotes and their smaller size suggests an independent origin[179,183].

The present data on the ribosomal RNAs, as may be seen from Table V, has given some useful pointers with regard to plant phylogeny. We must await, however, further detailed knowledge on the sequence of bases in some of the discrete fragments which are now being worked on before the biochemical aspects of this evolution can be properly commented on. The base sequences of fourteen different transfer RNAs have been compared mainly to those from bacteria and fungi[174]. Thirteen nucleotide residues at homologous sites are identical in all those examined and there were a further eight sites which were occupied either by a purine or a pyrimidine. These similarities point to a single evolutionary origin for all tRNAs. When homologous tRNAs for a single amino acid are compared, the results are even more striking: 77% of the 82 bases in yeast and rat serine tRNAs, and 82% of the 81 bases in yeast and wheat phenylalanine tRNAs are identical. Although these percentages take no account of chance, they are indeed remarkable, and again indicate the need for more data. For, although the sequencing of the various tRNAs has given us a great insight into the actual conservatism in the genetic code itself, there is insufficient data to discern any evolutionary trends. It is to be hoped that this will be remedied ere long.

5. Proteins

(a) Introduction

Comparisons of the properties of homologous proteins from different species with a view to deducing possible taxonomic and phylogenetic relationships have been made in several different ways. Not surprisingly the main methods used, serology[184,185] (hybridisation), gel electrophoresis[185–187], and determination of the sequence of amino acids (bases) in the protein[46,188–195], show a remarkable similarity to those already described for the nucleic acids. Although some of the drawbacks already described for these methods apply in general terms when they are applied to proteins, proteins have the advantage over the nucleic acids both because of their greater structural variability and the fact that most of them possess enzymic activity. These two factors allow for more refined and sensitive separation techniques to be applied and permit a much wider range of differences to be detected. Unfortunately, none of these techniques have so far been applied over a sufficiently wide range of plant taxa to enable firm conclusions to be drawn about the overall evolution of proteins in the Plant Kingdom. However, there will undoubtedly be bigger advances during the next decade.

(b) Systematic serology

The earliest methods used for comparing proteins were based on the serological techniques which came to the fore in the last decade of the nineteenth century as a result of increased knowledge of the human immune response to pathogens[84]. The immune response is due to the *de novo* synthesis of antibody proteins (immunoglobins) in the blood which possess structures (determinants) that allow the antibody to react and "neutralise" the agent (antigen) eliciting the response. A particular antibody will react only with the specific antigen and a few closely related substances, and this reaction can be used *in vitro* to detect homologies in any of a wide number of different classes of substances, including foreign proteins, which elicit an immune reaction[84,185]. In the production of antibodies for investigations of protein homology, the given, preferably highly purified, foreign protein is introduced intravenously into the test animal (usually a rabbit) at intervals over a period of a few weeks. Blood samples are then collected and the sera containing the antibodies are prepared[84,185]. There are several quantitative ways of measuring the reaction between the antibody (or properly the antisera) and the proteins under test, some of which (*e.g.* microcomplement fixation) are extremely sensitive, and qualitative techniques which are equally discriminatory have also been devised most of which involve observing the reaction in suitable gels[185].

Using earlier rather cruder methods, Metz and his collaborators in the 1920s produced spectacular phylogenetic trees of the whole Plant Kingdom, but his work was largely discredited when a rival school produced a completely different set of relationships[57]. Indeed it is only with the development of the more sensitive and reproducible techniques mentioned above, that serological methods have come to be accepted in taxonomic practice[57,184,185]. Even so, although very many useful relationships between various plant taxa have been obtained by serological studies[184], no one has attempted phylogenetic studies on the grand scale. This is in contrast with work on animals[196–198], and to a lesser extent, bacteria[199].

However, it should be noted that, as with DNA hybridisations, the nature of the reaction between the determinant sites on the immunoglobulin antibody and the active sites on the antigen is ill understood[199] and it is difficult therefore to decide what represents an important evolutionary difference. For example two quite different protein antigens could show equivalence because they possessed a similar sequence of amino acids somewhere in their chain which reacted with the determinant. Nevertheless, as we know more about the structure of antibodies and the biochemical nature of the immune reaction, this method, because of its high specificity and sensitivity, will undoubtedly be used far more widely.

(c) Electrophoretic separations

Proteins, because they contain varying proportions of basic, neutral and acidic amino acids, show wide differences in their net electric charge at any given pH value. Advantage can be taken of such differences to separate proteins from each other and isolate them in a pure state. On a large scale, it is usual to utilise weakly basic or acidic ion exchangers for separations, but on a small scale, such as is useful in comparative work, electrophoretic methods can be used[185, 200a]. In recent years the use of starch and acrylamide gels with defined pore sizes has allowed proteins to be separated on the basis of both size and charge. Coupled with sensitive histochemical methods for the detection of enzymes, this method has added much to our knowledge of protein polymorphism[186, 187, 200a]. However, unlike the work on rRNAs using such methods, which demonstrated the uniformity of size of certain fragments in the major phyla (see Table V and p. 160), there is no apparent uniformity in apparently homologous enzymes such as malic dehydrogenase from anything but the most closely related taxa[200a]. Indeed, in most individuals, several enzymes exist in a number of separable forms; sometimes these occur in the same organelle (isoenzymes). In other cases, however, proteins which are known to be different are not separated using this technique[200b] and this could be the rule rather than the exception in many cases. At the level of populations or intraspecific intergression, the gel electrophoresis of different enzymes is an extremely powerful tool; but it has not added much to our knowledge of phylogenetic relationships of plants in the wider sense.

(d) Amino acid sequences in proteins

In recent years, much weight has been given to the usefulness of comparative studies on the primary structure of proteins in deducing phylogenetic relationships[189–195]. Some workers have gone as far as to state that ordinary biological characters, including all simple chemical compounds, are much less suited to making evolutionary appraisals than are amino acid sequences of selected homologous proteins[190, 201]. Belief in the power of sequence analysis arose when the results on the very first proteins to be examined, the insulins from nine different mammalian sources, were compared and found to differ only at four sites among the 51 amino acid residues making up the two polypeptide chains[202]. This indicated a great conservatism in variation of the structure of this biologically important protein during the course of mammalian evolution. When more sequence data from animal proteins became available, especially those on the various cytochromes *c*[203] and hemoglobins[204], it was apparent that amino acid sequence data *per se* promised to be an important tool for accurately charting patristic relationships[188–192]. Moreover, since proteins, albeit imperfectly due to code degeneracy, reflect the sequence of bases in the structural DNA, the amino acid data could be used to read directly the evolutionary history of the genes.

The development of refined computer techniques became necessary for the analysis of the vast amount of sequence data and allowed the construction of phylogenetic trees which were in satisfactory agreement with paleontological investigations[190]. Several principles emerged from these results. One was that for any set of orthologous proteins[189], that is, proteins whose common evolutionary ancestor was represented by a single gene, the rate of amino acid substitution was linear if estimates from the fossil record of the times of divergence of the various faunal groups was taken to be correct[46,133,189,195,205]. A second was that the rate of substitution in different orthologous sets could vary widely; fibrinopeptides, for example, having one mutation fixed every million years which is a rate 30 times faster than that for the cytochromes *c*[46]. It may be noted that, in general, the most slowly changing sequences are found in proteins which one might expect, because of the need to conserve function, to remain the most invariant. As a corollary, it may be noted that within an individual set of orthologous proteins, certain amino acids are completely invariant (30 of the 104 amino acids in cytochrome *c* are of this type) and the fact that a large number of classes of enzymes are known which have near identical sequences at their active centre points to their evolution from a single primitive ancestral type[46,136,137,188–193,201]. The apparent constancy of amino acid substitution and the invariability of the active sites in many of the proteins so far examined has led to suggestions that all molecular evolution is neutral; that is, occurs as a result of random changes in the genome which are adaptively unselected because the altered protein has not changed in function[133,188,205]. This may be true for proteins like cytochrome *c* and hemoglobins, but it is very unwise to extrapolate from these examples that the whole genome or even the structural DNA undergoes only neutral changes during the course of evolution[206,207]. Furthermore, constancy of structure need not mean constancy of function as shown by the change in that of the thyroid hormones during evolution from amphibia to mammals[208]. If, as is believed, the majority of cognate enzymes arose by gene duplication followed by changes in one of the genomes, it is obvious that molecular evolution is far from neutral[136,137,193,209,210]. Indeed all present day biological characters which delineate taxa must result from an altered genome followed by selection. To understand these processes in molecular terms, one needs to look at the variations in the amino acid sequences of enzymes, such as the various dehydrogenases[211,212], which show adaptive changes in activity, and hence in the function of the metabolic sequences, in different organisms.

That molecular evolution is by no means constant is becoming clear from detailed studies of the more recently evolved animal's groups, and it is becoming accepted that the evolutionary change of proteins can be species as well as time dependent[213–215]. With these structures in mind, the results of recent extensive investigations of plant cytochromes *c* (almost entirely angiosperms) and the plant ferredoxins can be examined.

(e) Plant cytochrome c

In the last few years, Boulter and his colleagues have isolated and sequenced mitochondrial cytochrome *c* from twenty-one higher plant species ranging from *Ginkgo biloba* to *Cucurbita maxima*[216,217]. This enzyme functions as an electron carrier in the mitochondrial respiratory chain. The plant enzymes appear to be strictly homologous to those isolated from animals and fungi[218,219]. For example, they all contain the eleven residue sequence (70–80 inclusive on horse heart cytochrome *c* numbering) which is found invariant in all cytochromes *c* examined[220]. Furthermore, as in the animal enzyme, the heme group in the plant cytochromes *c* is attached through two cysteines, a histidine and a methionine in exactly the same equivalent positions, and there are similarly placed hydrophobic, aromatic and glycine residues along the chain. There are differences, however, in the number of basic and acidic residues; plant cytochromes *c* having on average seven less lysine residues (12 as against 19) and three less acidic residues (9 against 12) than has horse heat cytochrome *c*; the positions of the existing residues in the plant enzyme, however, correspond with those in animal cytochrome *c*[220]. Finally, the amino acid sequences of plant cytochromes *c* fit the projections of three-dimensional structure of horse heart and *Bonito* cytochromes *c*[219]. Since the enzymes also have an equivalent function in all eucaryotic organisms which have been examin-

References p. 287

ed, the sequence data can be properly used to determine the phylogeny of the species concerned.

The sequence data on the plant cytochromes *c*, were used to construct phylogenetic trees in two different ways; one by the ancestral sequence method[46,190] and secondly, by the numerical matrix method[190,221], using a flexible agglomerative strategy[222]. In both cases, the trees were derived without the use of any weighting factors and with the usual assumptions that molecular evolution has occurred with the minimum of mutational changes and that the rate of change has been relatively constant, at least over relatively long time scales[46,190,194]. A tree obtained with the flexible numerical matrix method is shown in tabular form in Table VI (refs. 216, 217). A variety of other trees constructed using this method by altering the variable parameter β from -0.4 to $+0.3$[222], all gave virtually the same pattern. Nor did it make any difference whether the amino acid changes or minimum mutational differences (the minimum number of nucleotide changes required to change from one cytochrome *c* to another[46]) were used to construct the tree. The alternative use of the matrix method of Fitch and Margoliash[221] also led to essentially the same results.

TABLE VI

Phylogenetic tree derived by the flexible numerical method[217,218]

Cytochrome c phylogeny	*Genus*	*Family*
	Brassica	Cruciferae
	Cucurbita	Cucurbitaceae
	Phaseolus	Leguminosae
	Sambucus	Caprifoliaceae
	Abutilon	Malvaceae
	Gossypium	
	Ricinus	Euphorbiaceae
	Lycopersicum	Solanaceae
	Sesamum	Pedalaceae
	Helianthus	Compositae
	Guizotia	
	Spinacia	Chenopodiaceae
	Fagopyrum	Polygonaceae

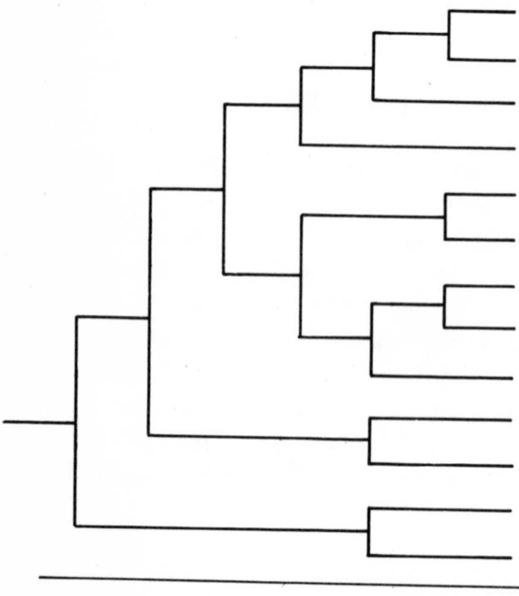

TABLE VII

Comparison of cytochrome *c* and classical angiosperm phylogeny

Cytochrome c phylogeny[a]	*Genus*	*Family*	*Order*[b]	*Sub-class*[b]
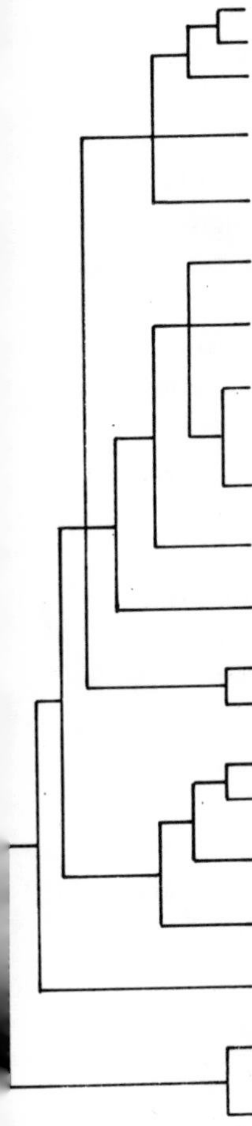	*Hordeum* *Triticum* *Zea*	Gramineae	Cyperales	Commelinidae
	Allium	Liliaceae	Liliales	Liliidae
	Arum	Araceae	Arales	Arecidae
	Nigella	Ranunculaceae	Ranunculales	Magnoliidae
	Tropaeoleum	Tropaeolaceae	Geraniales	Rosidae
	Cucurbita	Cucurbitaceae	Violales	Dilleniidae
	Brassica	Cruciferae	Capparales	Dilleniidae
	Phaseolus	Leguminosae	Rosales	Rosidae
	Sambucus	Caprifoliaceae	Dipsacales	Asteridae
	Helianthus *Guizotia*	Compositae	Asterales	Asteridae
	Abutilon *Gossypium*	Malvaceae	Malvales	Caryophyllidae
	Ricinus	Euphorbiaceae	Euphorbiales	Rosidae
	Sesamum	Pedaliaceae	Schrophulariales	Asteridae
	Lycopersicum	Solanaceae	Polemoniales	Asteridae
	Fagopyrum	Polygonaceae	Polygonales	Caryophyllidae
	Spinacia	Chenopodiaceae	Caryophyllales	Caryophyllidae

[a] Boulter[217,218]; tree constructed by the ancestral sequence method.
[b] Cronquist[56].

A tree obtained by the ancestral sequence method but with more data is shown in Table VII. It can be seen that there is little difference between the trees obtained by these two different methods. In the matrix tree (Table VI), the node leading to the Compositae comes before that giving rise to the Malvaceae 'group', whereas the opposite is true in the ancestral tree. The matrix tree includes the Solanaceae in the Malvaceae group, whereas it is clearly separated on the ancestral tree. However, the ancestral sequence method is preferred, since it shows an overall lower number of amino acid substitutions. The extent of parallel and back mutations (27% and 6% respectively) calculated from the plant cytochrome *c* data[217], is relatively the same (20% and 1%) as that from the animal data[233]. It appears likely, however, that this value may be too low[194].

The trees from the amino acid sequence data of cytochrome *c* from angiosperms (Tables VI and VII) certainly do not agree with the relationships proposed by several taxonomists mainly on the basis of morphological criteria[56, 58]. As mentioned earlier, (p. 151), the majority of present-day taxonomists favour the Magnoliidae (Ranalean complex) as the basic group from which other dicotyledons have evolved (see Fig. 3). Others, however, on chemical grounds have favoured an interpretation in which the Rosidae or Dilleniidae are at least coeval with the Magnoliidae[129, 130, 224–227]. Another generally accepted hypothesis is that polypetaly and polycarpy evolved prior to sympetaly and syncarpy, although it is possibly not true for all types of flowers[56, 58, 228, 229]. As can be seen, such views are certainly not in accord with the data shown in Table VII. One difficulty, however, is that there is much controversial evidence from the palynological[230] and macrofossil data[61] about the possible time of appearance of major modern plant taxa[62–64]; and it seems possible that the morphological characters on which major phylogenetic schemes are based may have evolved more than once. In any case, the molecular data implies that phylogenetic implications of the existing morphological evidence may need to be re-examined.

Using the plant sequence data on cytochrome *c*, Boulter has calculated the time of divergence of plants from other organisms (Table VIII)[216, 230a]. These results are generally in accordance with those published by other workers[46, 190]. It is interesting to note that the lines to the three kingdoms of eucaryotic organisms (animals, plants and fungi) appear to have diverged at dates earlier by a factor of two from the presumed origin of the

TABLE VIII

Times of Divergence of Various Classes of Organism based on Cytochrome *c* Data[216]

Groups compared	*Number of variant residues*	*Corrected time of divergence (m. yrs)*
Mammals↔Birds	9.9±1.7	280
Vertebrates↔Invertebrates	26.6±3.1	830
Animals↔Angiosperms	47.2±2.7	1860
Dicotyledons↔Monocotyledons	8.6±2.1	240
Angiosperms↔Gymnosperms	14.7±2.2	520
Angiosperms↔Fungi	52.3±2.4	2120

eucaryotic cells themselves[78]. Of course, the corrections which one should apply to such long time scales can only be a matter of conjecture and those used might be inexact for divergences as large as those found[194]. Nevertheless, it is interesting to note that the case of both the angiosperms→ gymnosperms and monocotyledons→dicotyledons, the time of diversion pre-dates the usually agreed origin of even the oldest of two groups concerned, as judged by the admittedly scanty paleological record[61,230]. It seems more likely, therefore, that plant cytochromes *c* have evolved at a faster rate than the animal enzymes. As pointed out above (p. 167) there is increasing evidence from animals that the rates of protein evolution are markedly species dependent[213–215] and there seems no reason why this should not be true for plants.

(*f*) *Ferredoxins*

Probably the most important group of proteins from the evolutionary point of view are the ferredoxins. These are non-heme iron-containing enzymes which act as electron carriers in a number of different biochemical processes[231–233]. About forty proteins of this type are now known, isolated from primitive bacteria to higher plants and animals. They can be divided into four groups depending on whether they contain 2, 4, 6 or 8 iron atoms and equivalent amounts of inorganic sulphur per molecule.

In each case the iron atoms are linked to the protein through the sulphur of cysteine residues, and the inorganic sulphur forms a link between the iron atoms (Fig. 4)[234]. They all have a low redox potential (−0.3 to −0.5 V at pH 7) and characteristic electron paramagnetic resonance spectra.

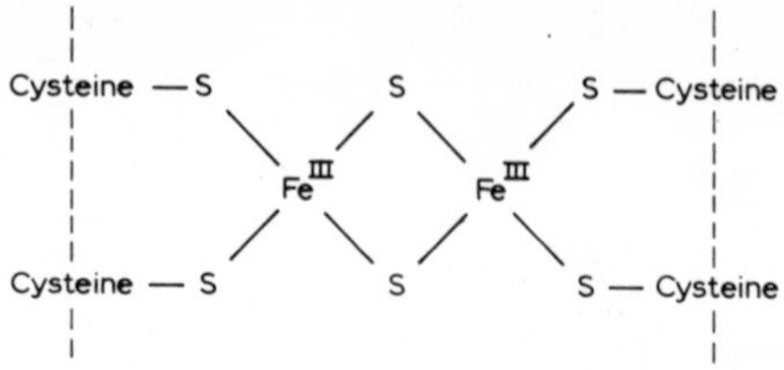

Cysteine — S S S — Cysteine
Fe^{III} Fe^{III}
Cysteine — S S S — Cysteine

Fig. 4. Structure of iron–ferredoxin bond. (One Fe^{III} is reduced to Fe^{II} on reduction).

The ferredoxins containing eight iron atoms occur only in anaerobic and photosynthetic bacteria. Unlike other ferredoxins, they transfer two electrons at a time, perhaps because the iron atoms are arranged in two separate clusters. The *Clostridia* ferredoxins are simple small proteins of 55 amino acids and contain no arginine, leucine, methionine or tryptophan which are commonly regarded as being relatively more advanced than other amino acids. Indeed 50 out of the 55 residues of the ferredoxin of *C. butyricum* contain the nine amino acids, glycine, alanine, valine, proline, glutamic and aspartic acids, serine, cysteine and isoleucine. The first six of these have been detected in the carbonaceous fraction from the Murchison meteorite[235] and all nine can be synthesized under primitive earth conditions[4,236]. Thus, the clostridial ferredoxins can be regarded as probably the most primitive electron carriers in living organisms.

Some of the four-iron ferredoxins also contain only 55 or so amino acids, and one part of the amino acid sequence of the ferredoxin from *Desulforibrio gigas* is similar to that from *Clostridia* and another part to segments of higher plant ferredoxins[232,233].

Among the two-iron ferredoxins, only one has fewer than 70 amino acids. The majority of the plant ferredoxins, including those from the blue-green alga *Microcystis*, have molecular weights of about 11 500 daltons and contain 95–98 amino acid residues. The sequences of these plant ferredoxins, and especially the position of the cysteine residues, show a fair degree of homology with those of the two-iron ferredoxins from *Pseudomonas putida* and beef adrenal glands, but they are not equivalent in their catalytic activity and not interchangeable. Arginine, leucine, methionine and tryptophan are present in the ferredoxin from *Microcystis*, so it shows considerable advancement on the *Clostridium* enzyme[232,233].

Based on the knowledge of the amino acid sequences so far known, the evolution of the ferredoxins can be represented as shown in Fig. 5.

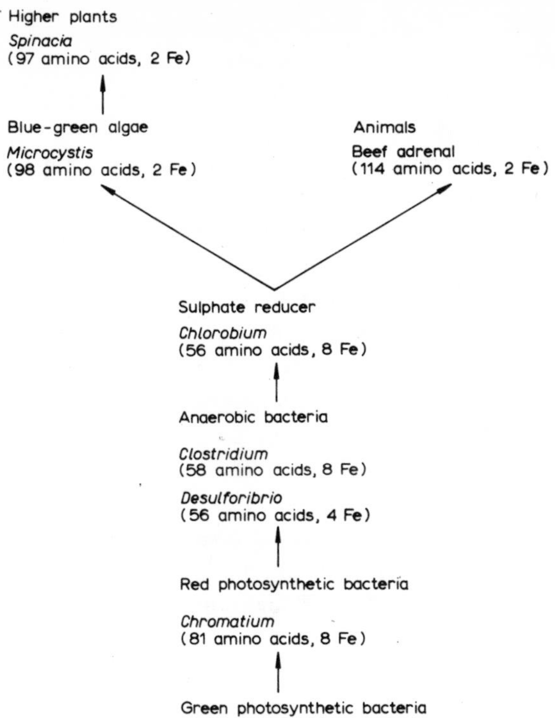

Fig. 5. The evolution of the ferredoxins[232].

It should be noted that there is some uncertainty about the conservation or loss of certain amino acid sequences in the ferredoxins and it may be found when more data are available that some will show examples of more than one origin. This would certainly be of great interest for it would indicate that evolutionary processes at the molecular level can be as complex as they appear to be at the morphological and secondary compound levels.

6. Metabolic pathways

If we accept that primitive procaryotes were present when the black chert of the Fig Tree series in South Africa was laid down about 3.2 billion years ago and that fairly advanced eucaryotes had evolved by the time the Bitter Springs formation was deposited about 1.0 billion years ago (Table III)[114–121], then it follows that the development of most of the common biochemical pathways which occur in all present-day organisms was probably completed by the end of the first half of the earth's history

References p. 287

about 2.0 billion years ago. The final elaboration of the nucleus, mitochondria, chloroplasts and other specialised eucaryotic organelles might not have been completed until 0.8 billion years ago when the oxygen levels had risen sufficiently to allow perfection of their function[61,78,92]. On the other hand, the intricate mechanisms of nuclear and cytoplasmic control necessary for the organisation and differentiation of multicellular forms obviously could not have advanced until such organisms appeared just before the start of the Cambrian (0.6 b.yrs ago). The evolution of such mechanisms, especially in relation to complex behavioural patterns in animals and secondary metabolism in plants, is undoubtedly still advancing.

The common biochemical pathways of metabolism referred to above include not only relatively simple sequences such as that leading to the utilisation of sugars for energy or the formation of the saturated fatty acids from acetate, but also to the intricate interlocking series of events involved in the synthesis of nucleic acids and proteins, including the elaboration of the ribosomes and associated transfer and messenger RNAs. It should be noted that in the cells of all present-day organisms from the most primitive bacterium to the most highly advanced plant or animal, each individual step in these catabolic or biosynthetic sequences is absolutely identical. In every case, each reaction involves catalysis by an enzyme which has a closely similar if not identical active centre tertiary structure and related sequence of amino acids to the corresponding enzyme in other organisms[46]. For example, NAD-linked dehydrogenases have roughly homologous sequences of amino acids at their active centres and only four types of active centre are known in all the proteolytic enzymes[206]: the main one has serine at the active site and is mainly confined to the proteases found in the Animal Kingdom and certain bacteria which can be distinguished from each other by the nature of the adjacent amino acids; the second, present in plant proteases and in those from Group A *Streptococci*, has cysteine at the active site; the third group consists mainly of exopeptidases which require a bivalent metal ion; and the last are the acid proteases, of which pepsin is an example. Furthermore, where necessary, the same enzymic co-factor for the transfer of hydrogen and phosphate and so on is utilised. Indeed it must be stressed that there is abolutely no variation throughout the entire living world in the structure and the mode of reaction of the common co-enzymes[237] such as NAD, NADP, FAD, FMN, ATP, UTP, Coenzyme A, biotin, lipoic acid, thiamine phosphate and pyridoxal. Often, of course, there is a variation in the rate at which individual reactions proceed in different organisms under a given set of conditions, and the enzymes involved may well have different pH and temperature optima, usually because of variations in their amino acid sequences. Even different cells of the same organism, as mentioned earlier, may contain two or more isomeric analogous enzymes some of which can be readily separated on account of their difference in charge or size by gel electrophoresis or related techniques[186,187,200a,b].

Does the possession of a common sequence of interlocking reactions in several organisms indicate a single evolutionary origin? Or does it merely show that there are only a few viable biochemical options open to the cell? In other words, are there just one or two solutions to the problem of, for example, reducing a carbonyl group or phosphorylating a hydroxyl group which can be encompassed within the energy limitations imposed by the cell? Again does the conservatism in enzyme structure only reflect that there are only a few ways possible in which a protein can be assembled in order to provide the right matrix to catalyse a given reaction by being able to bind the substrates and co-factors in the specific way which allows transfer of electrons or other moieties to readily take place? At first sight these questions are difficult to answer. But if the reason for biochemical similarity of primary metabolism in all present-day organisms were because of some innate unchangeable conservatism in the structure of the genetic code, one might expect very little organic evolution to have occurred at all! The reason then must lie in the paucity of solutions for carrying out the individual steps in the common pathways of metabolism. One might conclude that, if this were true, little insight would be obtained into evolutionary processes by comparing the biochemical mechanisms underlying the steps of primary metabolism in different phyla. But, while it is certain that the energy or co-factor requirements of an individual step will not have changed during evolution, the metabolic control mechanisms which affect it undoubtedly have done so.

Indeed, it is the development of such control mechanisms which distinguish evolutionarily advanced organisms from the more primitive ones. That such controls have an important evolutionary advantage is shown by experiments with strains of *Escherichia coli* which lack the ability to regulate the synthesis of proline. In competition with a wild strain the mutant did poorly both on a minimal medium and on one containing small amounts of proline, thus showing that it would not have normally survived[238]. Similar observations have been made on a mutant of *Bacillus*

References p. 287

subtilis in which the tryptophan pathway was derepressed[239].

Because there are several ways in which control of any metabolic pathway can be exercised and that the mechanisms of control probably did not evolve until relatively late, it is not surprising to find that there are often wide variations between one organism and another[168, 240–244]. It has been suggested that the evolutionary significance of a given pathway may often be more easily assessed by a comparison of the control mechanisms which are used by different organisms[245]. This might allow one to determine, for example, whether the pathway had a single evolutionary origin. The control of a single step in a sequence of metabolic reactions does, of course, affect all subsequent transformations in that sequence.

In the majority of cases control is effected either by repression of the synthesis of the enzyme catalysing the step in question or by directly inhibiting its activity[168, 241–244]. The former method is relatively slow to take effect and depends to a large extent on the rate of turnover of the enzyme. Furthermore, once synthesis is turned off, the pathway may take some time to re-establish. Direct inhibition is usually by either allosteric or competitive action by a later product in the metabolic sequence. This method of control can be very rapid and highly effective in the short term, but will rapidly be reversed as the inhibitor is itself further transformed or translocated. In several cases, inhibition is effected through a co-enzyme necessary for the activity of the enzyme; equally, lack of such co-enzymes often determines the rate of overall reactions of metabolic sequences. Where a pathway is linear in sequence, control is usually made through the first enzyme of the pathway[168, 241–244]. Where, however, branched pathways exist leading to two or more end-products, the situation can be very complex indeed and usually involves a great deal of variation between one organism and another (Fig. 6). This is especially true of eucaryotes where repression of enzyme synthesis may involve either control of template availability for the transcription of the genetic code to mRNA, or stability of the messenger, or modulation of translation as well as the control *via* the regulator–operator model as found in procaryotes. Unfortunately, however, little work has been done on the control mechanisms of the majority of metabolic pathways in the higher members of the Plant Kingdom[246], but the examples given below do indicate the types of variation which might be expected and their possible evolutionary significance.

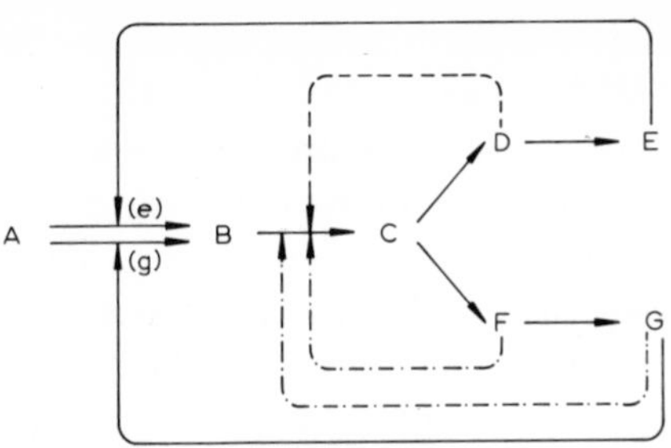

Fig. 6. Possible control mechanisms on branched pathways. ——, end-product feed-back inhibition; ---, sequential feed-back inhibition; -·-·-·-, concerted feed-back inhibition; A, B, etc., metabolites on the branched pathways; e and g are two separate enzymes catalysing the reaction A→B.

A relatively simple example of regulation by co-factors concerns citrate synthetase in bacteria[247]. This enzyme is the key to the entry of carbon into the tricarboxylic cycle and the two coenzymes NADH and AMP are involved in the overall operation of this sequence. It was found that neither cofactor had any effect on the citrate synthetase from Gram-positive bacteria such as *Bacillus cereus* or *Streptomyces somaliensis*. These organisms are regarded as being the most advanced group of Schizomycetes in having more complex intracytoplasmic membranous structures called mesozomes which contain the enzymes of the TCA cycle and are analogous to the eucaryotic mitochondria. All Gram-negative bacteria on the other hand contained citrate synthetase which is strongly inhibited by NADH. Within this group, the aerobic bacteria like *Azobacter vinilandii* and *Pseudomonas fluorescens*, had enzymes which were reactivated by AMP, whereas the enzymes in facultative anaerobes like *Escherichia coli* and *Salmonella anatum* were unaffected by the nucleoside. These latter organisms degrade glucose by the normal glycolytic pathway where presumably AMP (or ADP) act as inducers of key enzymes like phosphofructokinase or pyruvate kinase and are thus not required to control a later point in the overall catabolism of the sugars. The Gram-negative aerobes, on the other hand, catabolise glucose *via* 6-phosphogluconate which involves no feedback control from either AMP or ADP. This example indicates how divergence in the mechanism of control is related to the overall capabilities of the organism[247]. It should be noted that it is highly likely that the citrate synthetases from these different types of organism will probably show some variation in the amino acid sequence in or about their active centre. The significance of such variation on the ability of the organism to compete successfully with others needs further investigation.

A second example concerns the control of pathways leading to the degradation of the benzene ring in bacteria and fungi[240, 245, 248–250]. The most widely investigated group of organisms carrying out this sequence of reactions are the *Pseudomonas* and related Gram-negative aerobes[240, 245]. Most species of this genus degrade the aromatic ring by first hydroxylating it to give an *ortho*-dihydroxy grouping and then cleaving the ring between these two hydroxyl groups (Fig. 7). This oxidative degradation gives rise to *cis,cis*-muconic acid or its congeners depending on the substitution pattern of the aromatic ring. Another group of *Pseudomonas*

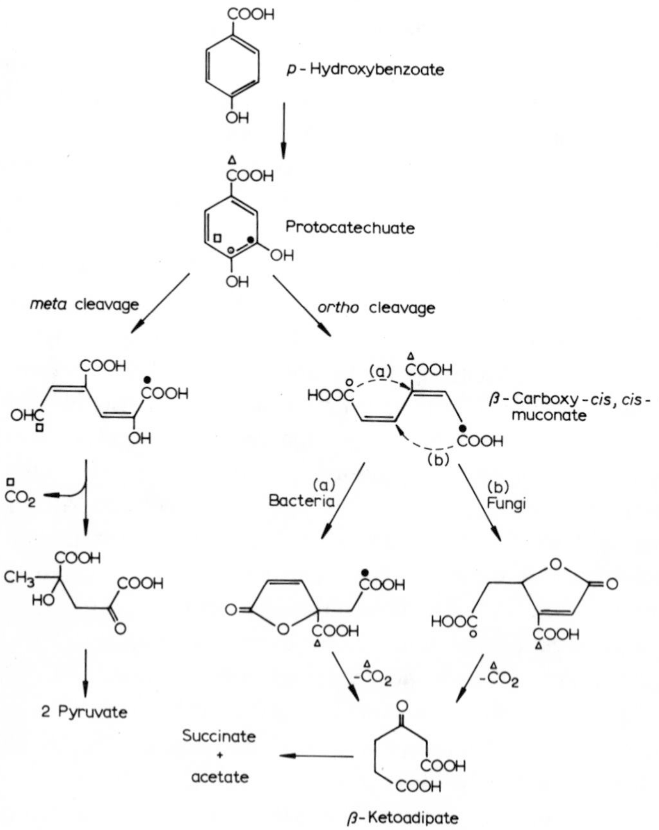

Fig. 7. Degradative aromatic pathways in bacteria and fungi[245, 248].

again hydroxylates the aromatic ring to the *ortho*-dihydroxy state, but then proceeds to cleave it in the so-called *meta* position to give a muconic acid semi-aldehyde. Among the organisms which use the main *ortho* pathway, there is a wide variation in the control mechanisms involved. In *Pseudomonas putida*, for example β-ketoadipate, which is the last intermediate between the degradation of both benzoic acid itself and *p*-hydroxybenzoic acid to compounds involved in primary metabolism (see Fig. 7), induces the formation of enzymes carrying out the last two steps of the latter pathway as well as the first common step between the enol lactone and β-keto-adipate itself. In *Acinetobacter (Moraxella) caloacetica*, on the other hand, the pathways from each initial substrate are controlled separately. Within strains of *Pseudomonas* using the same pathway and presumably the same inducers there are wide differences in the structure of the enzymes involved as shown by immunological studies. This indicates that the overall control of aromatic degradation depends on: (*a*) the reaction sequence followed; (*b*) the control mechanisms used for any given pathway; and (*c*) the possible variation between the homologous enzymes used for each individual step[240,245]. It is quite likely that such a state of affairs exists for each metabolic sequence in all living organisms.

In fungi, the sequence of degradative steps leading to the cleavage of the aromatic ring follows that found in the majority of bacteria as far as the formation of muconic acid[248–250]. However, whereas the bacteria convert the product from protocatechuic acid to the γ-carboxylactone, fungi form the isomeric β-carboxylactone (Fig. 7). Again there is a different induction control system in the various fungi studied. Whether this variation is followed in the degradation of the aromatic ring in other eucaryotes is at present unknown. Although it is obvious that higher plants can degrade aromatic rings, little is known of the mechanisms involved. Indeed, the only widely distributed metabolites resulting from aromatic rings fission are the betalains[251] (see p. 281), although the recent recognition of the biosynthetic origin of the major cyanogenetic glycoside from *Thalictrum aquilegifoleum*[252] as being from the ring opening of dihydroxyphenylalanine points to the possibility that the cleavage of aromatic rings in higher plants is probably widespread (Fig. 8). It is interesting to note that the biosynthesis of the betalains involves the *meta* cleavage of the 3,4-dihydroxy aromatic ring (L-DOPA), whereas the *Thalictrum* glycoside is formed by *ortho* cleavage of the presumed precursor, 3,4-dihydroxy mandelonitrile glucoside (Fig. 8).

References p. 287

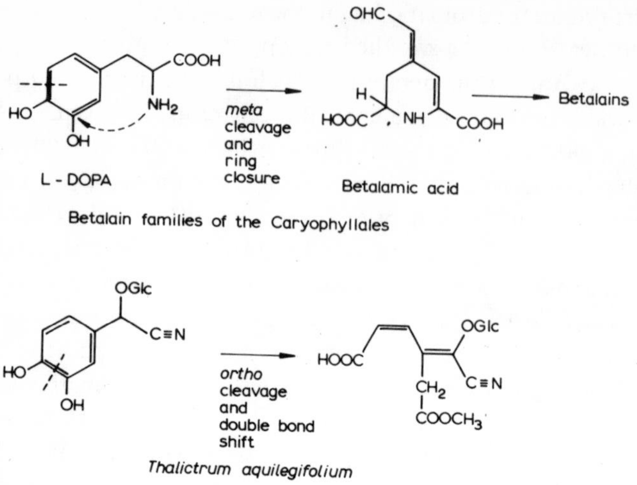

COOH
NH2
HO
OH
L - DOPA
meta
cleavage
and
ring
closure
OHC
H
HOOC
NH
COOH
Betalamic acid
Betalains
Betalain families of the Caryophyllales
OGlc
C≡N
HO
OH
ortho
cleavage
and
double bond
shift
HOOC
OGlc
CH2
C≡N
COOCH3
Thalictrum aquilegifolium

Fig. 8. Aromatic ring cleavage in Angiosperms[251,252].

Pyruvate
Acetate
Oxalo-
acetate
α-Keto-
glutarate
Aspartate
Homocitrate
Aspartate-
β-semialdehyde
CH2·COOH
(CH2)2
H2N - CH - COOH
α-Aminoadipic acid
(AAA)
H2N - CH - COOH
(CH2)3
H2N - CH COOH
Saccharopine
Diaminopimelic
acid
(DAP)
CH2NH2
(CH2)3
H2N-CH-COOH
Lysine

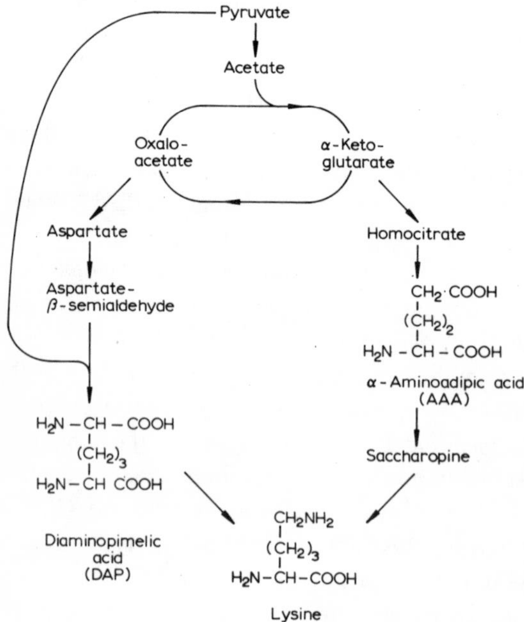

Fig. 9. Biosynthesis of lysine by the DAP and AAA pathways.

Another well investigated pathway is that leading to the amino acid lysine[95,253]. The majority of organisms which synthesise this diamino acid utilise a pathway which starts from L-aspartate and proceeds *via* α-ε-diaminopimelic acid (DAP). Most of the Fungi, however, with the exception of the Oomycetes and Hyphochytridiomycetes utilise a second pathway from homocitrate *via* α-aminoadipic acid (AAA) (Fig. 9, Table IX).

TABLE IX

Lysine Pathway in Different Taxa[a]

Diaminopimelic acid (DAP) pathway	Aminoadipic acid (AAA) pathway
Bacteria	
Cyanophyta (blue green algae)	
Chlorophyta (green algae)	Euglenophyta
Fungi: Oomycetes[b] Hyphochytridiomycetes[c]	Fungi: Basidiomycetes Ascomycetes Chytridiomycetes[d] Zygomycetes[e]
Tracheophyta: Filicinae Gymnospermae Angiospermae	

[a] See Fig. 9. The taxa are arranged according to Tables I and II. Data from Vogel[95,253].
[b–e] The lower fungi have either: biflagellate spores[b]; anteriorly uniflagellated spores[c]; posteriorly uniflagellated spores[d]; or no spores[e]. It should be noted that posteriorly uniflagellated spore formers in the Blastocladiales were once assigned to the Hyphochytridiomycetes and are now placed in the Chytridiomycetes, and that Oomycetes and Hyphochytridiomycetes are often removed from the Fungi proper[245].

It should be noted that some authorities exclude the Oomycetes and Hyphochytridiomycetes from the Fungi altogether and consider them as achlorophyllus algae[254]. This is supported by the fact that these two Sub-Divisions are also the only fungi with 'tinsel' flagellae, cellulose in the cell wall (see p. 201), and have (Oomycetes) a different sedimentation pattern of the enzymes of tryptophan biosynthesis[95]. It seems probable that on present evidence they are wrongly placed biochemically, and further data should be sought on this point since it may also throw important light on the origin of the saprophytic habit and hence of the fungi themselves. The exclusive distribution of the AAA pathway in the higher fungi listed in Table IX points more to the desirability of considering these organisms as a separate Kingdom[76]. Again, this is supported by the fact they are the only plants with chitin in the cell wall[95], and unlike bacteria and eucaryotic green plants, have tryptophan synthetases[255], and pre-aromatic enzymes (see below) which are inseparable into two separate enzymically active proteins.

As a final metabolic pathway, the one leading to the biosynthesis of phenylalanine, tyrosine and tryptophan is highly instructive with regard to the variation which has been observed in the underlying control mechanisms[246, 255–260]. The pathway has been found in all organisms which synthesise the aromatic amino acids and is especially important in higher plants, since it also leads to lignins, flavonoids, phenolic alkaloids and a number of other imporant aromatic secondary products[20, 37–41, 261] which are discussed later (p. 209).

The aromatic (shikimic acid) pathway involves the formation of the end-products in four distinct stages: first, the formation of the 7-carbon sugar acid, 3-deoxy-D-arabinoheptulosonic acid-7-phosphate (DHAP); next its cyclisation to 5-dehydroquinic acid; third, the addition of a 3-carbon side-chain; and finally the rearrangement and aromatisation to give the keto acids corresponding to phenylalanine and tyrosine (Fig. 10). Tryptophan synthesis involves a different sequence of reactions after the second stage which need not be outlined here.

In strains of *E. coli*, three separate DHAP synthetase isoenzymes exist[242, 256, 257]; one is inhibited and its synthesis repressed by relatively low concentrations of phenylalanine, a second is inhibited by tyrosine and the third is repressed (and in some strains only inhibited) by tryptophan. Similar mechanisms exist in other members of the Enterobacteriaceae. In *Bacillus*, however, inhibition of DHAP synthetases is mediated by the branch-point intermediates, chorismic acid and prephenic acid. Aerobic *Pseudomonas* possess a DHAP synthetase which is only sensitive to tyrosine whereas other members of the genus exhibit cumulative inhibition by phenylalanine and tyrosine. Among the fungi, similar variations in control were evident: *Neurospora crassa* and *Saccharomyces cerevisiae* (both Ascomycetes) are similar to *E. coli* in having three DHAP synthetase isoenzymes; *Sporobolomyces odorus* (Basidiomycetes), on the other hand, showed cumulative feedback inhibition[257].

In algae and in higher plants (*Pisum sativum* and *Phaseolus aureus*),

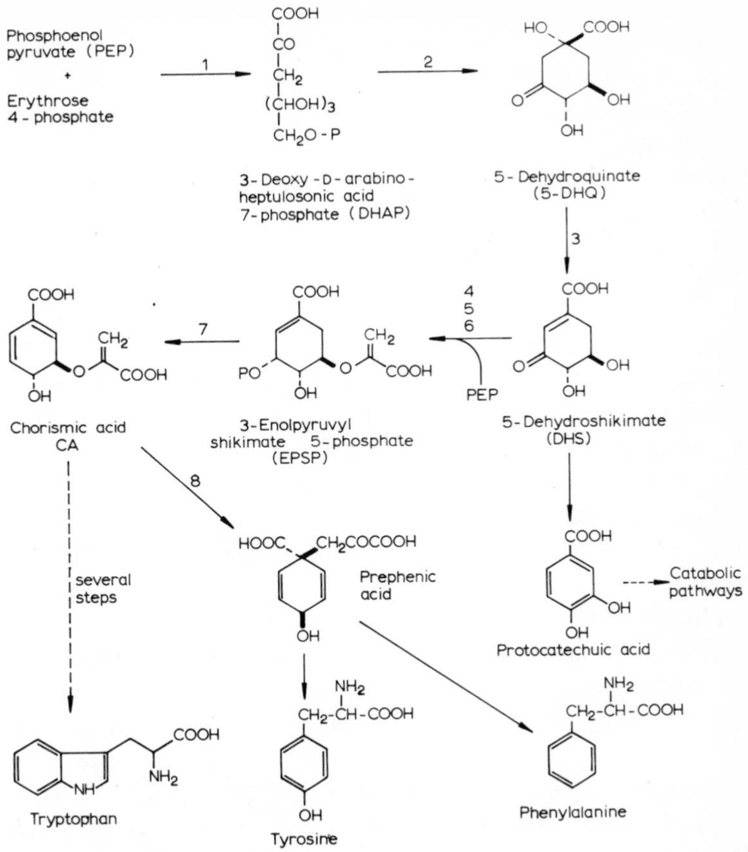

Fig. 10. Aromatic acid biosynthesis. The enzymes involved are: 1. DHAP synthetase; 2. DHQ synthetase; 3. DHQase; 4. DHS reductase; 5. Shikimic acid kinase; 6. EPSP synthetase; 7. CA synthetase; 8. CA mutase.

regulation of the shikimic acid pathway is effected in part by the inhibition of chorismate mutase by phenylalanine and tyrosine. Tryptophan, in this case, relieves the inhibition, and activates the enzyme when neither of the other two aromatic acids are present[246]. *P. aureus* possesses two isoenzymes, only one of which is subject to regulation by both amino acids, and the overall control in this plant is unresolved[262]. There are no reports of regulation *via* DHAP synthetase, but this obviously requires further investigation.

Besides the variation in the mode of enzyme regulation of the aromatic pathway in different plant taxa, there is also evidence of wide differences in the arrangement of the enzymes catalysing the five steps from the cyclisation of DHAP to the addition of pyruvate (PEP) to shikimate-5-phosphate (Fig. 10; enzymes 2–6) and the structural genes responsible for their synthesis[258–260]. In *E. coli* and five other bacteria examined, all five enzymes were readily separable by centrifugation on linear sucrose density gradients[258]. The individual enzymes from the different species, however, showed differences in their molecular weights. Furthermore, genetic analysis showed that the cistrons which encode each of the enzymes were not aggregated on the genome. In *N. crassa*, on the other hand, in spite of the similarity of its feedback regulation to *E. coli*, the five enzymes were found to be bound in an inseparable multi-enzyme complex of mol.wt. 220000, which are encoded by a single *arom* polycistronic cluster[259]. Other fungi, including 3 Zygomycetes, 2 Basidiomycetes and 2 Deuteromycetes, showed similar characteristics and three of the eight fungi examined had two dehydroquinase (DHQase) isoenzymes, one of which appeared to be part of a catabolic pathway *via* protocatechuic acid[259].

In photosynthetic organisms, from the blue-green alga *Anabaena variabilis*, the green alga *Chlamydomonas reinhardi*, the moss *Physcometrella patens* to three angiosperms, the enzymes were all separable except that DHQase and dehydroshikimic reductase were inseparable in the eucaryotes[260]. The enzymes in *Euglena gracilis*, on the other hand, are aggregated, but the mol.wt. of the cluster was about half that of the *N. crassa* complex[260]. Both aggregates appear to be composed of sub-units of about 60000 daltons.

It has been suggested that the evolutionary advantage of the *arom* aggregates in fungi is to channel the intermediates for the synthetic pathway and out of the potentially competitive catabolic pathway utilising dehydroquinic acid to form protocatechuic acid[258–260]. This may well be so since many higher plant cells on which these saprophytes live, contain sizeable amounts of quinate esters often produced in response to fungal invasion.

The other variations in control of the aromatic pathway do not show any clear evolutionary trend, and the data are too scanty at present to allow any generalisations to be made. If the biosynthesis of the higher-plant phenolic secondary products mentioned earlier is all channelled through phenylalanine (see pp. 259 and 274), it seems likely that more complex forms of feedback control must exist in such organisms. It is to be hoped that studies on this problem are currently in progress.

7. Photosynthesis

(a) Introduction

Photosynthesis may be defined as the light-driven production of both biochemical reducing power as a reduced pyridine nucleotide by removing electrons from a suitable weak donor such as water and the storage of this potential by the incorporation of reduced carbon dioxide into the metabolic pools, and high-energy chemical bonds, mainly as ATP[263–268]. Our knowledge of both the energetics of the conversion of light energy into biologically usable chemical energy and the various ways in which

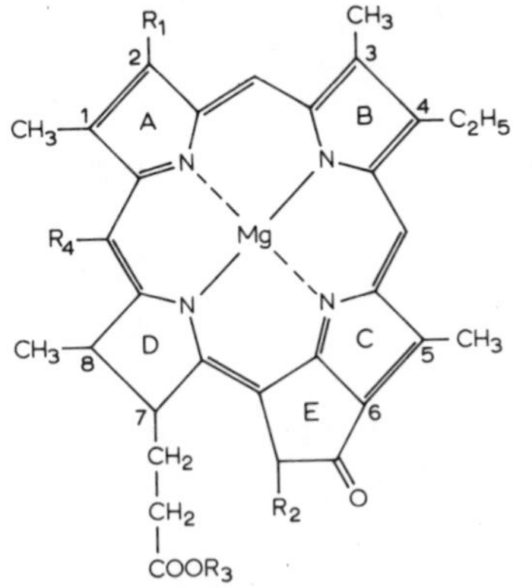

Fig. 11. The structures of plant and bacterial chlorophylls.

	Type	R_1	R_2	R_3	R_4	*Other differences*
Chlorophyll	*a*	$-CH{=}CH_2$	$-COOCH_3$	phytyl(C-20)	H	
	b	$-CH{=}CH_2$	$-COOCH_3$	phytyl	H	$-CHO$ at C-3 instead of CH_3
	c	$-CH{=}CH_2$	$-COOCH_3$	H	H	$\Delta C_7–C_8$
	d	$-CHO$	$-COOCH_3$	phytyl	H	
Bacteriochlorophyll[a]	*a*	$-COCH_3$	$-COOCH_3$	phytyl	H	3,4-dihydro
	b		unknown			
	c	$-CHOHCH_3$	H	farnesyl (C-15)	CH_3	
	d	$-CHOHCH_3$	H	farnesyl	H	

[a] Bacteriochlorophylls *c* and *d* are mixtures: only the main components are given here.

References p. 287

CO_2 is incorporated into the various reactive biochemical sequences in different organisms is still far from complete. However, certain generalisations can be made which have an evolutionary significance in the Plant Kingdom.

(b) Photosynthetic pigments

The primary photosynthetic light-acceptor in all organisms examined is one of the magnesium-containing porphyrins which are grouped together as chlorophylls (Fig. 11). In Cyanophyta and all eucaryotic green plants, the major light-accepting molecule is chlorophyll *a* which is present in relatively high concentrations in the chloroplasts and chromatophores of all these photosynthetic organisms. With the exception of the blue-green algae and the Chrysophyta which contain only this compound, all photosynthetic organisms contain other chlorophylls besides chlorophyll *a* (Table X). The photosynthetic purple bacteria possess chlorophylls in which the vinyl group at C-2 has been oxidised to an acetyl group[264, 269]. The introduction of this conjugated carbonyl group leads to absorbance of

TABLE X

Distribution of Chlorophylls in the Plant Kingdom

Division	*Chlorophyll*					*Bacteriochlorophyll*			
	a	*b*	*c*	*d*	*e*[a]	*a*	*b*	*c*	*d*
Bacteria									
Chlorobacteriaceae	−	−	−	−	−	−	−	+	+
Thiorhodaceae	−	−	−	−	−	+	+	−	−
Athiorhodaceae	−	−	−	−	−	+	+	−	−
Cyanophyta	+	−	−	−	−	−	−	−	−
Euglenophyta	+	+	−	−	−	−	−	−	−
Chlorophyta	+	+	−	−	−	−	−	−	−
Rhodophyta	+	−	−	±	−	−	−	−	−
Cryptophyta	+	−	+	−	−	−	−	−	−
Pyrrophyta	+	−	+	−	−	−	−	−	−
Chrysophyta									
Xanthophyceae	+	−	−	−	+	−	−	−	−
Chrysophyceae	+	−	−	−	−	−	−	−	−
Bacillariophyceae	+	−	+	−	−	−	−	−	−
Bryophyta	+	+	−	−	−	−	−	−	−
Tracheophyta	+	+	−	−	−	−	−	−	−

[a] Only found in one organism and may be identical to chlorophyll *c*.

light by the bacteriochlorophylls *a* and *b* at much longer wavelengths (820 mm and 1025 mm respectively)[16] than the higher plant chlorophylls (640–660 mm)[270] and this is undoubtedly advantageous to the purple bacteria which inhabit muddy environments where light of low wavelengths is more greatly scattered. The green sulphur bacteria possess chlorophylls with a farnesol rather than a phytol as the alcoholic moiety and also lack the carboxymethyl group at C-10 (Fig. 11).

(c) Biosynthesis of chlorophylls

As can be seen from Fig. 11, the variation in structure of the various chlorophylls are relatively minor in relation to the conservative retention of the porphyrin ring structure which they all possess as a result of having the same biosynthetic origin[37–41, 261, 270–272]. This same pathway is also used in all organisms examined for the synthesis of the porphyrin moiety in the heme group of the cytochromes, peroxidase and haemoglobin, and leads, by degradation of the porphyrin ring to the chromophores of the phycobilins (Fig. 12). The corrin moiety of vitamin B_{12} arises in a similar manner. The unusual feature of the pathway is that the four porphobilinogen molecules which give rise to the four pyrrol rings of the

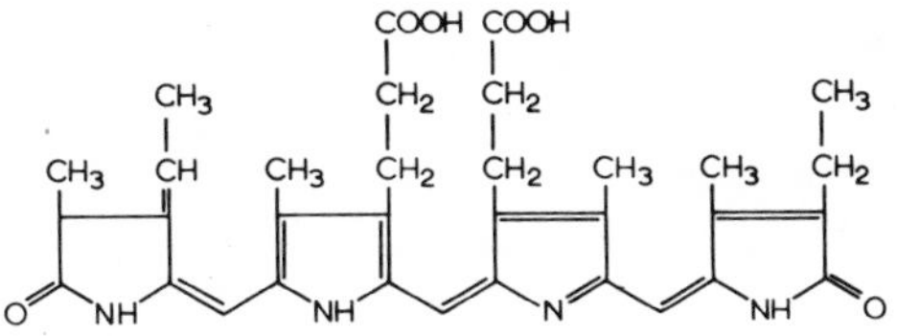

Phycocyanobilin

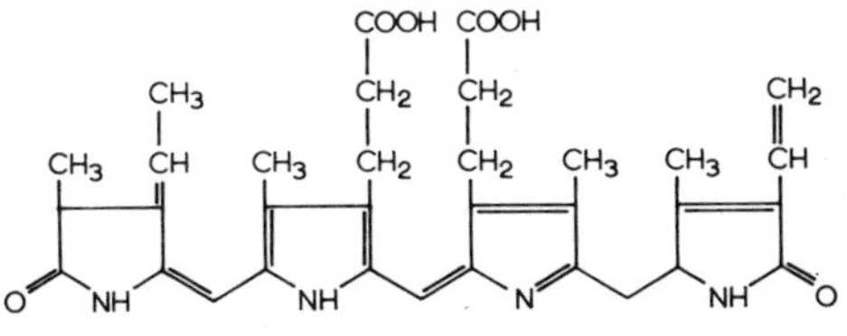

Phycoerythrobilin

Fig. 12. Chromophores of the phycobilins.

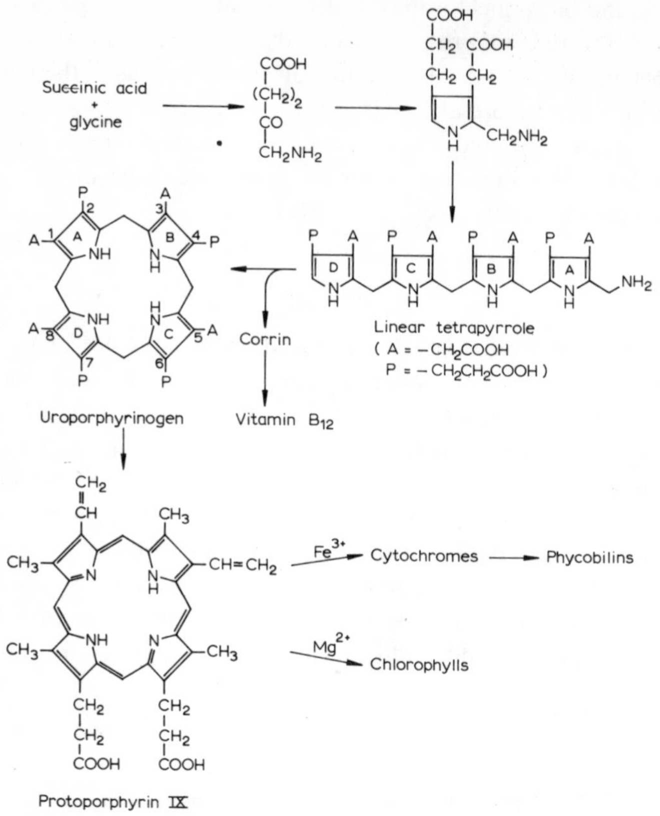

Fig. 13. Biosynthesis of chlorophyll and related porphyrins.

porphyrin are not all linked head to tail as might be expected, one (pyrrole ring D in uroporphyrinogen III, Fig. 13) being inserted back to front. This mode of synthesis gives two contingent propionyl units at C-6 and C-7 and two acetyls at C-1 and C-8, rather than the alternate arrangements of propionyl and acetyl groups found between the other rings. Since this feature is found in all biologically active porphyrin-containing compounds and in the corrin ring of vitamin B_{12} it must represent an extremely conservative and ancient character. Porphyrins corresponding to a completely head-to-tail mode of combination of the porphobilinogens are known, but none have any physiological role and their synthesis may represent a degradative pathway for the utilisation of excess porphobilinogen or other intermediates[37]. The mechanism whereby the pyrrole ring D is inserted into the uroporphyrinogen III molecule is still unknown, although evidence points to this ring (D) being the starter porphobilinogen molecule for the fabrication of the hypothetical acyclic tetrapyrrole precursor of the porphyrins, and that the isomerisation occurs during ring closure to ring A to form urophorphyrinogen III[271]. It seems likely that the same type of reaction occurs in corrin[271] biosynthesis, but in this case leading to a direct union of rings A and D. The sequence of events from uroporphyrinogen III to protoporphyrin IX involves decarboxylation of all four acetyl groups, oxidative decarboxylation of the two propionyl groups on rings A and B to give vinyl residues, and oxidation of the methane bridges to methine. Protoporphyrin IX is the branch point leading to heme and the chlorophylls. The route to the latter involves first the introduction of magnesium, followed by elaboration of the cyclopentane ring, E (Fig. 11), from the propionyl unit at C-6. The vinyl group at C-4 and the double bond in ring D are then both reduced to give chlorophyllide *a* and esterification of the latter with phytol gives chlorophyll *a*. It is believed that chlorophylls *b* and *d* as well as the bacteriochlorophylls *a* and *b* are derived directly from chlorophyll *a* itself, while bacteriochlorophylls *c* and *d* are formed from chlorophyllide *a*[271]. All these other pigments may thus be regarded as being biosynthetically more advanced than chlorophyll *a*, and it is obviously not possible to discern any real evolutionary significance from the variation in chlorophyll structure.

(d) Auxiliary pigments

The chloroplasts and chromatophores of all photosynthetic organisms contain, besides the auxiliary chlorophylls described above, other compounds which absorb visible light, but it is not known with any degree of certainty whether or not they play a role in photosynthesis, except in a very few cases. The main pigments found in the grana of the chloroplasts of all green plants are the four carotenoids β-carotene(VIII), lutein(IX), violaxanthin(X) and neoxanthin(XI). The function of these compounds in the chloroplast is equivocal and is discussed later (see p. 237). It seems highly likely, however, that the major carotenoid present in the Phaeophyta, fucoxanthin(V), does act as a photosynthetic light receptor in these algae.

References p. 287

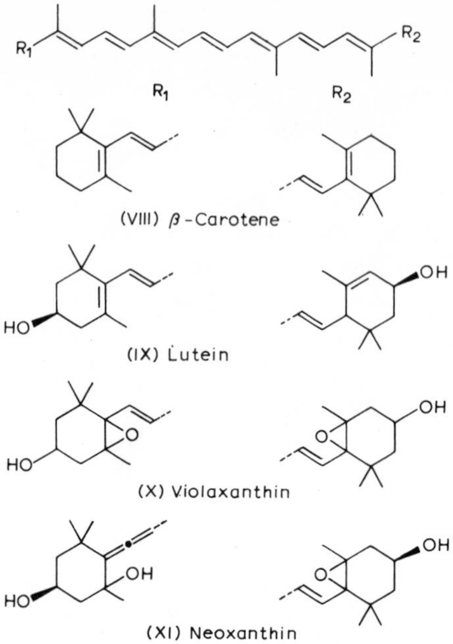

R1 R2

R1 R2

(VIII) β-Carotene

(IX) Lutein

(X) Violaxanthin

(XI) Neoxanthin

One other class of pigments which have been implicated as light-acceptors in photosynthesis are the phycobilins mentioned above[273–277]. These are chromoproteins which are only found in three Divisions of the Plant Kingdom, Cyanophyta, Rhodophyta and Cryptophyta (Table I). They possess, as their chromophoric group, either phycocyanobilin (bilitriene) or phycoerythrobilin. It has been proposed that these tetrapyrroles (Fig. 12) are derived by ring opening of a porphyrin of the heme type between the pyrrole rings A and B with the elimination of the methine carbon. However, unlike heme, they are bound to their proteins through covalent bonds involving the two carboxyl groups and this may indicate that they have a different biosynthetic origin. The two chromophores are not distributed evenly. The Cyanophyta and the more primitive sub-division of the Rhodophyta, the Bangiophyceae, accumulate phycocyanobilin-based biliproteins as their major pigments, whereas these are only weakly present in the more advanced Florideophyceae and most of the Cryptophyta[275]. These organisms instead accumulate biliproteins with the phycoerythrobilin chromophore which are antigenically dissimilar to those containing the bilitriene. The occurrence of phycobilins in the cryptomonads links these algae biochemically with the red algae and this association is paralleled by the nature of the carotenoids which the two divisions contain (see Table XVI). On other grounds, such as the number of flagella, however, the Cryptophyta are believed to be more closely associated with the Pyrrophyta and this is supported by the fact that both Divisions contain chlorophyll *c*. Thus, it has been suggested that the presence of the phycobilins in the Cryptophyta represents an example of parallel evolution. It seems more probable that the ability of the cryptomonads to accumulate and utilise the admittedly ancient phycobilin pigments represents an example of mosaic evolution, an ancient character being retained as several others advanced in the same taxon. It is probable that the accumulation of phycobilins in the chloroplasts of the three Divisions of the algae and their utilisation as auxiliary light acceptors is merely a reflection of the ecological niche occupied by the organisms rather than a relic of biochemical evolution. Indeed, recent evidence has shown that the important regulatory chromoprotein in higher plants, phytochrome[26], contains a phycobilin-like chromophore, and furthermore, is antigenically similar to several of the phycocyanobilin-containing biliproteins[278]. This strongly indicates that the ability to synthetise phycobilins has been retained throughout the evolution of the Plant Kingdom.

(e) Energy-transfer mechanisms

Although there is still much to be learned about the exact role and position of the various components of the photosynthetic apparatus in the overall energy-trapping systems, there is a general consensus about the broad outlines of the photosynthetic processes[264–266,279,280].

In present-day photosynthetic bacteria there are two independent photosynthetic systems[16,27,269]. The first photoact is responsible for the cyclic photophosphorylation of ATP and the second, which is non-cyclic, leads to the production of reducing power which is channelled *via* ferredoxin (p. 171) to NADPH (Fig. 14). Both of these primary photoreactions generate about 0.8 electron volts. The second reaction requires a donor to supply the electrons required for reduction, and, since ferredoxin has an E'_0 of -0.43, the weakest electron donor that can be utilised must have an E'_0 lower than $+0.37$. Thus the bacterial systems are unable to photolyse water (E'_0 $+0.82$) and have to rely instead on a stronger electron donor such as H_2S.

In higher plants, there are again two light-trapping systems both generat-

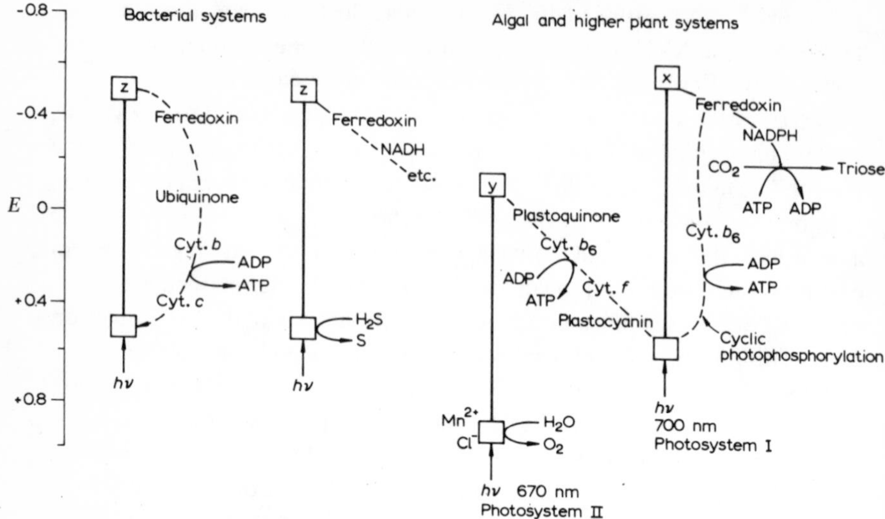

Fig. 14. Photosynthetic electron transfer in bacteria and in higher plants.

ing 0.8 electron volts, or possibly a little more, but in this case they are coupled together[177,266,281]. One of them is like the second bacterial system in so far as it boosts electrons from a weak donor into a strong reductant (E'_0 *ca.* -0.5) from which they are transferred to ferredoxin and thence to NADPH. The second photosystem (II) is also involved in the production of reducing power but at an intermediate level with an E'_0 of about 0. This means that it can utilise water as the initial electron donor and is thus the system responsible for the production of oxygen. The reduced intermediate formed by photosystem II passes on its electrons "downhill" generating some ATP on the way to produce the electron donor with an E'_0 *ca.* $+0.40$ for photosystem I (Fig. 14). Higher plants also carry out cyclic photophosphorylation of ATP. Most authorities believe this to be another facet of the activity of the normal photosystem I, with ATP being generated by electron flow through a cytochrome system which might or might not include ferredoxin (Fig. 14). It has also been postulated that, as in the case of bacteria, cyclic photophosphorylation is carried by a separate system not coupled with photosystem II[281]. It should be noted that three molecules of ATP and two of NADPH are required to reduce CO_2 to the sugar level of oxidation. Since the coupled system only produces one ATP for every NADPH, it is obvious that cyclic photophosphorylation is needed to ensure carbon utilisation with the minimum expenditure of energy. It is generally agreed that the final quantum trapping centre in photosystem I is a special form of chlorophyll *a* bound in the matrix in such a way that its red absorption band is changed from about 670 nm (the major long wavelength band of the pigment when bound in chloroplasts) to 700 nm[279]. Other chlorophyll *a* molecules and any effective accessory pigments present in the system pass their energy to the trapping centre. Photosystem II contains bound chlorphyll *a* absorbing at 670 nm, together with most of the auxiliary pigments, chlorophylls *b*, *c* or *d*, the phycobilins and the carotenoids, and requires both manganese possibly in the form of a porphyrin complex, and chloride ions for activity. This system, unlike photosystem I, can be readily inhibited by a variety of poisons and this has allowed the sequence of many of the various quinone and cytochrome electron transport intermediates which are found in the chloroplast to be determined (Fig. 14).

(f) The pathway of carbon in photosynthesis

Some experts regard photosynthesis as consisting of the generation of reducing power and ATP only. They exclude the incorporation of reduced carbon dioxide into the metabolic pools on the grounds that no light energy is required for either the carboxylation or subsequent reduction steps. Nevertheless, the requisite enzymes for these and related steps are intimately associated with the light-trapping system in the chloroplast and it seems highly likely that both processes evolved as a whole.

Most, if not all, photosynthetic organisms incorporate carbon dioxide in only one way and this is the carboxylation of ribulose 1,5-phosphate (RDP)[27,263–265,282]. This reaction gives an unstable intermediate which splits into two molecules of phosphoglyceric acid which is then reduced by NAPDH to phosphoglyceraldehyde. Part of the latter triose remains in the Calvin sugar cycle leading to the regeneration of RDP and the rest is channeled into the general sugar pool leading to the formation of sucrose and starch in higher plants[263] and to the accumulation of poly-β-hydroxybutyric acid in photosynthetic bacteria[16]. The enzyme catalysing the incorporation of carbon dioxide in this pathway, RDP carboxylase, has been found in all photosynthetic organisms examined from higher plants to bacteria, and also in some chemosynthetic autotrophic bacteria. It is not present in other bacteria.

References p. 287

Some tropical angiosperms, mainly monocots, contain, in addition to the Calvin cycle described above, a system which incorporates carbon dioxide by the carboxylation of phosphoenolpyruvate (PEP) to give the C_4 acid, oxaloacetate[282]. It is suggested that this type of incorporation is restricted to a certain type of chloroplast which is only found in plants having this pathway: other chloroplasts found in C_4-type plants fix carbon dioxide by the normal photosynthetic reduced-carbon pathway. However, the enzyme responsible for the formation of oxaloacetate, PEP carboxylase, is found in the chloroplasts of plants which do not apparently use the C_4 pathway and, indeed, in a wide variety of non-photosynthetic organisms including many bacteria and animals. The overall incorporation of carbon dioxide by the C_4 pathway involves the utilisation of two more molecules of ATP than does the Calvin cycle *via* ribulose 1,5-diphosphate and hence is much less efficient. It seems probable that the C_4 pathway may be the more ancient of the two and might be used by tropical plants as an auxiliary to the normal reductive cycle since it can make use of much lower concentrations of CO_2, the K_m for the gas for the PEP carboxylase ($1.5 \cdot 10^{-5}$) being about ten times lower than that of RDP carboxylase ($4.5 \cdot 10^{-4}$).

Some green and purple photosynthetic bacteria, in addition to possessing an active Calvin cycle, fix carbon by a reversal of the decarboxylation steps of the citric acid cycle[16,282]. These undoubtedly represent very ancient methods for the incorporation of CO_2 and are much less efficient than those mentioned above. A large number of other carboxylation reactions are known in living organisms, many of them biotin-dependent. Many are used for the activation of biosynthetic precursors or for anaplerotic reactions that lead to the replenishment of essential primary intermediates. None appear to be of any importance in overall carbon incorporation in photosynthesis.

(g) The evolution of photosynthesis

It is highly probable that the most primitive sequence in photosynthesis is that leading solely to the formation of high-energy chemical bonds. The cyclic reaction requires no electron donor and all the necessary electron transport intermediates can be combined in a relatively simple organelle. It can be assumed by their involvement in photosynthesis in contemporary organisms and their presence in the presumed primitive

Clostridia, that the ferredoxins (see p. 171) acted as the primary acceptors of the electrons from the reduced intermediate formed in the photoact in the primitive photosynthesisers. If the redox step was similar in magnitude to that found in present-day organisms as may be assumed by the probable involvement of a chlorophyll as a light acceptor there must have been a relatively strong cyclic electron donor like one of the cytochromes. As mentioned earlier, the primary chlorophyll was probably more like chlorophyll *a* than either bacteriochlorophylls *a* or *b* which represent a relative advancement in biosynthesis. Bearing in mind that the present-day green sulphur bacteria (*e.g. Chlorobium*) are the most primitive of the photosynthetic organisms, being obligate anaerobes, having the smallest chromatophores[16,269], and having more primitive ferredoxins[232], it is possible that a chlorophyll *a* with farnesol rather than phytol represented the most primitive light acceptor. Since the evolution of the cytochromes undoubtedly preceded photosynthesis, it can be presumed that the biosynthesis of the chlorophylls developed from the heme pathway. It is probable, that the phycobilins are equally ancient, but being less efficient for energy-trapping purposes were discarded early in the evolution of photosynthetic apparatus.

The next step in the evolution of photosynthesis was to divert the electron flow from the cyclic phosphorylation pathway to one producing reducing power in the form of reduced pyridine nucleotide using electrons from an electron donor like H_2S. This was perhaps expectable since ferredoxins take part in several electron-transfer reactions. It is probable, however, that this development was coupled with the evolution of an enzyme complex designed exclusively for the incorporation and reduction of carbon dioxide into triose or other metabolite. It is probable that at an earlier stage organisms used the reversal of citric acid cycle reactions as means of incorporating CO_2, but as pointed out above this is grossly inefficient and was never built into the chromatophore. It is possible, however, that the initial fixation reaction used was the carboxylation of phosphoenolpyruvate rather than ribulose-1,5-diphosphate although the presence of the RDP carboxylase in present-day chemosynthetic bacteria points to it possibly having evolved prior to photosynthesis.

The next major step in the evolution of photosynthesis was the development of an algal photosystem II and its coupling with photosystem I. It may be presumed that the original photosystem II might have utilised exclusively a second type of chlorophyll (*b* or *d*) so that the two photoacts were separated by having different absorption maxima. This may have been coupled with a Mn^{4+}/Mn^{3+} reaction centre to allow the use of weaker electron donors. Finally, a photosystem II evolved which used chlorophyll *a* almost exclusively. This development came about by evolving an ability to complex chlorophyll *a* in different ways to give pigments absorbing at 670, 680 or 700 nm. All the above developments are shown in Fig. 15.

Coupled light reactions I and II so that weaker electron donors could be used such as water.

↑

Evolution of ribulose-1,5-diphosphate carboxylase and the requisite coupled enzyme systems for the efficient incorporation of CO_2.

↑

Development of a second photosystem with H_2S as electron donor producing NADH and associated with PEP carboxylase and the requisite enzyme complex for the incorporation of CO_2.

↑

Evolution of a photosynthetic unit capable of transforming light energy to chemical energy by the cyclic phosphorylation of ATP *via* ferredoxin, with a chlorophyll *a*-like photo-acceptor perhaps with a farnesol side-chain.

↑

Chemotrophic organisms forming ATP and NADH from SO_4 to S transitions containing ferredoxin and cytochromes.

↑

Fig. 15. Evolution of Photosynthetic Systems.

8. Cell walls

(a) Introduction

Plant cells, with few exceptions, have a firm wall outside the plasma membrane surrounding the protoplasm which serves both as a protective and a structural element[16,24,25,95,283–286]. Animal cells, by contrast, have only a thin selectively permeable membrane and these differences have important consequences, especially during the early stages of development of higher organisms[87]. In some unicellular plants the cell wall consists of a sculptured silica or calcium carbonate sheath embedded in a polysaccharide matrix[81], but in the majority of cases the wall consists mainly of mixtures of several different polysaccharides associated with protein and lipids[283–286].

In all plants, the cell wall is formed during cell division, but there is little detailed information about the underlying biochemistry of the process in most unicellular organisms[24]. In higher plants, a plate of pectic acid-like material is deposited along the equator of the dividing cell, apparently being formed by the dictysomes of the Golgi apparatus which fuse together and incorporate elements of the endoplasmic reticulum[283]. The precise chemical nature of the plate material is not known, but since it later gives rise to the middle-lamella of the cell wall it is presumably a mixture of pectic acid-like polymers. Primary walls are then deposited on each side of this "middle-lamella". The primary walls mainly consist of cellulose fibrils which are deposited in a series of lamellae, the orientation of each successive layer being altered somewhat randomly to give the final wall a laminated appearance[283]. In most cases the cells have not reached their full size at this point and the walls must be flexible to allow for further growth[286]. These walls owe their extensibility and elasticity to the fact that the cellulose microfibrils are intially relatively short and are embedded in a matrix of hemicelluloses, pectin and other polysaccharides. It is also possible that the presence of proline-rich proteins in these primary walls may also be important[287]. In higher plants, when the cell has assumed its final size, secondary wall synthesis commences. This secondary wall again consists of cellulose fibrils, but now they are much larger and successive oriented layers are laid down at definite angles to increase the overall strength of the wall[283–285]. Subsequently, lignification may take place in certain cells to give added rigidity[24, 283]. The random phenylpropane polymer, lignin, is laid down in the middle lamella, primary and secondary wall, being deposited between the cellulose fibrils. The evolutionary significance of this polymer is described below (p. 205). In the epidermal cells of land plants, the outer wall contains a further feature, namely a cuticle which is often of considerable thickness[285]. Besides polysaccharides, this contains a number of lipids (see p. 221). Unfortunately, little is known about the variation in this layer in Bryophytes and Tracheophytes and no evolutionary deductions can be made.

(b) Structure of cell walls

As might be expected, there are large variations in the structure of the cell walls of multicellular and unicellular organisms and, among the latter, between pro- and eu-caryotes. In the bacteria, the inner wall consists of chains of a complex aminopolysaccharide cross-linked with a relatively short peptide usualy containing four amino acids, L- and D-alanine, D-glutamic acid and either L-lysine or diaminopimelic acid[25,285,288]. The polysaccharide chain consists of alternating units of *N*-acetylglucosamine linked β-1,6 to *N*-acetyl-3-*O*-lactylglucosamine, which is then joined β-1,4 to a further *N*-acetylglucosamine moiety and so on. The chains are linked *via* their lactyl carboxyl groups. The bag-like peptide glycan macromolecule (Fig. 16) so formed constitutes the major part of the cell wall of most of the Gram-positive bacteria and may be the factor which prevents egress of the dye in the Gram-staining procedure. These bacteria show greater variation in the linking peptide part of the polymer than do those organisms which give a negative Gram stain. Many have teichoic acids attached as an outer coat to their cell walls. These are glycerol or ribitol phosphate polymers containing a number of different amino acids and sugars which impart specific strain differences and confer some immunological advantage to the different pathogenic organisms. In Gram-negative bacteria, the peptideglycan layer is much thinner and is covered with a different antigenic coat containing several unusual sugars including keto-deoxyoctulonate, and colitose[25,84]. The bacterial wall thus serves three purposes: first, it imparts a rigidity to the cell; second, it protects the bacterium against attack by foreign organisms, especially fungi, by having a structure which contains a number of unusual D-amino acid links which

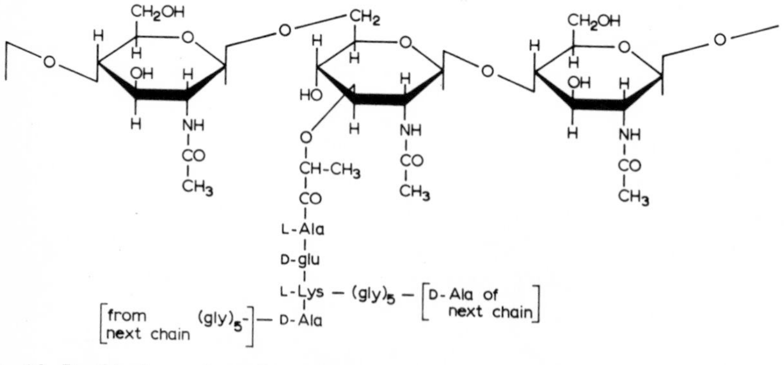

Fig. 16. Peptidoglycan from *Micrococcus lysodeikticus*. In all Gram-negative bacteria, L-lysine is replaced by diaminopimelic acid (1,5-diaminohexanoic acid), the D-glutamic acid free (d) carboxyl group is aminated, and the linking pentapeptide is either missing (in which case D-alanine links to the terminal amino group of L-lysine or DAP) or the glycine is replaced by alanine and threonine.

are not readily broken by normal hydrolases; and, third, the variation in the structure of the outer coat of the wall imparts a counter defense against antigens. However, since the immunogenic response is only found in vertebrates[84], this latter feature of the bacterial cell is perhaps a relatively late development in bacterial evolution.

All the bacteria, including the obligate parasites such as *Rickettsia* and the more highly advanced Actinomycetes have muramic acid in their cell wall. The blue-green algae also contain polysaccharides containing this sugar and, these are presumably peptide-linked in a similar way as in bacteria, since many yield diaminopimelic acid in large amounts on hydrolysis[16]. They also have an outer sheath of a complex hemicellulose polymer containing glucose and xylose. Cellulose (Fig. 17) is also found in the wall of some members of the Cyanophyta, but not as the sole structural element as is the case in higher plants[81]. It is interesting also to note that two genera of bacteria, *Acetobacter* and *Sarcinia* can also synthesise cellulose, but only as an excretory product[81].

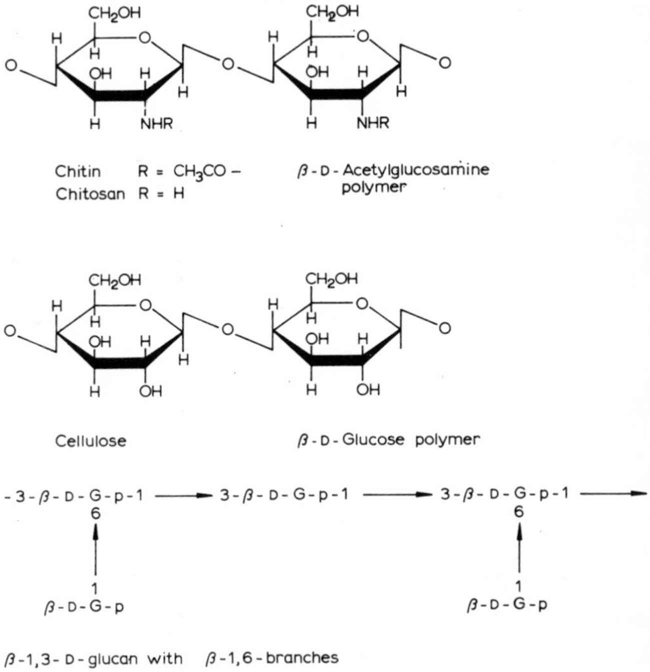

Chitin R = CH_3CO –
Chitosan R = H

β-D-Acetylglucosamine polymer

Cellulose

β-D-Glucose polymer

-3-β-D-G-p-1 ⟶ 3-β-D-G-p-1 ⟶ 3-β-D-G-p-1 ⟶

(6 ← 1 β-D-G-p branches on the first and third units)

β-1,3-D-glucan with β-1,6-branches

Fig. 17. Examples of plant polysaccharides.

The eucaryotic algae contain none of the amino-sugar containing polysaccharides found in the procaryotes. Many resemble the higher land plants in containing cellulose fibrils but others have large quantities of unique polysaccharides whose role in the cell wall, whether as a fibril or as part of the surrounding matrix is unknown[81,283–286]. Present knowledge is too fragmentary to allow any concrete evolutionary changes to be discerned, but the general pattern is usually similar. This is of one, or less usually two, homopolymer(s) or regularly formed heteropolymers occurring in crystalline or semi-crystalline aggregates or microfibrils which are oriented in a specific way in layers or laminates round the cell. These microfibrils are embedded in and encrusted with other non-crystalline heteropolymeric polysaccharides which are usually branched or acidic or both and often form the initial plate in dividing cells. This arrangement affords two major advantages to the organism. First it imparts reasonable strength without too much rigidity during the time of cell expansion[286,287] and yet affords the ability to impart more stiffness, if and where needed, in multicellular organisms. Secondly the heterogeneous nature of the surrounding polysaccharide matrix provides, as with the outer coat of bacteria, a reasonable first line of defence against invading pathogens which would need a battery of different glycan hydrolases to break down the numerous polysaccharide types present before gaining access to the cell.

In all photosynthetic land plants the pattern of wall formation appears to follow that found in angiosperms. Cellulose is the main strengthening polysaccharide and it is embedded in a matrix of hemicelluloses, usually branched and often specific to each taxa, based on heteropolymers containing galactose, mannose, rhamnose, xylose and arabinose[283–285,289]. In addition a number of acidic polysaccharides are found, such as pectin in the middle lamella, containing galacturonic acid. However, little of evolutionary significance can be discerned, except perhaps in the relative quantity of the various heteropolymers. Thus, angiosperms generally contain low quantities of mannose-containing polysaccharides, whereas other tracheophytes have larger amounts[290]. However, much more data is required before definite conclusions can be drawn.

The most interesting group from the point of view of variation in cell-wall composition are the fungi which have long been known to have chemically diverse cell walls[95,96,285]. Over fifty years ago, the aquatic Phycomycetes were divided into two groups on the basis of having either cellulose or the $\beta 1 \rightarrow 4$ polymer of *N*-acetylglucosamine, chitin (Fig. 17).

Physically, the majority of fungal cell walls are like those of other eucaryote members of the Plant Kingdom with interwoven partially crystalline fibrils of polysaccharides embedded in an amorphous matrix. The major polysaccharides found in the cell walls of fungi include cellulose (although the two taxa which contain it are often considered achlorophyllous algae[254]) and a branched β-1,3-D-glucan, chitin and chitosan (deacetylated chitin), a highly branched α-1,6-D-mannan and several glucurans.

The distribution of the most important of these polymers in the fungi is outlined in Table XI. It should be noted that cellulose, glucans and mannans occur in most other Divisions of the Plant Kingdom as described previously. Chitin, on the other hand, is apparently restricted to the fungi although widely present in the Animal Kingdom in the invertebrates and more primitive animal phyla including *Sarcodina*[291,292]. Chordates can synthesise a number of other polymers containing *N*-acetylglucosamine, but these usually also contain sulphated sugars or uronic acids[289], whereas arthropod chitin is basically similar to fungal chitin, although there are differences in the degree of acetylation and crystallinity[291,292]. There are reports that chitin occurs as a coating substance

TABLE XI

The Distribution of Different Wall Polysaccharides in the Fungi

Fungal group	*Polymer*				
	Chitin	*Chitosan*	*Cellulose*	*Glucan*	*Mannan*
Phycomycetes					
Oomycetes[a]			+	+	
Hypochytridiomycetes[a]	+		+		
Chytridiomycetes	+			+	
Zygomycetes	+	+		(+)[b]	
Ascomycetes					
Hemiascomycetidae				+	+
Euascomycetidae	+			+	
Basidiomycetes					
Heterobasidiomycetidae	+				+
Homobasidiomycetidae	+			+	

[a] Sometimes regarded as non-photosynthetic algae[254].
[b] Only in spore cell walls.

References p. 287

on the outside wall of certain green algae (*e.g. Cladophora*)[81], but further chemical investigation seems warranted since no other aminopolysaccharide is reported to occur in other eucaryotic photosynthetic plants.

The co-occurrence of chitin in fungi and the animal kingdom is often regarded as a case of convergent evolution. It seems more likely, however, that as suggested by Jeuniaux[292a] it is a remarkable example of the persistence of a primitive character. The ability to polymerise amino sugars, so well developed in the procaryotes, was perhaps retained by a primitive achlorophyllous eucaryote which gave rise to both the fungi and protozoa. The early chlorophyll-containing eucaryotes, on the other hand, presumably developed further the capacity to synthesise cellulose

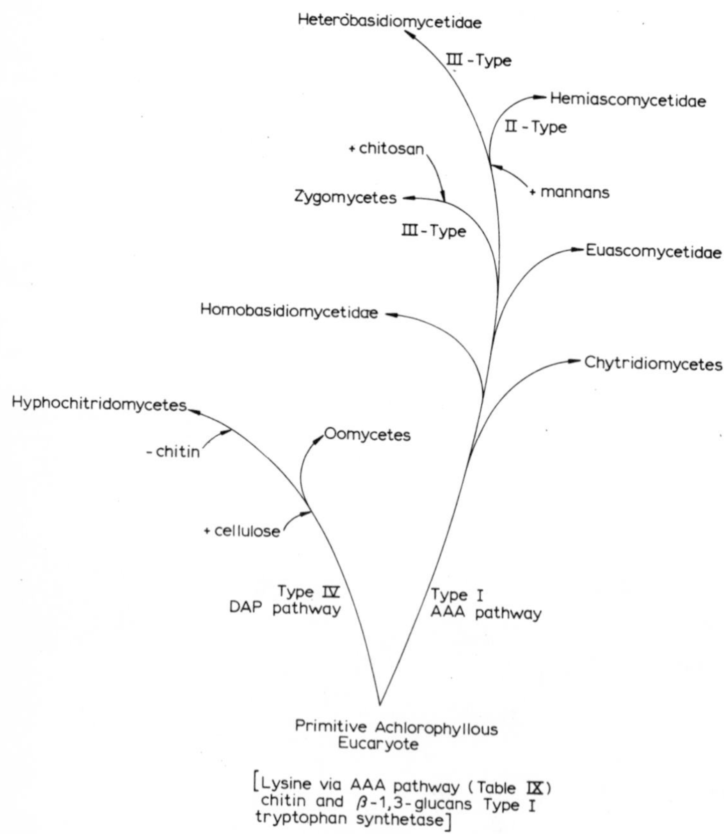

Fig. 18. Evolution of fungi. Type I–IV refers to types of tryptophan synthetase.

found in the Cyanophyta, and either lost or suppressed the ability to make aminopolysaccharides.

In the presumed phylogeny of the fungi, the majority of systematic mycologists agree that the evolutionary sequence proceeds from the Phycomycetes to the Basidiomycetes. On the basis of the presence of cellulose in the cell wall and other criteria Bartnicki–Garcia has suggested a separate origin for the Oomycetes and Hyphochytridiomycetes from the other groups given in Table XI[95,96,254]. There is obviously a strong biochemical link between the Chytridiomycetes and the higher mycelial fungi, Euascomycetidae and Homobasidiomycetidae, all of which have the chitin-β-1,3-glucan type of cell wall. The Zygomycetes, on the other hand, have no glucose polymer in the mycelial cell wall, although it is prominently present in the sporangiospore wall. This change in composition from mycelial to sporangial wall does not apparently occur in representative members of other fungal groups. In the higher fungi, the yeast-like forms diverge from the main group in having large amounts of the branched 1,6-mannan in their walls at the expense of either chitin or the glucan. The presence of this highly branched non-crystalline polysaccharide presumably increases the plasticity of the wall as well as forming a biochemically more inert barrier to invading pathogens.

The limited evidence from these cell-wall differences suggests that the fungi might have evolved as shown in Fig. 18 and this is supported from other evidence including the presence or absence of flagella and differences in the biosynthesis of tryptophan and lysine mentioned earlier (p. 181). However, it must be noted that, as in almost all other cases, exceptions have been found in the generalisations shown in Table XI. Two species of the genus *Ceratocystis* of the Euascomycetidae were reported to contain chitin and cellulose, while two others contained chitin alone[96].

One other feature of the fungal cell walls needs to be noted. Spore walls of many higher fungi contain, besides polysaccharides, encrusting dark-coloured polymers usually termed melanins. In several cases it has been

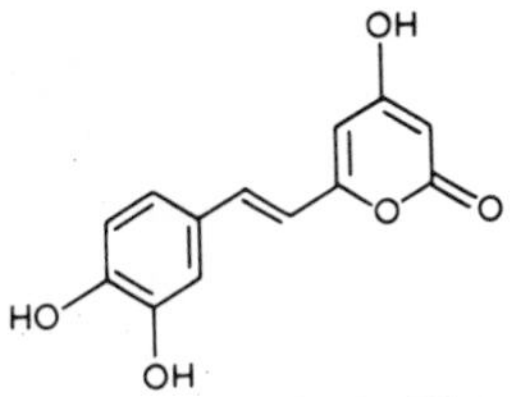

OH

HO

OH

O

O

(XII) Hispidin

References p. 287

shown that these compounds are non-nitrogenous and yield simple phenols and phenolic acids on alkaline degradation. In one case, *Polyporus hispida*, the precursor has been shown to be a dihydric phenol, hispidin (XII)[293]. It seems likely, therefore, that the fungi have the capacity to synthesise phenolic polymers presumably by an oxidative mechanism involving phenolase rather than peroxidase. This is analagous to the formation of true nitrogen-containing melanins by animals which involves the oxidation of 3,4-dihydroxyphenylalanine[294].

(c) Biosynthesis of polysaccharides

It is known that in higher plants cellulose is synthesised from guanosine diphosphate glucose (GDP-G) whereas in the cellulose-excreting bacterium *Acetobacter xylinum* the uridine analogue, UDP-G, is the starting point[295]. The use of UDP-G with enzyme preparations from higher plants leads to a β-1,3-glucan. It is possible, of course, that these nucleotide derivatives are not necessarily the glucosyl donors *in vivo*, for there is evidence in both higher plants and in bacteria that cell-wall synthesis involves the transfer of sugars attached to lipid. Nevertheless, it can be assumed that the mechanism for the synthesis of cellulose and the β-1,3-glucan are likely to be quite different. In any case, a separate branching enzyme, like the Q-enzyme in starch synthesis must be also present in those fungi which produce glucans. Chitin is synthesised in *Neurospora crassa* as in animals from UDP-acetyl-glucosamine, but it is not known whether chitosan requires a separate polymerase or if it is derived from chitin by deacetylation.

In higher plants, GDP-mannose gives rise to β-1,4-mannans, and it seems probable that a complex series of polymerases and branching enzymes are required for the fungal mannan. The possession of the polymers by the different fungal groups (Table XI) thus points to fundamental differences in polysaccharide synthesis.

(d) Evolutionary trends in cell-wall polysaccharides

There does not seem to be sufficient data to make any comments about the evolution of polysaccharide biosynthesis. Simple carbohydrates are known to be formed under primitive earth conditions[4], but no attention has been paid to polymerisation processes[296]. It seems likely, by analogy with the proteins, that linear and branched polymers could be formed from various hexoses under suitable conditions. However, the occurrence of peptidoglycans in the cell walls of the most primitive of living organisms suggests that mixed heteropolymers might have predominated. One can imagine that evolutionary development in cell-wall polymers started from these peptidoglycans which, with the cross-linking afforded by D-amino acids gave suitably flexible, though growth restricting, membrane protection. The next stage was the loss of the peptide cross-link and its replacement by a combination of hydrogen bonding to protein which allowed for cell expansion followed, when the cell had reached full size, by a cross-linking either by the formation of co-valent links ("tanning") by the quinones from simple phenolic compounds (*e.g.* protocatechuic acid or dopamine as in insects) or by calcification[297,298]. In both cases there would be an advantage for *partial* deacetylation of the chitin chain. It should be noted, however, that non-amino polysaccharides could not be cross-linked by "tanning". The use of non-amino non-crystalline heteropolysaccharides was obviously a much later step, since although ideal for cell expansion, the opportunity for strengthening the cell-wall by cross-linking, although less necessary in marine algae, was now absent. Thus, the evolution of oriented crystalline cellulose, coupled with that of acidic, calcium cross-linked, polysaccharides can be regarded as a necessary step for land plant evolution. If this were so, it perhaps required the primary evolution of carbohydrate polymerases which could handle GDP-rather than UDP-glucose so that proper control of orientation of the polymeric β-1,4-glucans could be made. It would obviously be of interest to examine the degree of polymerisation and orientation of cellulose chains in lower algae including the Cyanophyta.

One other feature, lignification, is discussed below, but it might be remarked that the presence of lignin in land plants seems at first sight to parallel the evolution of the use of phenolic polymerisation agents in fungal spores (hispidin, see p. 204) and in insects. However, the phenolic moieties in these last two cases almost certainly form cross-links with the aminopolysaccharide chains, whereas lignin only forms its three-dimensional structure by cross-linking with itself.

(e) Lignin

Although the carbohydrate components of the cell wall of land plants impart a good deal of rigidity and thus help to maintain the dimensions of the individual cell relatively constant, they give very little strength between cells and are insufficient in themselves to hold anything other than the smallest shoots of any plant in an erect position[63,64,81]. In some taxa, for example *Equisetum*, the polysaccharides are augmented by infiltration of the wall with silica, but this is too brittle itself to afford much extra strength[63]. The vast majority of higher land plants have overcome these difficulties by deposition of lignin in and between the cell walls[24,283–285]. Lignin is a three-dimensional heteropolymer formed by the co-polymerisation of one or more of the cinnamyl alcohols corresponding to the common cinnamic acids, *p*-coumaric (XIII), ferulic (XIV) and sinapic acids (XV)[299,300]. This accounts for the production of the similarly oriented benzaldehydes, *p*-hydroxybenzaldehyde, vanillin and syringaldehyde which lignins yield on mild oxidation. The cinnamyl alcohols, which are produced in the plant from the corresponding acids (see p. 258), undergo a one-electron oxidation catalysed by a peroxidase-like enzyme giving rise to a polymer containing covalent C to C and C to O links involving both the side-chain and the aromatic ring[300].

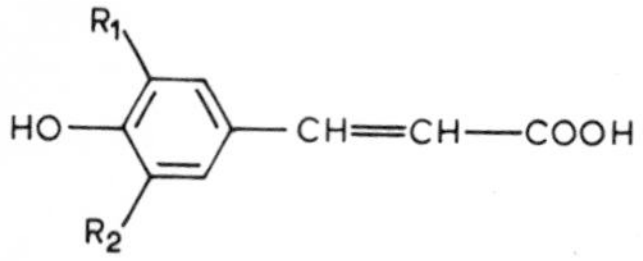

R_1, HO—(benzene ring)—CH=CH—COOH, R_2

(XIII) $R_1, R_2 = H$ *p*-Coumaric acid
(XIV) $R_1 = OCH_3, R_2 = H$ Ferulic acid
(XV) $R_1, R_2 = OCH_3$ Sinapic acid
(XVI) $R_1, R_2 = OH$ Caffeic acid
(XVII) $R_1 = H, R_2 = OCH_3$ Isoferulic acid

In most gymnosperms, lignin is almost entirely based on nuclei having the vanillin substitution pattern whereas in woody angiosperms there are also a considerable number of syringyl elements[301,302]. The lignin from herbaceous angiosperms contains, in addition, nuclei giving rise to *p*-hydroxybenzaldehyde. This latter aldehyde is also the major component produced by oxidation of the insoluble 'lignin' of the cell wall of the mosses *Sphagnum balticum* and the giant *Dawsonia* species[303]. This distribution of different substitution patterns in the aromatic moiety suggests that lignin may have evolved from a primitive polymer based solely on units yielding *p*-hydroxybenzaldehyde, *via* a vanillin-producing lignin, to one of the syringyl type. It is interesting to note that the oxidation products from the oldest fossil gymnosperm-lignin examined are also much richer in *p*-hydroxybenzaldehyde than those obtained from present day specimens[304]. It seems likely that a polymer formed entirely from *p*-hydroxyphenyl propane units might have different chemical and mechanical properties than one formed from precursors containing additional *O*-methoxy groups which would limit the formation of C–C bonds between the aromatic rings. However, angiosperm-like lignin containing syringyl units is present in *Selaginella* and *Isoetes* (Lycopsida) and the lignin from other lycopods yields syringic acid on hydrolysis, so that there is some doubt about the evolutionary significance of the methoxylation pattern[302].

The only multicellular land plants which do not contain lignin are the higher Fungi, and this is somewhat puzzling, since many of the higher Basidiomycetes have more or less all the enzymes necessary to elaborate lignin from phenylalanine. Thus they contain phenylalanine ammonia lyase which catalyses the deamination of L-phenylalanine to cinnamic acid[305] and has been postulated as a key enzyme in the biosynthesis of lignin and related phenolic compounds including the flavonoids (see p. 258); they also contain hydroxylases which can convert cinnamic acid to *p*-coumaric acid and finally to caffeic acid (XVI)[306]; *O*-methyl transferases for the methylation of phenolic hydroxyl groups[307]; carboxyl reductases which give rise to cinnamyl alcohol[308] and one-electron oxidases which can catalyse the polymerisation of many phenolic compounds. It might appear, therefore, that the Basidiomycetes had all the necessary enzyme systems to produce lignin. However, whereas the photosynthetic land plants have *O*-methyl transferases which catalyse the methylation of the *meta*-oriented hydroxyl groups (that is, caffeic acid to ferulic acid), the enzymes from fungi give rise to *para*-oriented methoxy groups like isoferulic acid (XVII), *p*-methoxycinnamic acid, or catalyse the methoxylation of phenolic hydroxy groups in polyketide metabolites[127]. The products formed are not likely to readily yield free radicals on oxidation and thus may not give lignin-like polymers. In any case, the ability to form lignin probably requires a highly ordered series of reactions which are all under strict control either at the precursor level or at the polymerisation site between the cells[283]. Although we know little about the phenolic compounds present in the eucaryotic algae, it seems likely that some may possess similar capabilities to the land plants especially since flavonoids have been reported in *Nitella hookeri* (Chlorophyta)[309].

However, it has been shown that the peroxidases from Rhodophyta and Phaeophyta, unlike those from higher plants, cannot polymerise lignin model precursors, so that these organisms would presumably be incapable of synthesising lignin[310]. Green algal peroxidase, on the other hand, was as active as the angiosperm enzymes, and it can only be presumed that lignification is controlled in the Chlorophyta either because of lack of a pathway to *p*-hydroxycinnamyl or vanillyl alcohol or the lack of peroxidase in the cell walls. It is interesting to note that the submerged parts of angiosperm water plants also almost lack lignin.

Further investigations on the metabolism of cinnamic acid by fungi and other non-lignified plants would be useful, therefore, in pointing the way to such control mechanisms and perhaps to the evolutionary changes which may have taken place in the organisation of present-day flora.

9. Secondary products

(a) Introduction

It was pointed out in the introduction to this Chapter that plants produce a far greater diversity of chemical compounds than do animals. The majority of these substances have no obvious role in the primary metabolism of the plant and their very heterogeneity has meant that no defined function could readily be seen for them. They are thus generally referred to as secondary products. This epithet is unfortunate, for it has led many workers to assume that such compounds are not involved in the dynamic pathways of metabolism, but merely exist as a means of ridding the cell of excess metabolites or excretory products derived from primary pathways, which otherwise would be deleterious to their functioning[22,23,127,311–313]. Another proposed role for secondary products is that their synthesis provides a way for the utilisation of certain primary metabolites and thus helps to keep in operation the primary production lines which are essential to cell multiplication, at times when some of these latter pathways might shut down altogether because external conditions were unfavourable for cell division or growth[311]. As a corollary to this hypothesis, it is suggested that the nature of the secondary compounds formed is relatively unimportant and this accounts for their conspicuous structural variation. Besides these two main proposals a number of equally unimportant roles have been suggested for secondary compounds ranging from methyl acceptors to redox participants[23].

None of these hypotheses take into account either the general complexity of the biosynthetic pathways which lead to the elaboration of secondary products or the fact that both their synthesis *and* their degradation is under strict regulatory control. If the role of secondary products was merely to detoxify excess primary metabolites or keep the wheels of primary energy production ticking over within the cell, there would be no need to synthesise more than one or two types of product from any given primary precursor, which would involve as few enzyme-catalysed steps as possible. It surely would not be necessary for example for *Streptomyces aureofaciens* to produce 7-chlorotetracycline (XVIII) to keep polyketide synthesis in trim or to use up excess malonate or acetate[314]. Moreover, it would undoubtedly be more productive for the organism to "use" its genome for closer control of primary metabolism than to synthesise a whole host of apparently useless products which would presumably need to be translocated to special storage pools. Finally, the fact that secondary compounds are turned over in the cell, sometimes quite rapidly, to produce primary metabolites and also are found in many different types of actively dividing tissue in higher plants, tells against hypotheses linking secondary metabolism either with utilisation of primary substances or with resting stages in cell division.

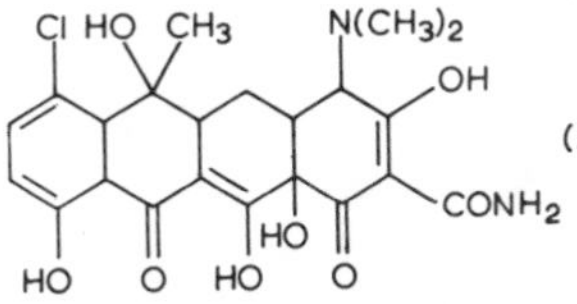

(XVIII) 7-Chlorotetracycline

There is now overwhelming evidence that the real function of secondary compounds is as repellents, inhibitors or attractants of other organisms[315–320]. Interactions dependent on the production of defined chemical substances occur between all types of living organisms (see Chapter 1), the most widely understood at present being those between higher plants and either insects or fungal and bacterial pathogens. Indeed, it is quite clear that it is because of their function that secondary products are present, like any other character, as part of the evolutionary strategy of the taxon in which they occur. They are thus much more likely to change as a result of selection than more conservative biochemical features such as those involved in primary metabolism. Their heterogeneity, far from being fortuitous, is a measure of the diversity of solutions which different organisms have adapted for co-existence in any given ecological niche. The above facts indicate the extreme importance of secondary products in biochemical evolution in plants and an examination of the various major classes of such compounds, their functions, distribution and development forms the subject of the rest of this Chapter.

(b) Types of secondary products

There is really no completely satisfactory way of defining a secondary product. In part, this is due to the fact that there has been a rapid erosion of the number of classes of compound which are thought to have no recognisable role in the economy of the organism. For example, ten or fifteen years ago monoterpenes would have been readily ascribed to the latter category, whereas now they are known to determine to a considerable extent the interaction between higher plants and insects as both repellents and attractants[321]. On the other hand, if secondary products were to be more narrowly defined as compounds which are not involved in primary metabolism they would include such substances as cellulose and chlorophyll which plainly have an important function in the plant. Nevertheless, it is probably better to define secondary products in this latter way, and to classify them into groups according to their biosynthetic origin from three or four primary metabolites. Such a classification brings together compounds whose function may sometimes be diverse, but whose biochemical evolution can be considered to be more or less identical. Using such a classification, we can consider secondary products in the following categories: complex polysaccharides arising from sugars in the primary pool; shikimate derived metabolites including flavonoids, cinnamic acids, lignin, aromatic alkaloids and the simple hydroxybenzoic acids; polyketides from acetate; unusual lipid constituents such as alkanes and acetylenes also from acetate; a wide variety of terpenoid constituents, including steroids and carotenoids, from mevalonic acid; and non-aromatic alkaloids from ornithine and lysine. The relation between these different classes of compounds and their primary precursors is shown in Fig. 19.

(c) Biosynthesis of secondary compounds

It should be noted that although some unique individual steps occur in the biosynthesis of secondary products, the majority have parallels in primary metabolism. For example, the route of biosynthesis and metabolism of the branched-chain amino acids valine and isoleucine, contain steps which are exactly analogous to those leading to mevalonic acid, and the biosynthesis

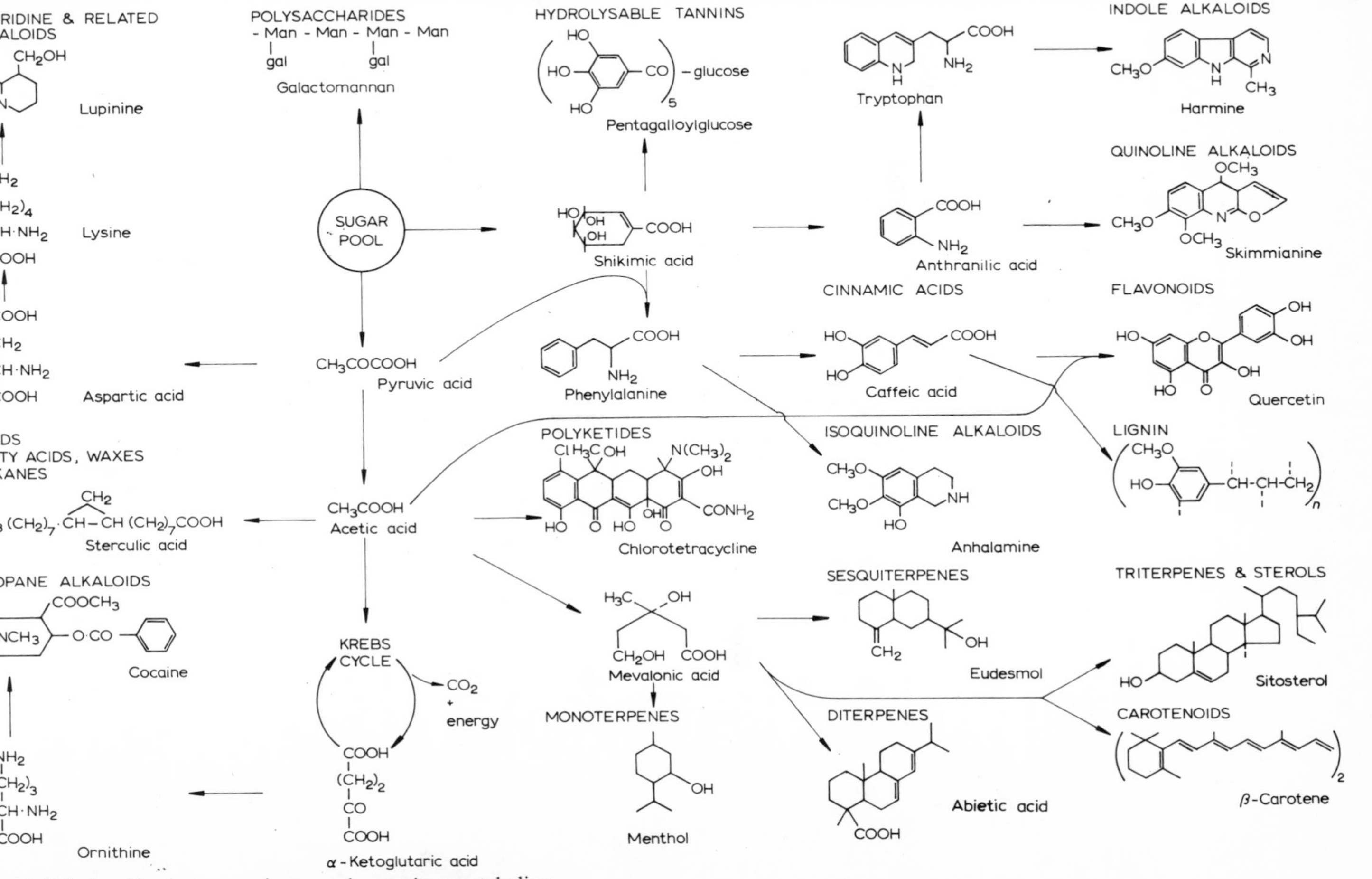

Fig. 19. Relationships between primary and secondary metabolism.

of the polyketides is akin to that of the common fatty acids (Fig. 19). Moreover, since many secondary products such as polyprenols, complex polysaccharides and so on are found in the most primitive of present-day organisms, many of the reactions which can be considered as belonging exclusively to secondary metabolism must be extremely primitive. Where investigations have been carried out, the pathways leading to the biosynthesis of secondary products have been shown to be subject to similar control mechanisms as those giving rise to primary metabolites (see p. 175). However, experiments on the biosynthesis of several different types of secondary compounds in higher plants have shown that this control is not so exact and many compounds which are not presumed to be obligate intermediates in a given pathway can often be incorporated into end-products or modified end-products.

For example, the dicotyledon, *Scutellaria galericulata* normally synthesises the 7-glucuronide of chrysin (XIX, R=H), the 2-phenyl ring of which is known to arise from that of phenylalanine (see p. 258). When *p*-fluorophenylalanine was fed to the plant, the abnormal 4′-fluorochrysin (XIX, R=F) was isolated[322]. Similarly, the alkaloid nicotine (XX,

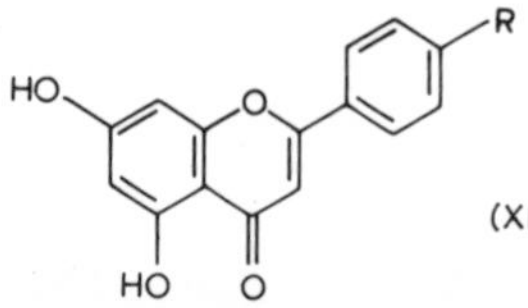

(XIX) R=H Chrysin
R=F 4′-Fluorochrysin

$R_1=R_2=H$) is known to arise from the condensation of nicotinic acid and 1-methyl-Δ′-pyrrolinium salt (see p. 272); when either the corresponding 1,3-dimethyl-Δ′-pyrrolinium salt[323] or 5-fluoronicotinic acid[324] was fed to *Nicotiana tabacum*, aberrant synthesis occurred leading to either 3′-methyl- (XX, $R_1=H$, $R_2=CH_3$) or 5-fluoronicotine (XX, $R_1=F$, $R_2=H$), neither of which are known in Nature. Another pertinent observation concerns the conversion of thebaine (XXI) to morphine (XXII) which is the normal route of biosynthesis of the latter alkaloid. This transformation can be easily observed in tissue cultures of the morphine-containing *Papaver somniferum* but expectably not in the morphine-less sister species *P. bracteatum*. However, the conversion is carried out surprisingly by tissue cultures of *Nicotiana tabacum*, a plant which as far as is known contains no morphine alkaloids at all[325]! It is possible in fact, that *N. tabacum* does contain small traces of morphine alkaloids since there are several cases where compounds have been found in plants in such small amounts as normally to go undetected. For example, the isoquinoline alkaloid, narcotine (XXIII) which is readily obtained from *P. somniferum* can be isolated from a number of common foodstuffs, where it occurs in quantities of less than 1 ppm, providing sufficiently large quantities are taken[326]. The same has been shown to be true of azetidine-2-carboxylic acid (XXIV), which was believed to be restricted to members of the Liliaceae, and one legume, *Delonix regia*. This amino acid was unexpectedly found in the commercial work-up of several tons of sugar beet (*Beta vulgaris*)[327,328]. There are

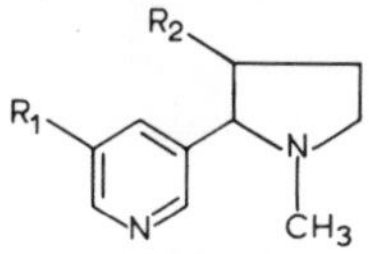

(XX) $R_1, R_2 = H$ Nicotine

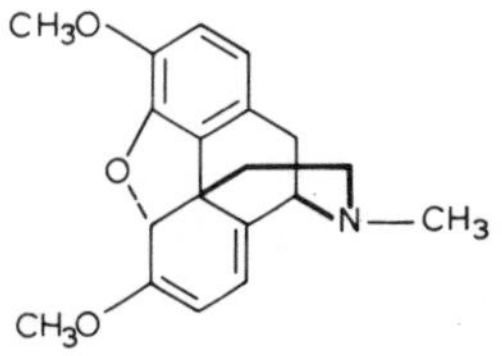

(XXI) Thebaine

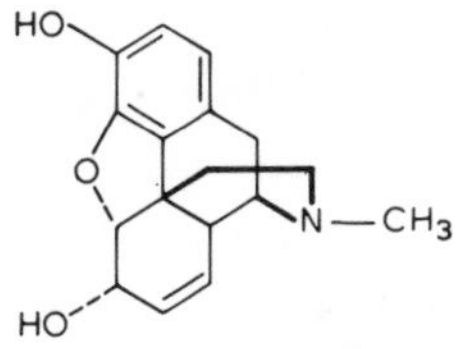

(XXII) Morphine

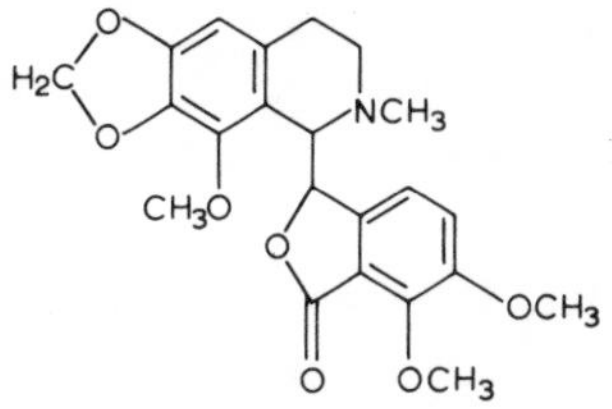

(XXIII) Narcotine

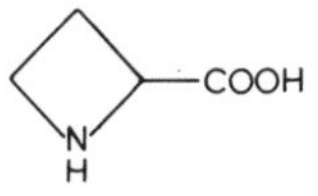

(XXIV) Azetidine-2-carboxylic acid

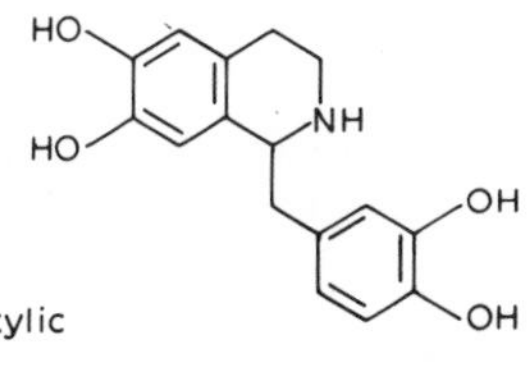

(XXV) Norlaudanosoline

References p. 287

numerous other examples reported in the literature, especially since the advent of ultrasensitive separation and detection methods such as gas-liquid chromatography.

These findings might be taken to indicate that all plants possess the necessary genes to synthesise all known compounds but that under normal circumstances the necessary biosynthetic pathways are totally repressed. It is certainly true that plants possess a much greater number of biosynthetic or degradative routes than are used for normal metabolism, since many are capable of transforming a large number of foreign compounds like herbicides to harmless detoxification products[329]. It is also true that the ability to synthesise and accumulate any complex metabolite implies the possibility of making a number of cognate compounds which do not occur in the plant, at any rate not in sufficient quantity to be detected[37–41]. Indeed, part of the argument presented in the section on cell walls (p. 207) implied that if an activity was present (*e.g.* an oxidative phenolic coupling reaction) in an organism, then a compound whose synthesis required a step involving such an activity (*e.g.* lignin) might be synthesised and therefore present. Since, as was pointed out above, the steps in secondary metabolism are similar to those in primary metabolism, one must expect transformations to take place and compounds to be synthesised by routes which do not normally occur as the major pathways in the organism in question. Two examples will suffice. In the early work on carotenoid biosynthesis, it was shown that leucine was a precursor of the pigments[330]. It is now known that leucine is converted into hydroxymethylglutaric acid and hence into mevalonic acid by a rather tortuous pathway, and although the reactions involved obviously can occur, no one would point to them as a main pathway to the terpenoids. Rather they show the catholicity of metabolic pathways in plants, and in the case given, perhaps a way of utilising excess leucine which otherwise would repress or inhibit the pathway to valine and isoleucine[246].

The second example is the demonstration that the typical plant alkaloid norlaudanosoline (XXV) can be formed from dopamine in rat-brain tissue[331]. This, obviously, is not a pathway which one could regard as normal in animals, although probably of high importance in both ethanol addiction and other syndromes[332].

It can be stated, therefore, with some degree of conviction that while all plants have the *ability* to synthesise a wide variety of organic compounds, most of these substances will not normally occur in the cell. Where they do, they will be present in insufficient concentration, either because of slow synthesis or rapid degradation, to have any effect on the overall metabolism of the plant or on its interaction with the environment. On the other hand, compounds which accumulate in relatively high concentration (greater than 0.1 μmole per g) are, or have been, of use to the plant and their rate of synthesis, and degradation is under strict control, which implies a single pathway for each. Such compounds can be regarded as true metabolites, and may undergo evolutionary change as pressures decree.

10. The lipids

(a) Introduction

The lipids comprise a relatively uniform group of compounds, all of which contain one or more alkyl chains containing from 12 to over 30 carbon atoms[36]. The main group of lipids, called acyl lipids, are all esters of fatty acids. They include, besides the true fats, which are esters of glycerol, phosphoglycerides like lecithin (XXVI), galactosyl glycerides (XXVII) and cognate derivatives containing other sugars, sphingolipids such as the cerebroside from wheat flour (XXVIII), estolides which form the resistant matrix of cutin (see below), and waxes. Other lipids are related to the

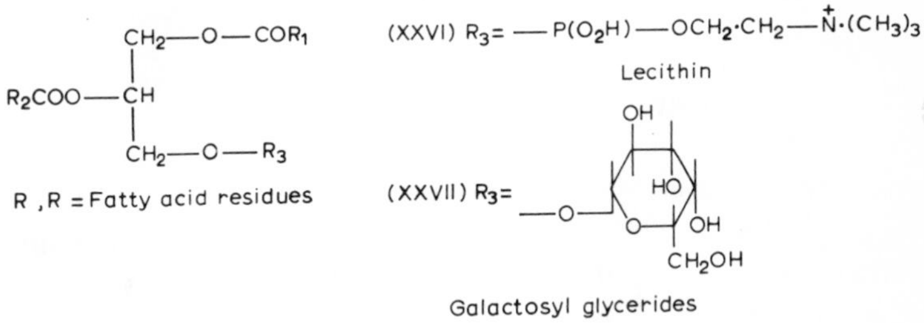

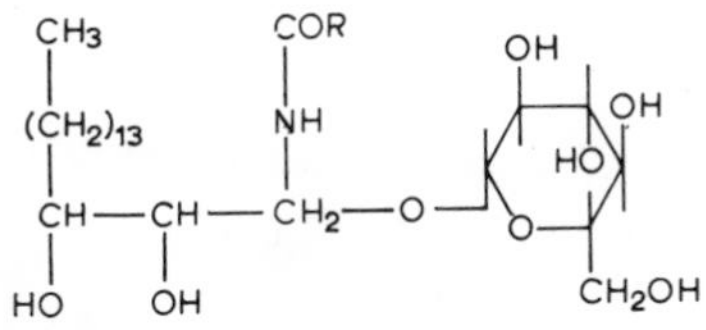

$CH_2—O—COR_1$
$R_2COO—CH$
$CH_2—O—R_3$

R ,R = Fatty acid residues

(XXVI) $R_3 = —P(O_2H)—OCH_2{\cdot}CH_2—\overset{+}{N}{\cdot}(CH_3)_3$

Lecithin

(XXVII) $R_3 =$ —O— (galactose: OH, HO, OH, CH_2OH)

Galactosyl glycerides

CH_3, $(CH_2)_{13}$, COR, NH, OH, HO, OH, CH_2OH
$CH—CH—CH_2—O—$
HO OH

(XXVIII) Cerebroside

References p. 287

fatty acids, including the alkanes, which occur in the surface layer of higher plant leaves, and the acetylenes, such as carlina oxide (XXIX) from the root of *Carlina acaulis*.

The major fatty acids in the angiosperms, in order of the amounts found in all tissues, are oleic acid (XXX), usually symbolised as 18:1 (9*c*) to show eighteen carbon atoms with one *cis* double bond at 9,10; linoleic acid, 18:2 (9*c*,12*c*), and palmitic acid, 16:0. α-Linolenic acid 18:3 (9*c*, 12*c*, 15*c*), stearic acid, 18:0, myristic acid, 14:0, and lauric acid, 12:0 are present in smaller amounts[36]. A large number of other saturated and unsaturated fatty acids occur in the Plant Kingdom, most having between 16 and 22 carbon atoms but many of these are unique to individual

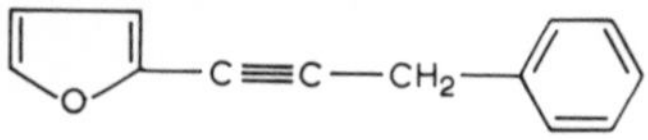

(XXIX) Carlina oxide

$CH_3—(CH_2)_7—CH\overset{cis}{=}CH—(CH_2)_7\,COOH$ (XXX) Oleic acid

families. Hydroxy and keto acids of varying chain length also are found and, especially in the bacteria[333], fatty acids with extra in-chain methyl groups or cyclopropane rings. The acetylenes found in plants usually contain from 13 to 17 carbon atoms[334,335], while most of the alkanes and alkyl alcohols, on the other hand, have between 25 and 35 carbon atoms[336]. A number of alkanes are known which contain a methyl group either at C-2 or C-3.

(*b*) *The biosynthesis of fatty acids and other lipids*

In all organisms, fatty acids are formed by the repeated addition of malonyl-CoA to an existing acyl chain with subsequent decarboxylation, followed by reduction of the original acyl-CO group to CH_2 before the process is repeated[36,337,338]. In this fashion, the chain grows by two carbon atoms each time. If the starter group for the acyl chain is acetate, the resulting products will have an even number of carbon atoms as is found in the normal fatty acids. Other starter acyl groups account for the occurrence of the less usual odd-chain or terminal α-methyl-substituted acids and related compounds.

The sequence of reactions which occurs during the addition of malonate involves the following succession of enzymes: a 3-ketoacyl synthetase, a 3-ketoacyl reductase (CO→CHOH), a 3-hydroxyacyl dehydrase (enoylhydratase, $CHOH \cdot CH_2 \rightarrow CH{=}CH$), and an enoyl reductase ($CH{=}CH \rightarrow CH_2 \cdot CH_2$). In addition, a non-enzymic protein component is also required, the acyl-carrier protein (ACP) to which, as its name suggests, the acyl group is attached before condensation and during the subsequent series of reactions. In higher plants, in photoautotrophic *Euglena gracilis* and a variety of bacteria including *Escherichia coli*, all these enzymes are separable. In *Saccharomyces cerevisiae*, *Mycobacterium phlei*, etiolated *Euglena* and a wide variety of animal tissues, on the other hand, the enzymes exist in a complex with a molecular weight of between 0.5 and 2 million[339]. It seems probable that even where the enzymes are individually isolable, they form a complex in the cell, otherwise one might expect a greater variety of intermediate products of fatty acid biosynthesis to be found in the free state. The chain length of the final product, usually 16 or 18, is probably determined by the specificity of the final thiolesterase which splits the fatty acid from the ACP[36]. Alternatively, the variation in the rate of addition of malonyl-ACP to the growing chain, which is dependent on synthetase activity, may control termination. There is little evidence that there is any variation in the biosynthesis of fatty acids between the most primitive and advanced organisms. Indeed, the only difference in the overall reactions which has been noted is that the fungal enoyl reductases require FMN, whereas those from *E. coli* are dependent on pyridine nucleotides[36].

In contrast to the uniformity of saturated fatty acid biosynthesis, there is a major variation in the mode of formation of unsaturated acids between certain bacteria on the one hand and all other organisms, including both plants and animals[36,337]. A number of anaerobic and photosynthetic bacteria synthesise the common C-16, C-18 mono-unsaturated acids by chain elongation from a shorter chain length enoyl-ACP derivate in which the double bond is not reduced before further condensation with malonyl units takes place. The elimination of water from the 3-hydroxy-acyl-ACP under normal circumstances yields a *trans* double bond. In those bacteria, which utilise this anaerobic route to unsaturated fatty acids, these *trans* acids are isomerised to the *cis* form which is not reduced by the enoyl reductase. Thus, in *E. coli*, *trans*-2-decenoate, which is reduced and had further C-2 units added would give palmitate and stearate, after isomerisation and subsequent elongation gives rise to palmitoleate and vaccenate (Fig. 20). This anaerobic pathway is found in some obligate aerobes such as *Pseudomonas fluorescens*. However, in most aerobic

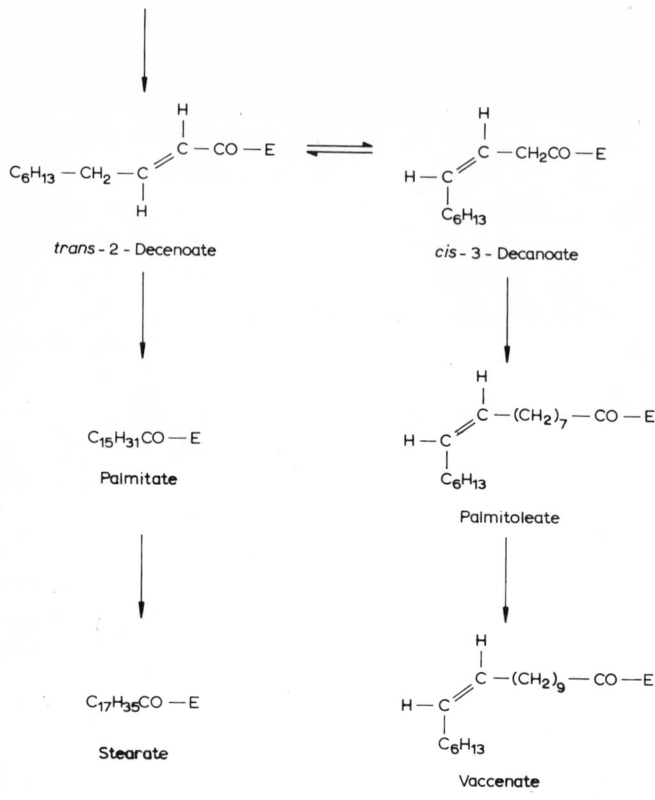

Fig. 20. Biosynthesis of unsaturated fatty acids in *Escherichia coli.*

bacteria, like *Micrococcus lysodeikticus* and *Corynebacterium diphtheriae*, all algae, fungi and higher plants examined, unsaturation arises by an aerobic pathway in which an existing saturated bond is dehydrogenated by a mixed function oxygenase which requires both oxygen and NADPH[36,337]. One atom of oxygen is reduced by the co-enzyme and the other provides for the oxidation of the substrate. Usually, a hydroxy intermediate would be expected in such reactions, at least transitorily, but no such derivative has been found during the biosynthesis of unsaturated fatty acids. Indeed, ricinoleic acid, 12-hydroxyoleic acid (12h–18:1–9*c*), which is formed from oleic acid (18:1–9*c*) in the presence of oxygen and reduced pyridine nucleotide, is not converted to linoleic acid (18:2–9*c*, 12*c*) by the relevant desaturase enzyme. There is some difficulty in demonstrating the desaturation step in higher plants because of the lack of a transferase to attach the higher saturated fatty acid-CoA esters (C-16 or C-18) to the carrier protein where the dehydrogenation takes place. In *Chlorella vulgaris* these transferases may be induced to form when the organisms are grown on a minimal medium under strong light[36].

In those bacteria which possess the anaerobic pathway, no poly-unsaturated fatty acids are found. In algae and higher plants these acids are synthesised by sequential desaturation in a similar manner as described above. The second double bond is usually introduced between the first and the methyl-end of the chain, with one methylene group in between. Thus α-linolenic acid is synthesised from oleic acid as follows: 18:0→18:1(9*c*)→18:2(9*c*, 12*c*)→18:3(9*c*, 12*c*, 15*c*). In some algae, however, a double bond can be introduced on the carboxyl side. Thus arachidonic acid is synthesised in algae from linoleic acid (18:2–9*c*, 12*c*) *via* γ-linolenic acid (18:3–6*c*, 9*c*, 12*c*) followed by chain-lengthening to the corresponding C-20 acid and the introduction of a fourth double bond to yield the final product (20:4–5*c*, 8*c*, 11*c*, 14*c*). In *Euglena*, on the other hand, this C-20 acid is synthesised by an initial chain extension from α-linoleic acid to 20:2 (11*c*, 14*c*) and thence by two successive desaturations on the carboxyl side of the existing double bonds. The ability of certain plants to desaturate monenoic acids on the carboxyl side of the existing double bond is paralleled in animals who can only desaturate acids in this way and thus cannot synthesise α-linoleic acid at all[36].

Conjugated ethylenic acids are relatively rare and probably arise in a different way from the normal dienoic and trienoic acids since in all cases they contain at least one *trans* double bond[36]. Although it is possible that hydroxy compounds are intermediates in their synthesis, no definite evidence exists that this is the case. The conjugated hydroxydienes which are found, may well be end-products of a reaction sequence as in the case of ricinoleic acid mentioned above.

There is also little data on the biosynthesis of the acetylenic fatty acids[340]. It is known that oleic acid (18:1–9*c*) can give rise directly to crepenynic acid (18:2–9*c*, 12*a*) in *Crepis rubra*, but neither α-linoleic acid (18:2–9*c*, 12*c*), its 12-*trans* isomer, nor vernolic acid (*cis*-12,13-epoxyoleic acid) all of which occur in the plant, were effective as precursors. The conversion of oleic acid to the acetylenic acid was shown to require oxygen and is probably thus catalysed by a mixed function oxidase as described above. The hydroxylation of oleic acid to ricinoleic acid in *Ricinus communis* as mentioned earlier requires oxygen and NADPH.

References p. 287

In ergot, *Claviceps purpurea*, however, this hydroxy acid is formed by hydration of the 12,13-double bond of α-linoleic acid. Whether this represents a universal difference between fungi and higher plants is not known.

The biogenesis of fatty acids with methyl groups either at the carboxyl or distal end of the chain usually involves changes in the precursors used during synthesis[341]. For iso- and anteiso-acids, which have methyl groups on the C-atom 2-or 3-atoms distant from the methyl- end and which occur widely in bacteria, 3-methylbutyrate or 2-methylbutyrate, derived from leucine and isoleucine, act as 'starter' acids instead of acetate. Similarly, the substitution of methylmalonate, from the carboxylation of propionate, instead of malonate itself in the terminal step of synthesis leads to a 2-methyl-substituted fatty acid. Repetition of this step would give rise to the 2,4,6-trimethyl-substituted acids, well-known in the mycobacteria[333]. In-chain methylation and the formation of cyclopropane groups in bacteria and higher plants involve the addition of an extra carbon atom with participation of *S*-adenosyl methionine.

Alcohols, ketones, alkanes and acetylenes related to the fatty acids are all undoubtedly derived from them by reduction, desaturation, decarboxylation and other appropriate reactions.

(c) The function of lipids

The most important function of the acyl lipids in living organisms is their role in the biological membranes[342–344]. All membranes contain such lipids, usually galactosyl- or phospho-glycerides, although sulpholipids are found in chloroplasts. Various models to explain the function of lipids in the membranes have been proposed most of which assign lipids both a structural and a metabolic role[36, 342–344]. It seems highly probable, for example, on the basis of their amphiphilic nature that acyl-lipids play some part in the transport of metabolites across membrane boundaries. However, the suggestion that the polyunsaturated acids found in high concentration in chloroplasts play some role in electron transfer in those organelles seems to be unlikely, for certain Cyanophyta contain only monoenoic acids. As might be expected, the membrane function of lipids does not appear to vary much with the evolutionary status of the organism, and differences in the nature of the fatty acid moieties associated with the various organelles between the angiosperms and certain lower plants, do not appear to be of any significance. The other role of acyl lipids is in the energy reserves in the seeds of higher plants[345]. Here the efficiency of the various triglycerides as energy stores does not seem to be affected by the nature of the acylating fatty acids present. One might speculate that the incorporation of unusual fatty acids into seed fats such as the cyclopropane compounds malvalic (XXXI) and sterculic acids (XXXII) in the Malvaceae, might inhibit their utilisation by bacteria and fungi, but it should be noted that the majority of plant seeds contain only the more common fatty acids[346].

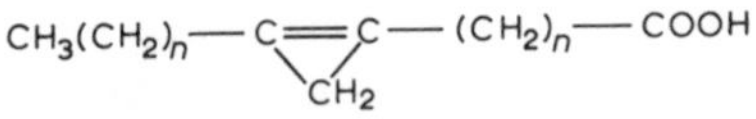

$$CH_3(CH_2)_n{-}\underset{\diagdown\,CH_2\,\diagup}{C{=}C}{-}(CH_2)_n{-}COOH$$

(XXXI) $n=6$ Malvalic acid
(XXXII) $n=7$ Sterculic acid

The function of some of the other lipid components is somewhat clearer. All land plants need to conserve water, and the formation of a water-resistant yet transparent coating semi-permeable to gases to cover the leaf surfaces is obviously of the highest importance. Such a coat should preferably also be resistant to attack by micro-organisms, be tough enough to withstand the rigours of climate and abrasion from insects and the like, and yet be sufficiently flexible to allow for plant growth and movement. This set of parameters is ideally met by the cutin and wax layer found on the leaves of land plants[336,347]. Cutin which is a three-dimensional polymer of poly-hydroxy fatty acids such as 9,16,18-trihydroxyoctadecanoic acid and 10,16-dihydroxyhexadecanoic acid, formed by intra-ester links, is apparently finally elaborated on the surface of the leaf. The primary precursors, which arise from the more highly unsaturated fatty acids by the action of lipoxygenase and subsequent partial reduction, are transported across the epidermal cell wall aided by the action of one of the phospholipids[347]. The same is true of the long-chain alcohols and acids (C-20 to C-35) which unite to form the waxes, and the associated alkanes and other lipid residues. The waxes crystallise out to form regular and distinctive molecular aggregates whose patterns are readily seen with a scanning electron microscope[336,348]. These give a marked "roughness" on the surface at the cellular level which is apparently important in the mechanism of self incompatibility, and probably acts as a barrier to the approach of potentially dangerous bacteria and fungi. In other plants, as in the case of the pitcher plant *Nepenthes*, the orientation of the waxy layer as a series of slippery plates effectively helps to trap insects[36].

The acetylenic compounds present a different case. Those obtained from the fungi, mainly from Basidiomycetes, were first discovered because the organisms in question were known to produce potent antibiotics[334,335]. Subsequently the compounds responsible for this property were shown to be acetylenes. It seems safe to assume therefore that most, if not all, such substances will show similar activity, and their occurrence in the roots of several members of the Compositae and in the seeds of plants from other families of angiosperms are of advantage to the organisms concerned. Unfortunately knowledge of the distribution of these compounds is rather poor because of their lability and difficulty of detection unless they possess a conjugated chromophore[335].

(d) Distribution of lipids in the Plant Kingdom

As mentioned earlier, the major fatty acids present overall in the angiosperms are oleic acid, linoleic acid and palmitic acid[36,346]. In leaves, however, the amount of oleic is much reduced and replaced by linolenic acid which makes up about half of the total fatty acids present (Table XII). Much of this acid appears to be associated with the chloroplast membrane where it may have more than a structural function[349]. Linolenic acid is also the predominant acid in other tracheophytes. In lower plants, however, the fatty acid composition of the cells is often quite different from that found in the angiosperms[350–353]. In the Cyanophyta, for example, few polyunsaturated acids are present and several species, such as *Anacystis nidulans*, contain none at all (Table XII). The Chlorophyta also contain much less linolenic acid than do the Tracheophyta: *Scenedesmus* contains instead a relatively large amount of the rare 16:4 acid, while *Nitella* and the marine algae *Codium* contain the corresponding C-20 analogues. The Chrysophyta are distinguished by having little or no C-18 acids of any sort, and instead have high amounts of palmitoleic acid (16:1) and the penta-unsaturated C-20 acid. This latter acid is also present in large amounts in the Rhodophyta and to a lesser extent in the Phaeophyta and *Euglena*. Although the 20:5 acid is found in the moss *Hypnum cupressiforme*, it is not found in any other land plant. The corresponding tetra-unsaturated C-20 acid, which is found in all Divisions of the eucaryotic algae[350], is also present in *Equisetum*, ferns and gymnosperms but not

TABLE XII

Fatty acid composition of various plant species

Species	*Fatty acids as % of total*[a]								
	16:0	*16:1*	*16:3*	*18:1*	*18:2*	*18:3*	*20:4*	*20:5*	*Other*
yanophyta									
Anabaena flos-aquae	39.5	5.5		5.2	36.5	10.7			
Spirulina platensis	43.4	9.7		5.0	12.4	21.4[b]			
Anacystis nidulans	47.0	38.8		10.0					
hlorophyta									
Chlorella vulgaris	26.0	8.4			33.5	20.1			
Scenedesmus obliquus	35.4			7.8	6.4	29.5			15.0% of 16:4
Nitella translucens	21.4				31.2	17.0	6.3		
Codium fragile[c]	28.1		12.3	10.8	5.5	27.2	3.0	1.9	3.2% of 22:0
hrysophyta									
Navicula pelliculosa	9.1	30.8	18.3		6.2	3.9	4.5	14.5	
Cyclotella cryptica[c]	15.7	34.8	7.8					18.8	
Phaeodactylum tricornutum[c]	15.8	31.1	7.3					25.8	
uglenophyta									
Euglena gracilis[d]	13.9	6.0		10.0	3.6	31.5	3.3	2.3	16.0% of 16:4
haeophyta[c]									
Fucus serratus	26.4	2.2		18.7	9.0	6.0	10.0	8.1	6.0% of 18:4
Fucus vesiculosus	25.6	1.6		17.2	7.4	8.2	10.1	7.6	6.0% of 18:4
hodophyta[c]									
Rhodomenia subfusca	29.1	4.9		14.9	1.4	1.1	14.0	24.1	
Placamium coccineum	26.6	5.7		5.6	3.0	0.4	11.5	21.6	6.6% of 20:3
ryophyta									
Hypnum cupressiforme	13.5	5.1	1.9	7.3	19.7	23.5	11.7	7.2	4.6% of 22:0
Sphagnum cymbifolium	20.6	1.4	2.2	25.6	19.5	13.2	10.2		3.0% of 20:3
phenopsida									
Equisetum arvense	28.2	4.4	9.4	4.0	11.5	43.0	2.1		
ilicinae									
Matteuccia struthiopteris	17.6	3.0	4.1	3.2	6.8	54.2	8.2		
Onoclea sensibilis	21.1	6.8	5.3	4.8	5.4	47.9	8.0		
ymnospermae									
Ginkgo biloba	19.3	1.5	4.9	4.8	8.2	48.2	3.4		
ngiospermae									
Antirrhinum majus	13.4	1.3		1.8	17.7	51.9			8.6% of 18:4
Vicia faba	11.7	6.9		3.4	14.3	56.4			

[a] Major acids only.
[b] γ-Linolenic acid (6*c*, 9*c*, 12*c*).
[c] Marine forms.
[d] Dependent on the growth conditions.

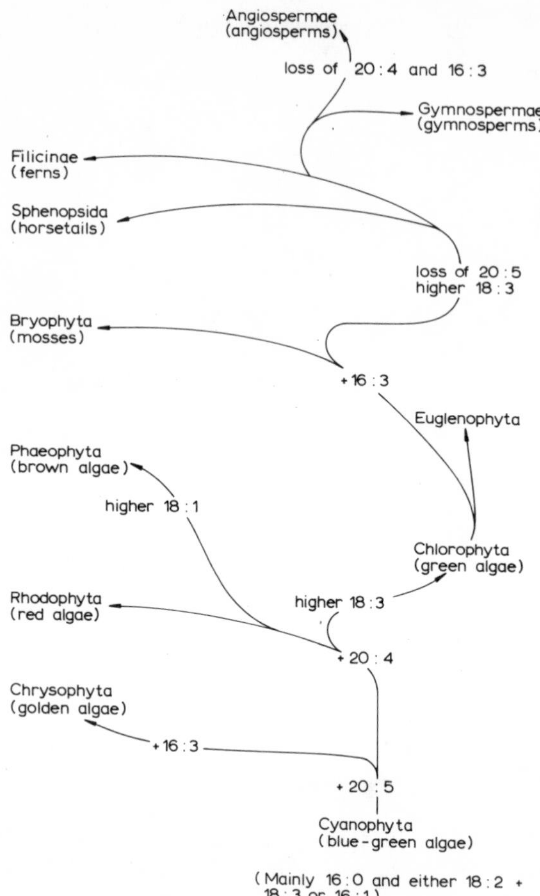

Fig. 21. The evolution of fatty acids in plants.

in the angiosperm species so far examined. These considerations lead to the possible evolution of the fatty acids as shown in Fig. 21. It is, of course, equally possible that the substitution of the C-20 polyunsaturated acids for linolenic acid in the algae is related to their marine or fresh-water habitat. It is noticeable that the marine forms of the algae have appreciable amounts of these compounds which are absent from the two fresh water Chlorophyceae[36]. The only stonewort examined *(Nitella)* also contains an appreciable quantity of the 20:4 acid.

In real evolutionary terms, however, the possession of arachidonic acid

(20:4) rather than linolenic acid[352] probably only involves a change in the specificity of chain-termination reactions of the fatty acid synthetases. The desaturation reactions are probably similar in each case. However, the differences noted earlier in the synthesis of the C-20 acid between algae and *Euglena* (see p. 219), may indicate that a totally different control mechanism exists in the different groups of plants which are capable of synthesising it. More is required before these questions can be satisfactorily answered.

The distribution of the wax alcohols and acids and of the associated alkanes are too imperfectly known to allow any phylogenetic conclusions to be made[347,354]. An interesting case, however, concerns the distribution of the C-21 alkene with six double bonds in the algae[353]. This hydrocarbon which is absent from higher plants, was first found in diatoms (Chrysophyta). A more extensive survey showed that it was absent from photosynthetic bacteria and blue-green algae, but present in all the eucaryotic algae except the Rhodophyta. This finding strengthens the suggestion that a close relation exists between the red algae and the blue-greens. It forms up to 90% of the hydrocarbon fraction of the species in the other algal Divisions, although it is only weakly present in the Euglenophyta and the Chlorophyta. Its presence in the Cryptophyta suggests these may be less closely related to the red algae than had previously been supposed (see pp. 146, 190). In the Chrysophyta, the absence of the hydrocarbon from all members of the Xanthophyceae (Heterokontae) whereas it is richly present in the other two families is of great interest for it points to a more definite split in the Division than had been earlier suggested. Obviously more needs to be known about the distribution and function of this highly unsaturated compound before any firm conclusions can be drawn.

As mentioned earlier the acetylenic compounds are found predominantly in the Basidiomycetes and the Compositae. In the fungi these compounds are restricted to the Agaricaceae (12 genera) of the Agaricales and to the Polyporaceae (6 genera) and one genus each of the Thelophoraceae and the Hydnaceae[334]. These last three families are all in the Polyporales. The length of the carbon chain varies from C-8 to C-13 with a preponderance of C-9 and C-10. A variety of end-groups have been found from alcohols, aldehydes, acids and so on. However, no heterocyclic compounds (other than epoxides) nor acetylenes containing an aromatic ring have been detected in the fungi. In contrast, well over fifty acetylenes with such features of chain length C-9 to C-18 have been isolated from the Compositae[334, 335]. Many are substances with thiophene rings although furan- and pyran-containing compounds have also been isolated. The only other families of angiosperms which have a number of acetylene-producing genera, are the Umbelliferae and the Araliaceae although acetylenic acids occur in the Santalaceae, Olacaceae and in one genus each of the Simarubaceae, Malvaceae and Sterculiaceae. With the exception of the last two families, the rest are grouped together in the Subclass Rosidae of the Dictyledonae. The Malvaceae and Sterculiaceae are close together in the Malvales in the related Subclass, Dilleniidae. It seems obvious from their distribution in the Compositae and the more recent orders of the other Subclasses that the ability to desaturate the enoic fatty acids and further transpose them to the mixture of acetylenes found is a relatively advanced one[340]. The presence of relatively simple acetylenes in fungi of generally shorter chain length may be perhaps regarded as an example of parallel evolution. It should be noted, however, that half a dozen compounds are known throughout the Plant Kingdom which have a single acetylenic bond and many more may await detection, so conclusions regarding the evolution of this feature must necessarily be tentative.

11. Terpenoid compounds

(a) Introduction

The terpenoids constitute a large class of compounds all of which arise from the condensation of two or more branched C-5 (isoprenoid) units[21, 33, 34, 355–357]. They are normally divided into mono- (C_{10}), sesqui- (C_{15}), di- (C_{20}) and tri- (C_{30}) terpenes, carotenoids (C_{40}–C_{50}), and polyisoprenoids, including the terpenoid quinones, bactoprenols and the high molecular weight polymers like rubber. Isoprene moieties are often found as *O*- or *C*-substituents in a number of different naturally occurring phenolic compounds such as coumarins and flavonoids[358], and, in addition, are found in a combined form in a number of alkaloids, especially those derived from tryptophan[22, 30–32, 127]. It should also be noted that the triterpenes, by the loss of methyl and other groups and the introduction of other features, give rise to a number of important derivatives including steroids, sapogenins, cardenolides, alkaloids and other classes of physiologically active compounds.

All present-day organisms have the capacity to synthesise one or more groups of terpenoid compound, so there is no doubt that they are the most ubiquitous group of diverse low-molecular substances which originate from a single biosynthetic pathway[359]. Nevertheless it is only the seed-bearing plants which are capable of synthesising the whole range of these compounds. Animals generally possess only a limited ability to make terpenoids and many taxa are incapable of producing those compounds essential to their metabolism. For example, insects rely absolutely on their diet for sterols[360] and no animals can synthesise carotenoids although several phyla utilise them in protective colouring or as part of their visual apparatus[361].

(b) Terpenoid biosynthesis

The major pathway of terpenoid biosynthesis in plants follows the pattern *via* mevalonic acid established in animals[362–364]. The individual steps are stereospecific and the formation of new double bonds involves the retention of designated hydrogen atoms of the primary precursor, depending on the geometry of the molecules formed. The initial branching reaction in which acetyl-CoA condenses with acetoacetyl-CoA to give β-hydroxy-β-methyl-glutaric acid yields, in all cases, only one isomer and must therefore be regarded as evolutionarily stable (Fig. 22). The same is true for the formation of isopentenyl pyrophosphate, its isomerisation to dimethyl-allyl pyrophosphate and the condensation of these two isoprenoid precursors to give geranyl pyrophosphate, the C-10 intermediate. The subsequent formation of the C-15 and C-20 building stones of the other major groups of terpenoids involves the addition of a further molecule of isopentenyl pyrophosphate to the growing isoprenoid in a similar manner as in the initial condensation reaction[362–364]. At each step, the new double bond which is formed is *trans* (except in the case of the biosynthesis of rubber and other polyprenols where it is *cis*) and it is the pro-*S* hydrogen atom from the original C-4 position of the mevalonic acid that is lost. Again, therefore, we may conclude that all these reactions and the enzymes which catalyse them are ancient and conservative. There is, however, little information about the overall control of the chain-lengthening system or of the termination steps. Since most lower organisms do not make mono- or sesquiterpenes, it must be assumed that the reactions from mevalonic acid to the immediate precursors of these groups

References p. 287

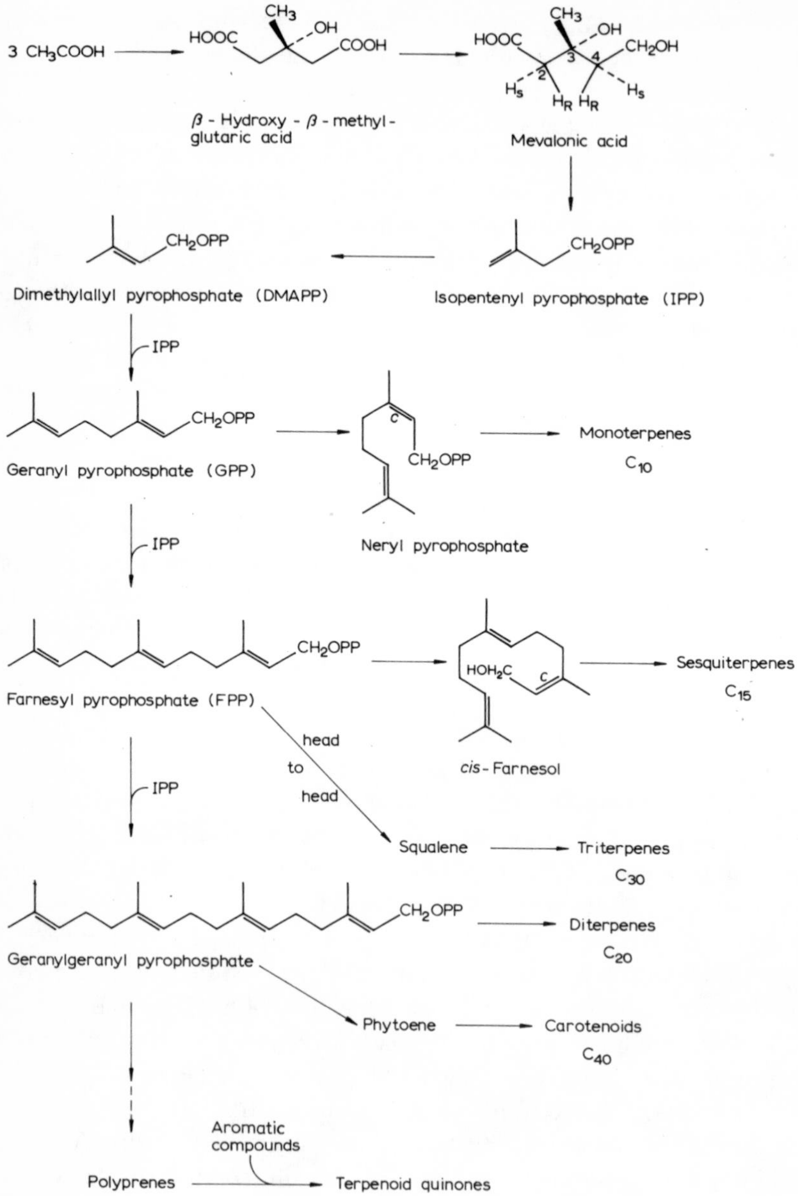

3 CH_3COOH → β-Hydroxy-β-methyl-glutaric acid → Mevalonic acid → Isopentenyl pyrophosphate (IPP) → Dimethylallyl pyrophosphate (DMAPP)

DMAPP —IPP→ Geranyl pyrophosphate (GPP) → Neryl pyrophosphate → Monoterpenes C_{10}

GPP —IPP→ Farnesyl pyrophosphate (FPP) → *cis*-Farnesol → Sesquiterpenes C_{15}

FPP —head to head→ Squalene → Triterpenes C_{30}

FPP —IPP→ Geranylgeranyl pyrophosphate → Diterpenes C_{20}

Geranylgeranyl pyrophosphate → Phytoene → Carotenoids C_{40}

Geranylgeranyl pyrophosphate ⇢ Polyprenes —Aromatic compounds→ Terpenoid quinones

Fig. 22. Biosynthesis of terpenoid compounds.

of compounds and their release from the enzyme matrix are intimately connected. Indeed, there is evidence that tightly bound dimethylallyl pyrophosphate precursors which do not readily label with added [^{14}C]-mevalonic acid are involved in monoterpene biosynthesis in higher plants[364–366]. It also seems likely that an essential step in the formation of cyclic compounds of both classes, and perhaps of diterpenoids also, is the isomerisation of the terminal *trans* double bond to the *cis* form. It is possible that this transformation curtails further chain elongation and effects release of the precursor from the enzyme complex. Thus, neryl pyrophosphate, the *cis* congener of the C-10 geranyl pyrophosphate appears to be the precursor of the cyclic monoterpenes[367,368] while *trans-cis*-farnesol apparently occupies the same role for the cyclic sesquiterpenes[369,370]. However there is little definite evidence for the further transformations which have been proposed to account for the origin of the different groups of these substances by a series of reactions involving carbonium ion intermediates[367–370].

It seems possible that in some cases the oxygen atom of the acyclic precursor may be retained since many mono- and sesqui-terpenoid alcohols and ketones are known with the O atom in the corresponding position to nerol and farnesol.

The diterpenes are also restricted mainly to the higher Divisions of the Plant Kingdom and the biosynthesis of the more widespread polycyclic forms involves appropriate ring formation from the terminal double-bond intermediate, geranyl–linolyl pyrophosphate[371]. The gibberellins (see p. 235) are formed by cyclisation of the primary diterpenoid precursor to give kaurene (XXXIII) which, after oxidation and ring contraction yields gibberellin A_4[372]. The other congeneric compounds are formed from this substance by further oxidation. In all organisms the route to the triterpenoids involves the tail-to-tail condensation of two molecules of farnesyl pyrophosphate *via* a cyclopropane intermediate[363,364] and subsequent reduction with NADPH to give squalene. This compound,

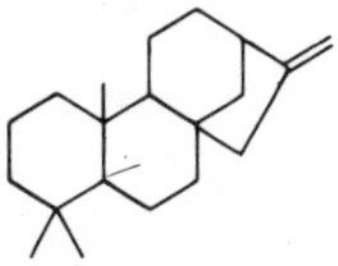

(XXXIII) Kaurene

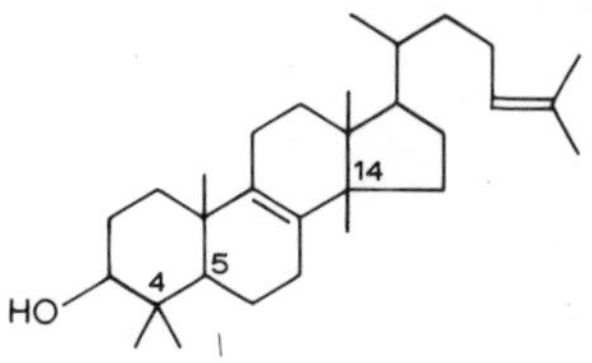

HO 4 5 14

(XXXIV) Lanosterol

partly because of the reduced central bond, can assume a large number of stereochemical configurations each of which, on cyclisation, leads to one of the various tetra- and penta-cyclic triterpenoids such as the steroid precursors and the triterpene acids.

In animals and in fungi, the first observable intermediate on the pathway to the sterols is lanosterol (XXXIV) which is formed by a concerted cyclisation reaction[362–364], which appears to be initiated by the oxidation of squalene to the 2,3-epoxide. This cyclisation is followed by migration of the two methyl groups at C-8 and C-14 to C-14 and C-13 respectively. These migrations may take place *via* cyclopropane intermediates (see cycloartenol below) and involve transfer of the corresponding hydrogen atoms at C-13 and C-20. The common animal sterol, cholesterol, is derived from lanosterol by the loss of the three methyl groups at C-4 and C-14, a shift of the Δ^8 double bond to Δ^5 (*via* Δ^7 and $\Delta^{5,7}$) and reduction of the side-chain double bond at C-24.

The sterols in higher plants and in algae are formed somewhat differently[363,364,373]. Although squalene epoxide is again the precursor and is folded in the same way as in the biosynthesis of lanosterol, cyclisation is accompanied by the elimination of a hydrogen from the methyl group at C-10, rather than that at the ring junction at C-9, leading to the formation of cycloartenol (XXXV). Most plant sterols contain a methyl (or methylene) or an ethyl (or ethylidene) group at C-24. For example, the commonest of plant sterols, sitosterol (previously called β-sitosterol), is 24-ethylcholesterol. The extra alkyl groups all arise from *S*-adenosylmethionine, alkylation taking place before elimination of the methyl groups at C-4 and C-14. 24-Methylenecycloartenol (XXXVI) thus appears to be an obligate intermediate in the biosynthesis of the majority of the plant sterols. The presence of this compound along with cycloartenol itself in Rhodophyta, Phaeophyta and marine Chlorophyta strongly suggests that all photosynthetic organisms use the same pathway for the biosynthesis of their sterols.

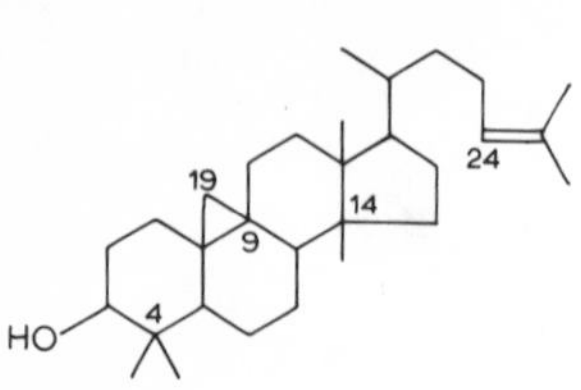

HO 4 9 19 14 24

(XXXV) Cycloartenol

HO CH2

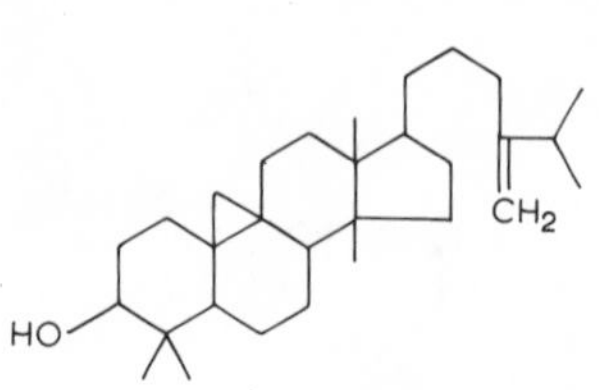

(XXXVI) 24-Methylene cycloartenol

Although fungi differ from other plants in the biosynthesis of sterols, the predominant compound they produce, ergosterol, possesses a methyl group at C-24 like many of the plant sterols. Again, this is derived from methionine *via* a methylene intermediate and alkylation appears to occur prior to the elimination of the extra methyl groups except, perhaps, in yeasts[363,364,374]. Ergosterol also contains a double bond at C-22 as found in a number of green plant sterols such as stigmasterol. However, whereas this bond is formed in fungi by the elimination of the pro-*S* hydrogens of the mevalonic acid precursor, in algae, and presumably in higher green plants, it is the pro-*R* hydrogens which are lost. Many more species need to be investigated before one can conclude that this is a consistent difference between fungi and other plants. A comparison of the existing data on sterol biosynthesis is shown in Table XIII[373,374].

TABLE XIII

Variation of Sterol Biosynthesis in Animals, Fungi and Higher Plants[374]

Feature	*Animals*	*Fungi*	*Higher Plants*
Main sterols	Cholesterol (C_{27})	Ergosterol (C_{28})	Sitosterol (C_{29})
C_{30} cyclic precursor	Lanosterol	Lanosterol	Cycloartenol
Order of methyl group removed	14, 4α, 4β	14, 4α, 4β	varied
24-Alkylation	Absent	Methyl, rarely ethyl	Ethyl and methyl
Double bond (Δ^{22} *trans*)	Absent	By loss of 2- pro-*S* H	By loss of 2 pro-*R* H
Pentacyclic triterpenes	Absent	Rare	Common

The formation of the various pentacyclic triterpenes from squalene results from different patterns of folding of the precursor. The simplest of these triterpenes are those based on hopane (XXXVII) since no methyl or other migrations are involved in the cyclisation. The other classes, such as α-amyrin and taraxasterol are derived by a series of methyl group transformations probably again involving cyclopropane-like intermediates[362–364].

The biosynthesis of the tetraterpenoids, the carotenoids, is dealt with later (see p. 240). The polyprenols contain up to 24 isoprenoid residues, some with *cis* and others with *trans* double bonds[375]. In the dolichols, which are found in animals and in fungi, the terminal residue attached

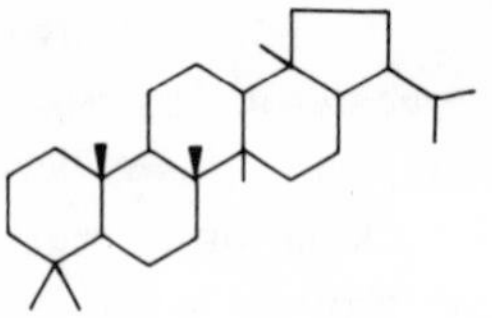

(XXXVII) Hopane

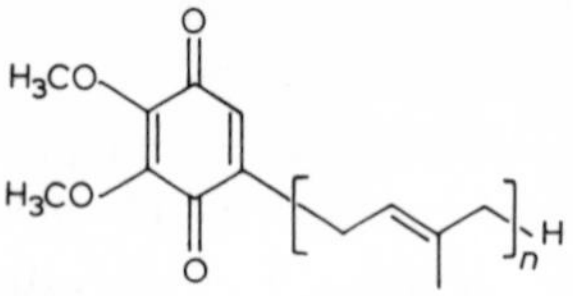

(XXXVIII) Ubiquinone
$n = 6-10$

to the hydroxyl group is reduced. These compounds are all synthesised by the condensation of isopentenyl pyrophosphate, pro-*S* or pro-*R* hydrogen being eliminated to form the *trans* and *cis* bonds respectively[375]. The biologically active quinones have polyprenyl residues which are all *trans* or reduced and are apparently all synthesised in the same way[376]. A polyprenyl group is added to the aromatic nucleus which serves as the precursor of the quinone moiety. For ubiquinones (XXXVIII), the aromatic precursor is *p*-hydroxybenzoic acid which after addition of the polyprenyl group is decarboxylated, oxidised, and *C*- and *O*-methylated. No difference has been found in the mode of biosynthesis in bacteria, fungi, algae and in animals. However, there may be some differences in higher plants since the intermediate polyprenylphenols have not been found. In the case of the plastoquinones, and tocopherols (*e.g.* IV), on the other hand, containing two *C*-methyl groups, one of these is derived from the γ-C-atom of the side-chain of phenylalanine or tyrosine. The ring system in prenylquinones based on naphthalene appear to originate from shikimic acid and a C-3 intermediate of as yet undetermined structure. In all cases, the polyprenyl side-chain is apparently added prior to final elaboration of the quinone moiety.

(c) The function of terpenoids

Terpenoid compounds are not only the most varied groups of natural products in both structure and distribution, but possess the widest range of functions of any group of compounds found in living organisms[359,360,362]. Indeed, it is apparent that their structural variation and range of function are highly interrelated. In both plants and animals, terpenes act as regulators of reproduction, growth and development, as part of electron-transport chains and cell-transport mechanisms, as important constituents of membranes and as attractants or repellents for a variety of other organisms[360,362]. Thus they are capable of eliciting biological responses not only within the cell, but at the level of multicellular organisation and between different organisms. Some, like the steroidal alkaloids, the cardiac glycosides and certain fungal triterpenoids, are extremely toxic to a wide variety of organisms from bacteria to mammals. Indeed the compound batrachotoxin, from the Colombian arrow frog *Phyllobates aurotaenia*, is the most potent neurotoxic agent known[362]. The terpenes are thus of immense importance in the total ecosystem.

The ability of terpenoids to fulfil these many roles lies in the multiplicity of structural forms which each group of terpenoids can assume. For example, there are over one hundred and fifty different monoterpenes known, which vary one from another in the position and number of double bonds, the number of rings and the nature and position of oxygen substitution[377]. Correspondingly fewer numbers of higher terpenes have been isolated in relation to possible forms, perhaps because of certain constraints on structure.

In higher plants, the monoterpenes appear to fulfil at least two roles. The first, and by far the most important, is the attraction of insect pollinators or scavengers and the repulsion of less desirous forms. Scavengers are those insects which help to protect the plant by destroying colonies of pathogenic or saprophytic micro-organisms and lesser, but more dangerous, forms of insect life, such as termites, miners and borers. The overall combination of monoterpenes which the plant contains appears to be more important than the possession of any one compound[360]. Insects are attracted or repelled by a whole variety of monoterpenoids and since most plants contain a complex mixture of such compounds, the overall effect cannot really be ascribed to any single one. For example, at least eight monoterpenes are involved in the complex colonisation of pine bark by *Dendroctonus* species, some only affecting female beetles, others only the male[378,379]. In a similar way, there is accumulating evidence of the synergistic action of the monoterpene sex and alarm pheromones produced by the insects themselves. It should also be noted that most plant species which accumulate large quantities of monoterpenes are unattractive as food for large herbivorous animals and for birds[380]. For example, deer apparently will only eat the young leaves of gymnosperm in which large amounts of terpenoids have not yet accumulated[381].

The second role of monoterpenes in higher plants is their effect as germination inhibitors of the seeds of other species. For example, the seed of many grasses in the Californian chapparal are inhibited from germination by the monoterpenes released or leached from the leaves of *Salvia leucophylla* and *Artemesia californica* resulting in the formation of bare patches round stands of these shrubs[382]. Although other factors, such as rodent foraging range[383], root competition and so on undoubtedly play a part, several experiments have shown that the main cause is most likely the released monoterpenes. There are many similar observations about the inhibition of germination and plant growth by other terpenoid-accumulating species such as hemlock, *Tsuga canadensis*, and *Eucalyptus*[382].

The sesquiterpenes exhibit a wider range of biological activity than do the monoterpenes[369]. They include plant and animal hormones, fungal, bacterial and animal toxins, as well as feeding deterrents. The most important from the viewpoint of higher plants, is the widely distributed hormone abscisic acid (XXXIX) which regulates leaf dormancy and abscission and inhibits seedling growth[384]. A number of other sesquiterpenoid acids and lactones, which have a more limited distribution in angiosperms, have similar growth inhibitory activity and it is likely that many others await detection. Another class of plant hormones is represented by sirenin (p. 247) which is an attractant of male gametes produced by female strains of the water mould *Allomyces*[385]. Again it seems likely that many more such compounds remain to be discovered, both in fungi and higher plants.

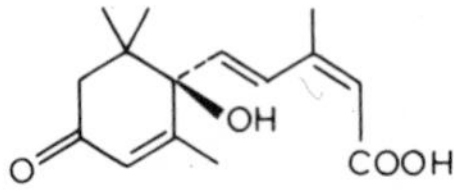

(XXXIX) Abscisic acid

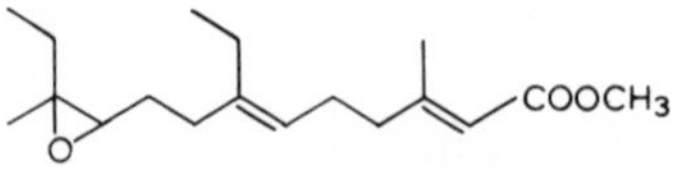

(XL) Juvenile hormone

In the Animal Kingdom, the most important sesquiterpene is the insect juvenile hormone (XL) which controls the maintenance of larval character during successive moults of insects until full growth is achieved[386,387]. This compound contains one extra carbon atom on two of the existing side methyl groups, but many gymnosperms (and some angiosperms) synthesise analogues which, though less active, are sufficiently effective to prevent metamorphosis of insects to the sexually mature adult stage and thus prevent breeding.

There are a large number of toxic sesquiterpenes. Several trees accumulate potent fungicides in their heartwoods and others produce such compounds on infection[369]. For example, the white and sweet potato synthesise the phytoalexins, rishitin (XLI) and ipomeamarone (XLII) respectively, when infected by the rotting fungi to which they are susceptible. Fungi themselves also produce potent sesquiterpene antibiotics which are not only bacteriostatic but are toxic to other fungal species[127]. Many higher plants produce C_{15} compounds which are toxic to higher animals including fish, while other sesquiterpenes deter phytophagous insects[369, 388].

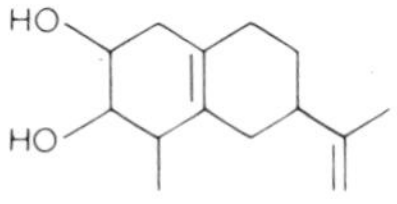

(XLI) Rishitin

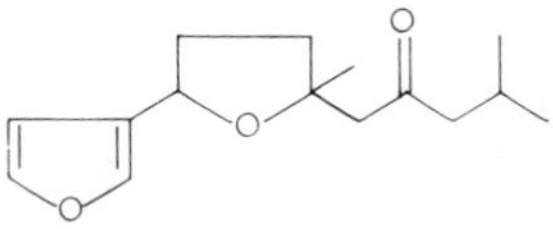

(XLII) Ipomeamarone

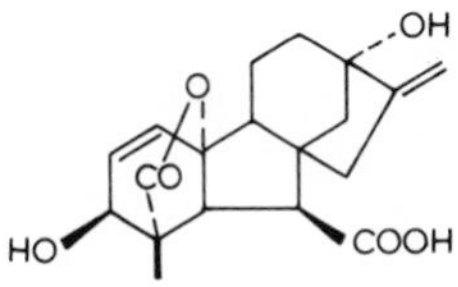

(XLIII) Gibberellic acid (A_3)

In the case of diterpenes, the most important compounds in higher plants are the gibberellins (XLIII), which act as growth substances[372, 389]. However, a large number of diterpene acids exist in conifers and it is highly likely that, as in case of the sesquiterpenes, these have a fungistatic action. The fungi themselves also synthesise a number of diterpenes, including the gibberellins, and a series of modified diterpenes, the trisporic acids, are reproductive agents in *Blakeslea* and *Mucor*[374].

The triterpenoids and their derivatives are probably the most important of all the classes of isoprenoid derivatives. There is little evidence for their

References p. 287

accumulation in all but a few bacteria, but they are found in Cyanophyta[390] and numerous eucaryotic algae[391,392]. In these more ancient Divisions of the Plant Kingdom, they probably only function as part of membrane structures for, as with the terpenoids so far discussed, it is only in the land plants that other functions are clearly expressed[362]. In the Animal Kingdom, on the other hand, triterpenoids are found in quite primitive phyla as toxins which presumably deter the ingestion of the organisms which possess them by their predators. In higher animals, of course, the steroids and their congeners play a large part in control mechanisms of one sort or another[362]. There are few reports of similar activities in plants, although certain fungi apparently require sterols to promote sexual reproduction. It is likely that the sterols found in plants may play some small role in the reproductive process, but their main function may be as defensive substances, especially in relation to the insects[386,387,393].

Most of the steroidal alkaloids and the corresponding cardenolides are extremely toxic to a wide variety of animals. Many plants also synthesise a second insect hormone, ecdysterone (XLIV) or its congeners[393]. This

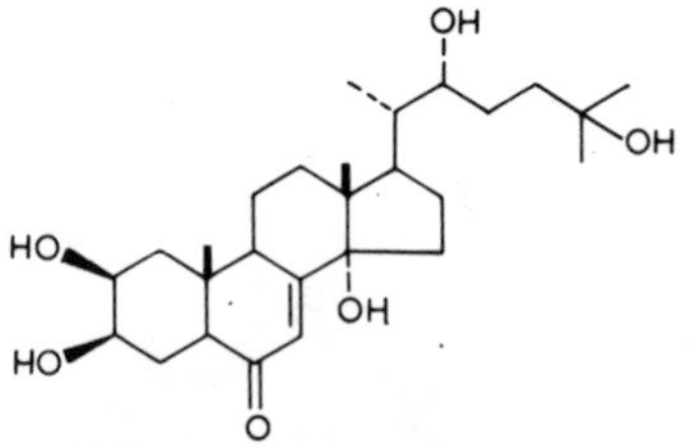

(XLIV) Ecdysterone

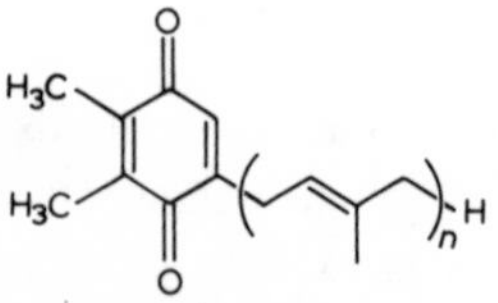

(XLV) Plastoquinone
$n = 8$ or 9

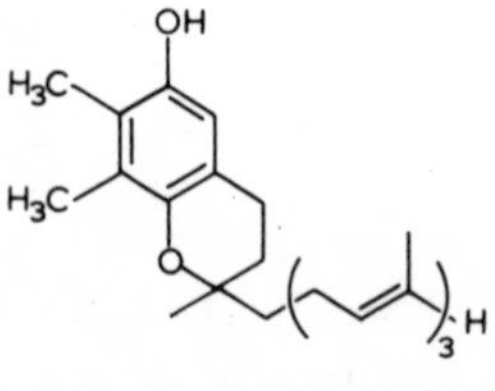

(XLVI) γ-Tocopherol

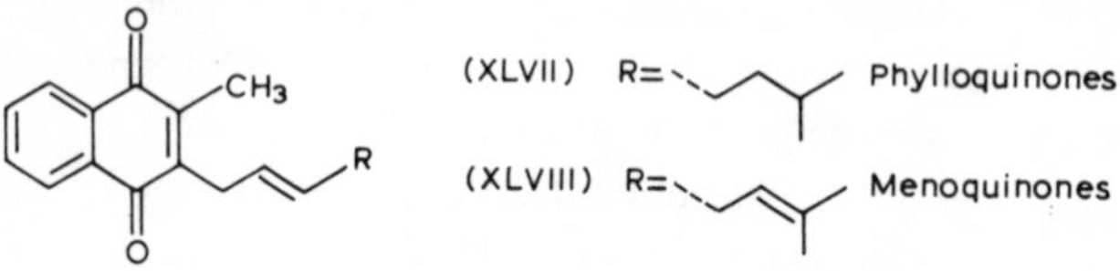

(XLVII) R= Phylloquinones

(XLVIII) R= Menoquinones

steroid controls moulting, the casting off of one hardened exoskeleton and the development of the next, between one larval stage and another during growth of insects. The ability of plants to synthesise the hormone, often in very large quantities, is apparently of advantage, for such species are not noticeably attacked by phytophagous insects.

The distribution of triterpenoids in the Plant Kingdom[394, 395] shows that the more primitive plants, like the bacteria, blue-green algae and the lichens, have hydrocarbons of the hopane type[106] in which no methyl groups have been displaced from their original positions in the C_{30} acyclic precursor. However, there does not seem to be any correlation between the distribution of other triterpenes and their degree of biosynthetic elaboration in the higher plants.

The most important class of tetraterpenoids, the carotenoids, is discussed separately below, and the final classes of terpenoid to be considered here are the terpenoid quinones and polyprenols. The quinones all contain an aromatic ring with a terpenoid side-chain having usually 6–10 isoprenoid units. In all cases, they appear to be actively involved in electron transport[396]. Their distribution (Table XIV) suggests that different classes may occupy the same role in diverse organisms, but their ubiquitous distribution indicates that they are required in all actively metabolic cells.

TABLE XIV

Distribution of Terpenoid Quinones

	Quinone				
	Ubi-quinones (XXXVIII)	*Plasto-quinones (XLV)*	*Toco-pherols (XLVI)*	*Phyllo-quinones (XLVII)*	*Meno-quinones (XLVIII)*
Procaryotes					
Bacteria					
Gram positive	−	−	−	−	+
Gram negative	+	−	−	−	+
Blue-green algae	−	+	(+)	+	−
Eucaryotes					
Fungi	+	−	−	−	−
Green algae	+	+	+	+	−
Higher plants	+	+	+	+	−
Animals	+	−	+	−	+

There is now good evidence that the polyprenols in bacteria are involved in the transport and assembly of the *N*-acetylmuramic acid peptides which constitute the cell wall (see p. 198)[373]. In higher plants, it appears likely that the prenols are important in membrane function especially in the mitochondria. Whether this indicates a structural or active role is not clear, so little can be said about their possible changing function during the course of evolution.

(d) The distribution of terpenes

The distributions of the various classes of terpenoid compounds in the Plant Kingdom is set out in Tables XIV and XV. It can be seen that while steroids, carotenoids and the terpenoid quinones are universal in plants, the relatively simpler mono-, sesqui- and di-terpenes have not been found except in isolated instances in Divisions lower than the Tracheophytes. Unfortunately, as indicated above, too little is known about the control mechanisms of biosynthesis of the terpenoids to allow more than speculation as to the reason for this state of affairs. It seems highly probable that the most primitive terpenoid was a linear polymer containing a terminal OH and 6–20 isoprene units which could be used to transport polar metabolites across lipid membranes and later be added to

TABLE XV

The Distribution of Terpenes in Plants[a]

Division	*Terpenoid*[b]				*Sterols* C_{21}–C_{27}	*Carotenoids*[c] C_{40}	*Quinones* $(C_5)_n$
	Mono- C_{10}	*Sesqui-* C_{15}	*Di-* C_{20}	*Tri-* C_{30}			
Bacteria				(+)	(+)?	+	+
Cyanophyta				(+)	(+)	+	+
Rhodophyta	(+)	(+)			+	+	+
Phaeophyta		(+)			+	+	+
Chlorophyta	(+)				+	+	+
Fungi	(+)	(+)	+	+	+	+	+
Bryophyta		+		+	+	+	+
Tracheophyta	+	+	+	+	+	+	+

[a] Many individual species do not contain representatives of each class of terpenoid shown.
[b] (+), means reported in only one or two species.
c See also Table XIV (p. 237).

a variety of aromatic nuclei so that these could act as electron-transport substances through organelle membranes. It was not until control mechanisms were evolved, probably in the late-Precambrian, that it was possible to divert this polyisoprene synthesis to the production of other terpenoid types. The first change was, perhaps, the development of an enzyme system for the head-to-head condensation of geranyl-geranyl pyrophosphate to give phytoene[397] and thence the carotenoids (see p. 240). A further development of this type of enzyme system involved condensation of farnesyl pyrophosphate to a cyclopropane intermediate followed by reduction which led to squalene which could be folded and cyclised to give the various triterpenoid and steroidal compounds of such importance to the Animal Kingdom. The ability to synthesise mono- and sesqui- and di-terpenes probably, as suggested earlier, awaited enzymes to catalyse the isomerisation of the terminal bond of the precursor from *trans* to *cis*. (Fig. 22).

Further elaboration of each class of terpenoid compound then took place depending on selection. Thus it seems likely that the immense number of mono-terpenes found in the present seed-bearing plants as against their virtual absence in ferns, must have arisen with the evolution of the present classes of insect pollinators. Conversely, the presence of large amounts of ecdysone-like compounds (XLIV) in ferns perhaps shows that the early Carboniferous arthropods had steroidal moulting hormones of this type.

12. The carotenoids

(a) Introduction

The carotenoids are mainly tetraterpenoid compounds (C_{40}) and they are widely distributed in Nature[398,399]. Four of them, β-carotene (VIII), lutein (IX), violaxanthin (X) and neoxanthin (XI), co-occur in the chloroplasts of all higher plants. The photosynthetic organelles of lower organisms also usually contain one or more of these four compounds although they are not necessarily the major carotenoid present. The carotenoids are also found in non-photosynthetic tissues. They occur predominately as the yellow, orange and red colouring matters in the flowers and fruits of higher plants. They are also found in fungi, in aerobic bacteria and, as pigments, in a wide range of animals from insects and lobsters to birds and mammals[361].

Carotenoids are, however, synthesised only by members of the Plant Kingdom[400–402]. Animals obtain their carotenoid pigments ultimately from plants but, with the exception of mammals, usually transform them after ingestion by oxidation or by complexing them with proteins to yield the unique colouring matters which are used for display and concealment.

(b) The biosynthesis of carotenoids

In plants, the biosynthesis of carotenoids takes place in the plastids or chromatophores. The initial precursor is geranylgeranyl pyrophosphate (GGPP, Fig. 22) which undergoes a "tail to tail" condensation to yield phytoene (Fig. 23)[397]. It should be noted that unlike the condensation of farnesyl pyrophosphate to give squalene and the C_{30} terpenoids, the formation of phytoene involves the introduction of a *trans* double bond at the junction. This constrains the folding of the resulting C_{40} molecule and thus leads to acyclic products rather than polycyclic ones. It is also possible that the precursor molecules (GGPP or its precursors) undergo some desaturation prior to condensation [401] which would further hinder the possibility of ring formation which is so characteristic of most other classes of terpenoid compound.

Regardless of the organism in which it is synthesised, GGPP is an all-*trans* compound. The same is true of almost all naturally occurring carotenoids. Although 15-*cis* phytoene has been isolated from carrot oil, it may not have been part of the normal biosynthetic pathway. Thus the all-*trans* compound was obtained from the bacterium *Flavobacterium dehydrogenans*. In tomato fruits and those of *Physalis alkenlengii* it has been shown that the biosynthesis of phytoene involves the loss of the pro-5*S* hydrogen atoms of mevalonic acid and thus the bond is likely to be *trans* in these cases also. The formation of the C_{40} molecule from GGPP expectably does not require any NADPH as does squalene from farnesyl pyrophosphate. The only absolute need is for the presence of sulphydryl reagents and for Mg^{2+} (or Mn^{2+}) when cell-free systems are used[400,401].

Phytoene, which has nine double bonds, undergoes a series of desaturations at conjugated centres (positions 1, 2 and 3 in Fig. 23) to yield phytofluene, ζ-carotene and neurosporene respectively. These desaturations involve, as may be expected, an overall *trans* loss of hydrogens from the adjoining carbon atoms. FAD is an absolute requirement in all cases studied and it seems possible that the reactions proceed in a like manner to those giving rise to the unsaturated fatty acids (see p. 216).

Neurosporene can either undergo a further dehydrogenation (at 4) to give lycopene, the tomato skin pigment, or undergo one of two cyclisation reactions which yield a six-membered ring at one of the terminal positions. One cyclisation gives rise to an α-type ring with a 4,5-double bond and the other gives a β-ring with a 5,6-bond. Again, FAD is an absolute requirement and NADP when also present appeared to lead to better yields of β-carotene in spinach leaf preparations. The monocyclic carotenoids formed from neurosporene in this way, α- and β-zeacarotene, can then be further desaturated (at 4) to give δ- and γ-carotene respectively. The latter compounds may also arise by cyclisation from lycopene (Fig. 23).

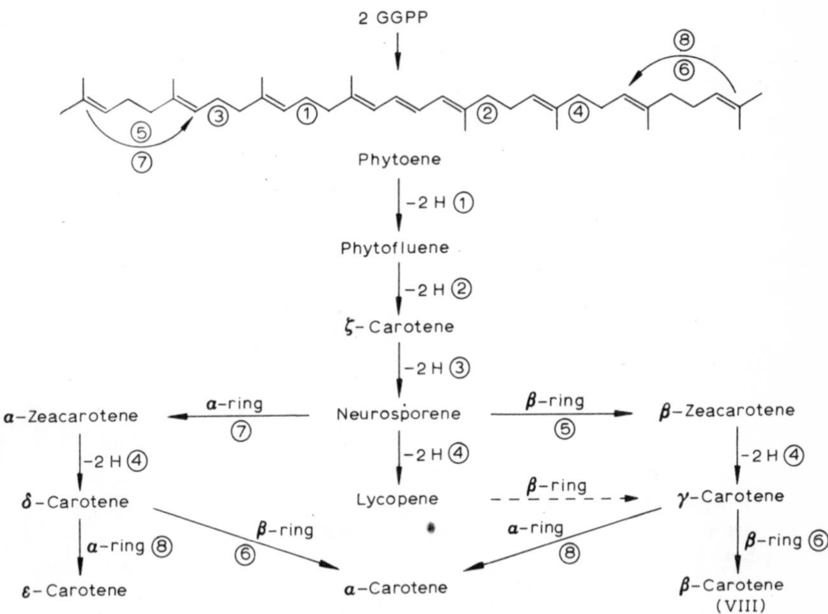

Fig. 23. Biosynthesis of carotenoids.

On further α- or β-cyclisation they give rise to the rare ε-carotene and the more common α- and β-carotenes respectively.

Many of the naturally occurring carotenoids contain hydroxyl groups at C-3 in one or both rings. Open-chain carotenoids having hydroxyl groups in equivalent positions do not occur, the hydroxylycopenes which are found in the photosynthetic bacteria being substituted in the 1 and 1′ positions as in spirilloxanthin (IL). The normal xanthophylls of higher plants arise by the action of a mixed function oxidase with NADPH (and possibly FAD), the hydroxy group coming from molecular oxygen rather than from water. The stereochemistry of the hydroxylation has been demonstrated to involve the elimination of the pro-5*R* hydrogen at the 3-position in both bacteria and higher plants, showing that homologous enzymes are involved[400,401].

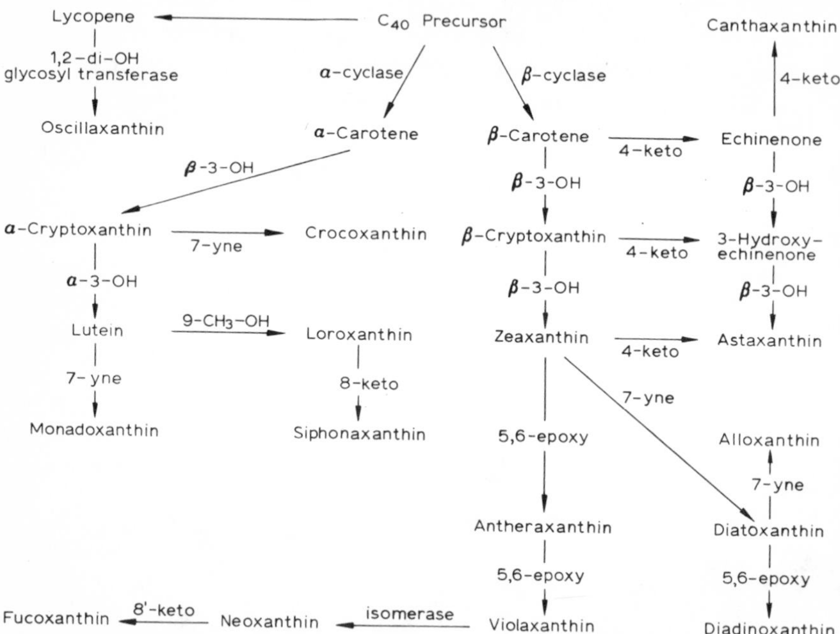

Fig. 24. Biosynthetic relationships of carotenoids. The enzymes catalysing the individual biosynthetic steps are abbreviated as follows: α- and β-cyclase—the enzyme catalysing the formation of α- and β-six-membered terminal rings as in α-carotene and β-carotene respectively: α- and β-3-OH, the hydroxylase for inserting an hydroxyl group in the α- or β-ring; 1,2-di-OH, enzyme for the formation of 1,2-dihydroxylycopene, possibly *via* the epoxide: glycosyl transferase, for the formation of 2-*O*-glycosides from the dihydroxy lycopene, presumably from the diphosphonucleotide sugar: 9-CH_3-OH, for the hydroxylation of the methyl group at C-9 (to β-ring) to yield a CH_2OH group: 4-keto, the system introducing a carbonyl group at C-4 of the β-ring, presumably acting *via* hydroxylation followed by subsequent dehydrogenation: 8-keto, the introduction of an (enolised) hydroxyl group at C-8 to the β-ring: 5,6-epoxy, the reversible epoxidation of the 5,6 double bond in a β-ring: isomerase, for the isomerisation of the 5,6-epoxide to a 5-hydroxy-7,8-diene: 7-yne, the desaturation of the double bond at C-7 to the β-ring to give a triple bond; it is possible, but unlikely, that this involves dehydration of the 8-(enol)-hydroxy compound.

The xanthophylls are transformed into the corresponding 5,6-epoxides in chloroplasts when the latter are strongly illuminated in the presence of oxygen. The reverse process takes place in the dark, but it is not generally agreed that these reactions have anything to do with photosynthesis[399]. In this way, the 3,3′-dihydroxy-β-carotene, zeaxanthin, is transformed into di-epoxy derivative violaxanthin (X). Under the action of an isomerase, this latter compound gives rise to neoxanthin (XI), but the actual nature of the reaction or the co-factors required are not known.

The hydroxylation of β-carotene at the 4-position probably involves a mixed function oxidase, but the formation of the 3,4-dihydroxy grouping present in 3,4,4′-trihydroxy-β-carotene from *Chlorococcum wimmeri* and other green algae, may involve an intermediate 3,4-epoxide. Desaturation of the 7,8-double bond in zeaxanthin to give the acetylenic compound diatoxanthin (VI) may be direct as has been suggested for the biosynthesis of crepenynic acid from linoleic acid, or it may involve the sequence *via* the allene neoxanthin (XI) which then rearranges with the elimination of the hydroxyl group to give the acetylenic bond in the 7,8-position. In-chain oxidations which give rise to the 8-keto carotenoids like fucoxanthin (V) may also involve an epoxy intermediate. In this case, the 5,6-epoxide might rearrange to give the 5,8-furan oxide (a reaction which occurs readily *in vitro*) and this could readily give rise to the 8-carbonyl group. This possibility seems more plausible than by the hydroxylation of the corresponding acetylenic compound. No acetylenic compounds were detected in *Codium fragile* which synthesises the 8-keto compound siphonoxanthin (L)[399]. The hydroxylation of the methyl group at C-9 of the latter compound presumably precedes the introduction of the carbonyl group (see Fig. 24).

(c) The function of carotenoids

Although β-carotene and other carotenoids are intimately associated with the chloroplast in all green plants and with the chromatophores in the photosynthetic bacteria, it appears unlikely that they play any part in photosynthesis in higher plants[399]. It is generally agreed, however, that fucoxanthin (V), the major carotenoid in the chloroplasts of brown algae (Phaeophyta), is implicated in the process, and transfers energy in photosystem II (see p. 189)[403]. From fluorescence measurements it has been concluded that β-carotene is involved in photosystem I in blue-green and red algae while in some green algae it may be involved in both systems[399]. None of the xanthophylls which are also present in the chloroplasts appear to be involved.

As mentioned earlier, there is evidence that the formation of the 5,6-epoxy group in violaxanthin is catalysed by light in the chloroplast[399]. This reaction may happen with other epoxides such as diadinoxanthin and is possibly coupled to the electron-transport chain between reduced plastoquinone and oxidised chlorophyll a_1, perhaps utilising some of the ATP produced by photosystem II. The importance of this reaction is not clear and is obviously not the only way in which epoxides are produced since they are present in high concentration in flower petals which contain no chloroplasts.

A perhaps more important function for carotenoids in photosynthetic tissues, especially bacteria, is that they protect cells against destruction in the presence of light and oxygen by what is termed photodynamic sensitisation[399]. Carotenoid-free mutants of *Rhodopseudomonas spheroides* and *Chlorella* have been shown to be readily killed under these circumstances and the protection afforded by carotenoids is probably due to their being able to neutralise singlet oxygen which is presumed to be responsible for the effect. It seems likely that carotenoids have a similar effect in the chloroplasts of higher plants; at least no plant has been found with carotenoid-free chloroplasts[359,399].

Carotenoids have also been implicated in some of the phototropic and phototaxis effects in higher plants. Much of the evidence lies in the observations of action spectra of these effects but the mode of action of the pigments is unknown[404]. The presence of specific carotenoids in the eye spot of the *Euglena*, however, certainly indicates their probable importance[399]. Carotenoids have been suggested to be important in reproductive processes in fungi. Here it seems more likely that their association with other terpenoid compounds which are the true sex hormones, is the reason for their accumulation (see p. 247)[374].

In non-photosynthetic tissues, carotenoids are of course important purely as colouring matters[395,405]. This is true of their presence in both animals and birds as well as in plants[361]. Most of the carotenoids in flower petals are yellow in colour and this has been shown to be most attractive to Lepidoptera. In fruits, on the other hand, more highly conjugated red or orange-red carotenoids occur which are more easily seen by birds[1]. By far the most important suggestion which has been made about the function of carotenoids in non-photosynthetic tissues is their involvement in the formation of spore and pollen exines[406]. This role, which has far-reaching phylogenetic implications is dealt with in a separate section (see p. 252).

(d) The distribution of carotenoids

In higher plants, as was pointed out earlier, the chloroplasts all contain β-carotene (VIII), lutein (IX), violaxanthin (X) and neoxanthin (XI), often with varying amounts of α-carotene and several xanthophylls[398,405]. It has been suggested that this fact strongly supports the premise that the Bryophyta and Tracheophyta were derived from a common ancestor, possibly in the Chlorophyta. However, it can equally be used as evidence

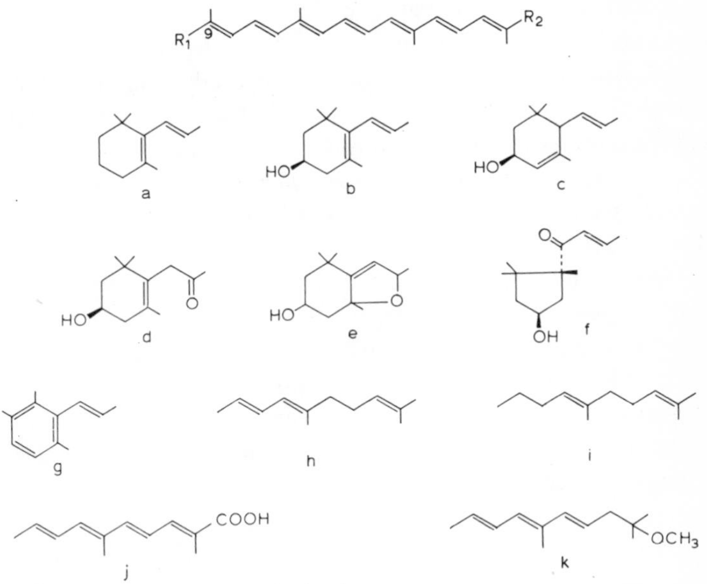

(IL)	Spirilloxanthin	$R_1=R_2=k$
(L)	Siphonoxanthin	$R_1=d$ and 9-CH_2OH, $R_2=c$
(LI)	Auroxanthin	$R_1=R_2=e$
(LII)	Rubixanthin	$R_1=b$ $R_2=h$
(LIII)	Capsanthin	$R_1=b$ $R_2=f$
(LIV)	Capsarubin	$R_1=R_2=f$
(LV)	Spheroidene	$R_1=k$ $R_2=i$
(LVI)	Chlorobactene	$R_1=g$ $R_2=h$
(LVII)	Torularhodin	$R_1=a$ $R_2=j$

that the chloroplasts of all eucaryotes were derived from a cyanophytic endosymbiont. Unfortunately, this latter suggestion falls down when the higher algae are considered (see below)[399]. The main variation in carotenoid pattern found in higher plants is found in flowers and fruits, where the concentration of pigments can rise to relatively large amounts[405]. Even so, there are few unique compounds. Flowers usually contain large amounts of xanthophylls, especially the 5,8-epoxides like auroxanthin (LI), although some hydrocarbons are found in orange coloured petals. There are a number of petal carotenoids which are very restricted in their distribution. For example, eschscholtzxanthin, a *retro*-zeaxanthin in which the double-bond system has been altered. Similar considerations apply to fruit pigments. Some fruits synthesise acyclic compounds like lycopene, others have a gamut of normal chloroplastic components, while yet others contain certain unique carotenoids like rubixanthin (LII) which occurs only in fruits of *Rosa* species. *Capsicum annuum* contains the keto-carotenoids capsanthin (LIII) and capsorubin (LIV) whose biosynthesis obviously involves an extra-oxidative step from β-carotene than does the production of the bulk of the xanthophylls[405]. In general, however, the biosynthesis of the unique compounds in flowers and fruits involves no unusual steps. Lycoxanthin, for example, which occurs in *Solanum dulcamara* is presumably formed from lycopene by an oxidative hydroxylation in a similar way to α-cryptoxanthin (Fig. 24)[401].

Bacteria contain few of the carotenoids found in higher plants. β-Carotene and other cyclic carotenoids are almost unknown and there are none of the hydroxy and epoxy compounds typical of higher plant chloroplasts[407–409]. Certain Gram-negative non-photosynthetic bacteria, like *Flavobacterium dehydrogenans*, contain carotenoids with one or two extra isoprenoid groups attached to the terminal ring system, such as decaprenoxanthin. Over ten of these C_{45} and C_{50} compounds are now known, including compounds with α-type rings and some based on hydroxylycopene[402,407–409]. In the photosynthetic purple bacteria, Athiorhodaceae, several acyclic carotenoids are known, all derived from lycopene or reduced lycopene by desaturation and hydroxylation usually followed by *O*-methylation. Examples are spheroidene (LV) from *Rhodopseudomonas spheroides* and spirilloxanthin (IL) from *Rhodospirillum rubrum*. Some species of purple bacteria are also able to effect oxidation of in-chain methyl groups to both alcohol and to aldehyde as in rhodopinal. Several other bacteria can form glycosides of the introduced hydroxyl groups. Such compounds are rare in higher organisms. Members of the Thiorhodaceae and Chlorobacteriaceae contain, in addition, carotenoids with aromaticised terminal rings, such as chlorobactene (LVI) from *Chlorobium* species[407–409]. Nothing is known about the biosynthesis of these aromatic rings but the fact that one of the two methyl groups from the C-1 of the presumed alicyclic precursor has been transfered to C-2 suggests a cyclopropane intermediate.

Fungi, unlike the bacteria, contain mainly carotenoid hydrocarbons such as lycopene (Fig. 23) and α- and γ-carotene. However, β-carotene, which is universal in green plants, is absent from many Fungi and the common xanthophylls, lutein, violaxanthin and neoxanthin have never been detected. Some fungi, however, do have the ability to hydroxylate β-carotene to give β-cryptoxanthin and zeaxanthin (Fig. 24). There are a few novel compounds in the Fungi, but some contain some novel acidic pigments like torularhodin (LVII) from *Rhodotorula gracilis*, which is obviously derived from γ-carotene[359]. One interesting feature is the fact that often widely different amounts of carotenoids are present in dimorphic + and − mating types. As mentioned earlier, however, there is no real evidence that such differences are significant or that carotenoids are involved in other sexual processes. It seems more likely that the presence or absence of carotenoids reflects the formation of terpenoid sex hormones such as trisporic acid (LVIII)[374], from *Blakeslea trispora* and *Mucor mucedo*, and sirenin (LIX) from various members of the Allomyces[385].

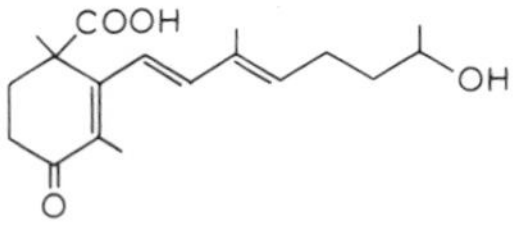

COOH
OH
O
(LVIII) Trisporic acid

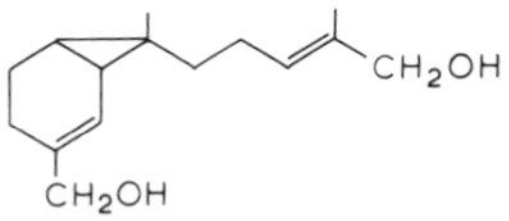

CH_2OH
CH_2OH
(LIX) Sirenin

As has been pointed out on several occasions by Goodwin, it is the algae, including the Cyanophyta, which show the largest variation in carotenoid distribution and moreover, variations which have an apparent phylogenetic significance[359]. The major algal carotenoids and their biosynthetic interrelationships are shown in Fig. 25. The individual steps outlined in the scheme indicate that thirteen different types of enzyme-catalysed reactions occur; α- and β-cyclisations, hydroxylations at C-3 of both α- and β-rings, 5,6-epoxidation, 5,6-epoxide isomerisations, hydroxylations at the methyl groups at C-4, C-8 and C-9 and, in open-chain compounds, at C-1 and C-2, desaturation of the double bond to give an acetylenic

TABLE XVI

The Distribution of Biosynthetically Significant Carotenoids in the Algae.

Division	*Compound present*[a]							
	Hydroxy-lycopene glycoside	*α-Carotene*	*Viola-xanthin*	*Neo-xanthin*	*Fuco-xanthin*	*Diadino-xanthin or allo-xanthin*	*Loro-xanthin or siphono-xanthin*	*Echin-enone*
Cyanophyta	+							+
Euglenophyta				+		+		(+)
Chlorophyta		+	+	+			+	(+)[b]
Rhodophyta		+	(+)					
Cryptophyta		+				+		
Pyrrophyta					+	+		
Phaeophyta					+	(+)		
Chrysophyta		(+)	+		+	+		

[a] All divisions have β-carotene and one or more carotenoids having a 3-hydroxy group: + means common or major compound and (+) means found rarely in the species examined.

[b] Extra plastid only.

bond at C-7, glycosyl transferase reactions, and the oxidation of the C-3 and C-4 hydroxyl groups to carbonyls. It is highly likely that each individual reaction of the same type, such as the hydroxylation of the C-3 of β-rings is catalysed by a separate enzyme and that even in closely related taxa, the enzymes catalysing the same step will show minor variations in primary structure. However, the major algal divisions show a remarkable uniformity of reaction sequences. For example, the Phaeophyta contain no carotenoids with an α-type ring system. The distribution in the algae of all the compounds shown in Fig. 24 has been given in several earlier publications[359]. For the present purpose it is only necessary to consider the distribution of the biosynthetically significant carotenoids. That is, of compounds whose presence in a given taxa indicates the possession of certain key types of enzymic activities. Such compounds are listed in Table XVI, and the implications of their distribution in relation to the presence of enzymes concerned in carotenoid biosynthesis is shown in Table XVII.

On the basis of such data, Goodwin[359] has suggested that the algal Divisions are interrelated as shown in Fig. 25. This is a modification of earlier schemes which he had put forward. In the scheme shown, the green bacteria, rather than the Cyanophyta, are suggested to give rise to the Euglenophyta (although this may be in error since it is apparently

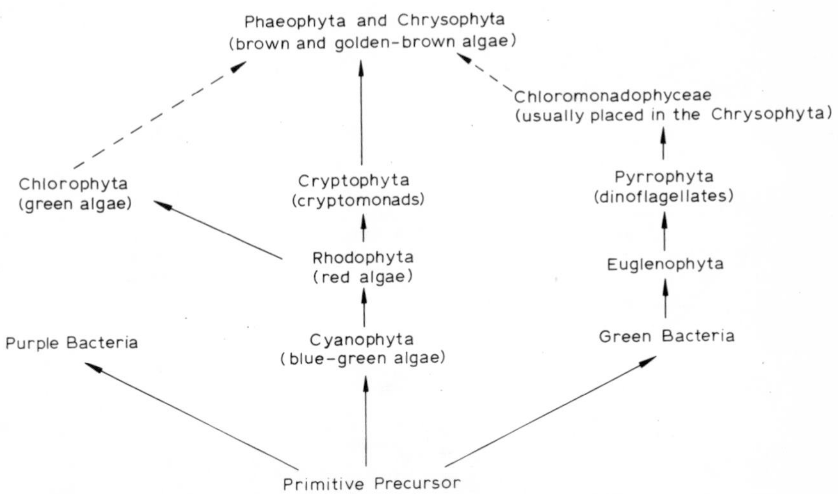

Fig. 25. Carotenoid evolution in algae according to Goodwin[359].

TABLE XVII

Distribution of the Enzymes of Carotenoid Biosynthesis in the Algae

Division	Enzyme[a]							
	α-Cyclase	*Hydroxylase*	*Keto-ase*		*5,6-Epoxydase*	*Epoxyisomerase*	*7-Desaturase*	*Other*
			4-	8-				
Cyanophyta		+[b]	+					Glycosidase
Euglenophyta			+		+	+	+	
Chlorophyta								
Siphonales	+	+[c]		+	+	+		
Others	+	+[c]	+[d]		+	+		5,8-Epoxydase
Rhodophyta	+				(+)			
Cryptophyta	+						+	
Pyrrophyta				+	+	+	+	
Phaeophyta				+	+	+	(+)	
Chrysophyta	(+)			+	+	+	+	

[a] All divisions have a α-cyclase and a 3-hydroxylase.
[b] Hydroxylating lycopene at C-1 and C-2 (γ-carotene numbering).
[c] Hydroxylating the methyl group attached to C-9: only present in a few genera.
[d] Extra-plastid carotenoids only.

at variance with the accompanying text), the latter group lead to the Pyrrophyta, which together with the Chlorophyta give rise to the Phaeophyta and Chrysophyta (which are not separated) and the Chrysophyta are now believed to be related phylogenetically to the Chlorophyta.

Consideration of the data given in Table XVII and the affinities shown in Fig. 25, however, show a number of anomalies in the proposed evolutionary scheme. For example, if the Rhodophyta were ancestral to the Chlorophyta, the latter needed to re-acquire the enzyme systems for the production of the 4-ketocarotenoids lost from the Cyanophyta. Similarly the typical α-cyclase type enzymes of the Rhodophyta and Chlorophyta must be lost if there is a connection between the latter and the Phaeophyta. The relation between the Cryptophyta and the Phaeophyta also seems obscure, since the typical Cryptophyta enzyme system producing α-carotene is lost and a wide number of others gained. It appears likely

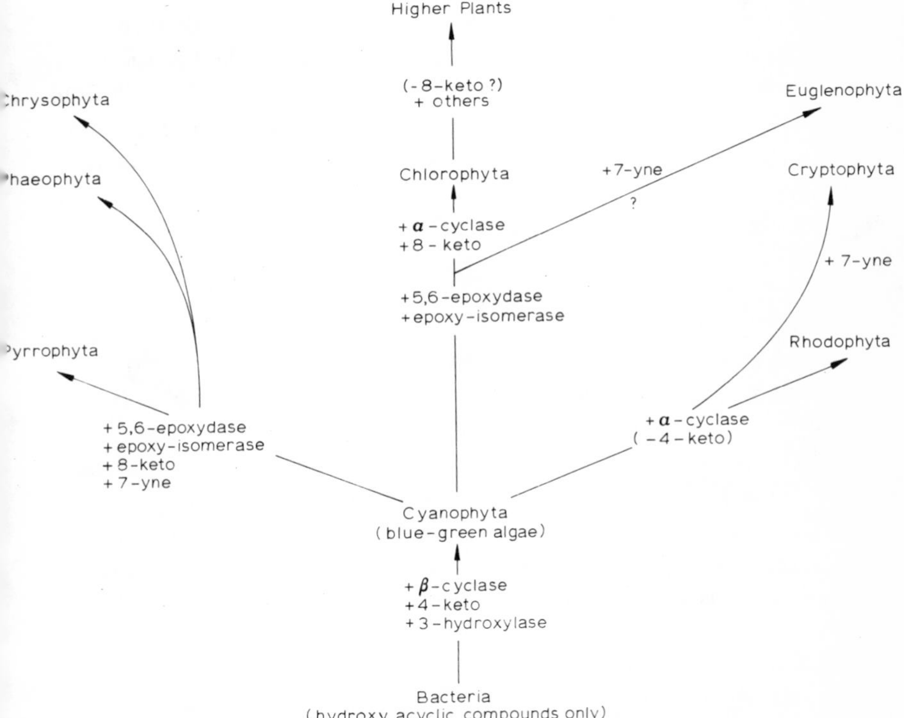

Fig. 26. Evolution of steps in carotenoid biosynthesis in the Plant Kingdom. For enzyme abbreviations, see Fig. 24.

References p. 287

that such difficulties would be more or less overcome if the interrelationships between the major algal divisions outlined earlier were adopted (Fig. 1). This leads to the scheme shown in Fig. 26 and comparison with the enzyme distribution shown in Table XVII, indicates that this grouping is more natural as far as the evolution of the algal carotenoids is concerned. The Rhodophyta and Cryptophyta form one line from the presumed Cyanophycean ancestors. These divisions have acquired only the α-cyclase activity and have lost the ability to hydroxylate lycopene and other acyclic compounds, which presumably was derived from the bacteria and is exclusive to procaryotes, and to form 4-ketocarotenoids. The Chlorophyta form another branch, which might lead to the Euglenophyta, which have retained the capability to synthesise echinenone-like compounds, but have in addition acquired a number of more advanced eucaryotic features such as epoxidation and epoxy-isomerisation, besides sharing with the red algae α-cyclase activity. The range of carotenoid-associated enzyme activities in the Chlorophyta is like that in land plants and suggests a close link between the green algae and the Bryophyta and the Tracheophyta. The Pyrrophyta, Phaeophyta and Chrysophyta form a third branch from the procaryote ancestor sharing in common the ability to desaturate the double bond at C-7 to give acetylenic carotenoids and the systems to produce fucoxanthin (V).

(e) Spore and pollen walls

The external walls of the pollen and spores of most present day higher plants show a unique diversity of morphological structure and are extremely resistant to attack by bacteria and fungi[406,410–415]. The walls are not readily oxidised and are more or less impenetrable to water. As a result, both spores and pollen have been preserved in the soil for remarkably long periods, almost intact. Indeed, much of our knowledge regarding Angiosperm phylogeny rests on the recognition of characteristic pollen remains in the fossil record as far back as the early Cretaceous[230]. Fossil remains of gymnosperm pollens are found much earlier, being prominent in Carboniferous coal deposits, while spores of ferns, lycopods and bryophytes occur along with the earliest record of land plants[230]. It is recognised that a number of these early fossil spores may not have arisen from land plants at all but are algal in origin. For example, many of the so-called hystrichospheres of the fossil genus *Tasmanites* found in several localities throughout the world in deposits from 230 to 250 million years of age, appear more akin to the spores of the dinoflagellates (Chrysophyta, see Table I)[230,414–416]. Other fossil spores may be related to the non-motile resting stage of several algae like the Prasinophyceae (Chlorophyta) but these suggestions need to be confirmed by microchemical and other analyses[416].

The chemistry of the exine (cell wall) of angiosperm pollen has attracted attention since the early nineteenth century, and the chemically resistant residue left after extracting pollen with ethanol and dilute alkali was called pollenin by John in 1814[411,417]. Zetzsche described the corresponding residue from pteridophyte spores, sporinin, and later recognising the similarity of the two, coined the term sporopollenin to cover both. This term refers to the material of the outer coat of both spores and pollen freed from polysaccharide by treatment with strong mineral acid[417]. It was readily established that sporopollenin was a highly unsaturated polymer containing hydroxyl and *C*-methyl groups which, on ozonisation, yielded a mixture of simple straight and branched mono- and di-carboxylic acids. These results, however, gave no real clue as to the essential structure of the polymer or of its mode of formation[411,417].

A plausible solution to this problem has been recently advanced by Shaw and his collaborators. They have shown that carotenoids and the fatty acid esters of xanthophylls, present as the dominant lipid component in the developing pollen grains of many plants, can be oxidatively polymerised in the presence of ionic catalysts to give a product with many properties in common with sporopollenin. These include molecular formula, the proportion of dicarboxylic and branched and straight chain mono-acids produced on ozonisation, I.R. spectra, the products formed by pyrolysis on gas chromatograms, and the pattern of phenolic acids produced on alkali fusion[411,417].

Comparison of the products prepared by oxidation of various pure carotenoids showed that, depending on the starting material, the synthetic polymers gave the same range of molecular formulae as natural sporopollenins from a variety of sources (Table XVIII). The I.R. spectra, pyrolysis gas chromatograms and alkali fusion and ozonolysis products were also similar. It is striking, for example, that the sporopollenin from the sexual spores of *Mucor mucedo* corresponds in every respect to the polymer from β-carotene since this is the almost exclusive carotenoid produced by the mated strains of this fungus[417].

TABLE XVIII

Composition of Sporopollenin from Various Sources[a]

Source of sporopollenin	*% in cell wall*	*Composition on a $C_{90}H_xO_y$ basis*[d]	
		X	*Y*
Fossil			
Tasmanites punctatus (250 million yrs old)	—	133	11
Chlorophyta[b]			
Chara corallina	41	117	28
Fungi			
Aspergillus niger	44	115	10
Mucor mucedo[c]	45	130	33
Lycopsida			
Lycopodium clavatum	93	144	27
Selaginella kraussina	91	124	18
Sphenopsida			
Equisetum arvense	92	144	31
Gymnospermae			
Pinus sylvestris	80	158	44
Picea excelsa	74	148	38
Angiospermae			
Dicotyledonae			
Chenopodium alba	72	144	35
Corylus avellana	76	138	22
Fagus sylvatica	72	144	35
Phoenix dactylifera	—	150	23
Quercus robur	81	144	33
Rumex acetosa	38	144	37
Monocotyledonae			
Festuca rubra	79	138	29
Lilium henryi	63	142	36
Narcissus pseudonarcissus	79	143	30
Phleum pratense	56	146	36
Carotenoid polymers			
Mixed *Lilium henryi*	—	148	38
β-Carotene	—	132	30
Zeaxanthin palmitate	—	135	24
Zeaxanthin	—	156	40
Echinenone	—	115	11

[a] Data from refs. 411, 417, 418.
[b] *Prasinocladus marinus* and *Pithophora oedogonia* have only cellulose in the spore wall.
[c] Asexual spores produce no sporopollenin.
[d] *E.g. Tasmanites* sporopollenin has composition $C_{90}H_{133}O_{11}$.

Brooks and Shaw have also examined sporopollenin from a number of fossil spores and the degradation products of insoluble organic matter in several Pre-cambrian sedimentary rocks[418]. In all cases they claimed that the materials were chemically identical with modern pollen, algal or fungal exines. In later papers they report the examination of the insoluble material in the carbonaceous chondrites such as the Murray and Orgueil meteorites. Again they claim that the material is identical to sporopollenin using the criteria outlined earlier, and on this basis, Shaw suggested that life on Earth may well have had an extraterrestrial origin[411].

Although this latter claim is undoubtedly exaggerated, it does seem likely that the carotenoids are involved in the formation of sporopollenin, at least in higher plants. However, this may not be the whole story. Cutin-like estolides (see p.221) may also be universally present in exines and aromatic melanin-like polymers are probably incorporated into the sporopollenins of some fungal spores. For example, labelled acetate was found to be readily incorporated into the exine of the pollen of *Gerbera jamesonii* (Compositae) whereas no activity was obtained using ^{14}C-labelled mevalonate[417]. Although this evidence points to a straight-chain lipid being involved in the formation of sporopollenin, it is possible that there is a membrane barrier which is impermeable to mevalonic acid like that found in the chloroplast[399]. Experiments with *Cucurbita pepo* showed that in this case neither ^{14}C acetate or palmitate were incorporated into the pollen exine to any great extent. Nevertheless, when carotenoids were isolated from the pollen after feeding radioactive palmitate and re-introduced into the anthers, it was found that the sporopollenin was highly labelled. This seems to point to the probability that it is only the fatty acid of the carotenoids which is incorporated, the latter merely only acting as a carrier. Certainly the data from molecular formulae, I.R. spectra and pyrolysis gas chromatography do not permit us to answer the question with certainty[411,417]. Furthermore, it seems highly likely that a proportion of the *p*-hydroxybenzoic and, especially, protocatechuic acids which are produced on alkaline fusion of sporopollenin may more easily arise either from bound forms of the acids which are known to be present in pollen[419,420] or from the equally ubiquitous flavonoids[412,419]. The latter are present in a non-extractable form in some angiosperm pollens[420]. It is interesting to note that these two acids are absent from the meteorite material[418] although this should, according to Brooks, be more highly aromatised than are present day pollens.

If we accept Shaw's premise that sporopollenin is more or less invariant throughout the Plant Kingdom and has been the same since the dawn of life (and perhaps before!) then there can be no possibility of discerning any phylogeny in the biochemical sense. The sporopollenins from lower plants do appear, however, to be more unsaturated as judged by their molecular formulae than those from higher plants (Table XVIII). The spread of analyses is too broad and the data too scanty to draw any firm conclusions. Nevertheless, the variation in the range of carotenoids found in the Plant Kingdom points to expectable differences in the polymers which may be formed. It also seems possible that there will be a variation in the accessory compounds present. As pointed out elsewhere (see p. 262), if one accepts the fact that the hydroxybenzoic acids perform a bacteriostatic role in plants and along with the flavonoids act as U.V. screens for plant cells, then we expect to find differences in the spore coats of marine and land plants in respect of these constituents. Unfortunately the data available at present does not allow this possibility to be examined further.

13. The flavonoids and related compounds

(a) Introduction

The flavonoids are a diverse group of phenolic compounds widely distributed in tracheophytes, the structures of which are based on 2-phenylchromone or flavone (LX)[20,421,422]. The individual classes of flavonoid compounds (catechins, leucoanthocyanins, flavanones, flavones, flavonols and anthocyanidins) differ from each other only in the state of oxidation of the heterocyclic ring. Isomeric with and interconvertible into the flavanones are the chalcones in which the heterocyclic ring has been opened, and another variant is represented by the aurones which have a 5- rather than a 6-membered oxygen-containing ring. Within classes, compounds vary depending on the number and orientation of phenolic hydroxyl groups on the aromatic rings, on the degree of *O*- and *C*-alkylation or prenylation, and on the number and nature of sugars attached to the hydroxy groups or directly to the aromatic nucleus as *O*- or *C*-glycosides respectively[421,422].

In addition to the flavonoids proper there are two classes of compound which have a similar biosynthetic origin, but in which the phenyl side group is at C-3 or C-4 of the chromone nucleus known as iso- and

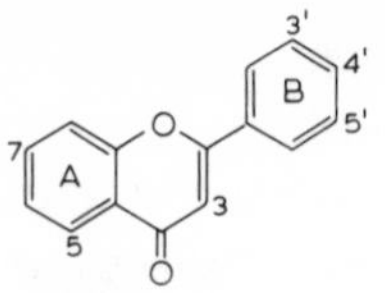

(LX) Flavone

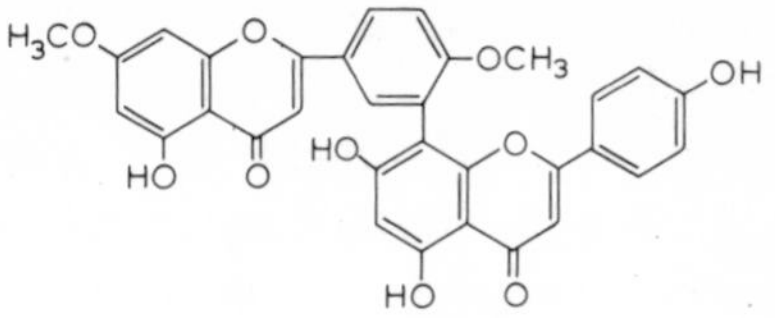

(LXI) Ginkgetin

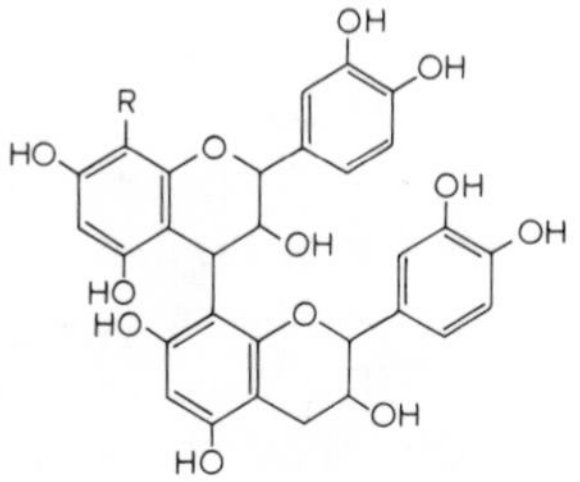

(LXII) Leucocyanidin
R=H dimer
R= further flavans, polymer

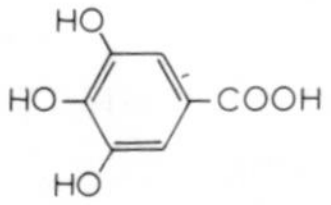

(LXIII) Gallic acid

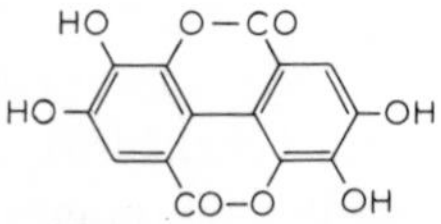

(LXIV) Ellagic acid

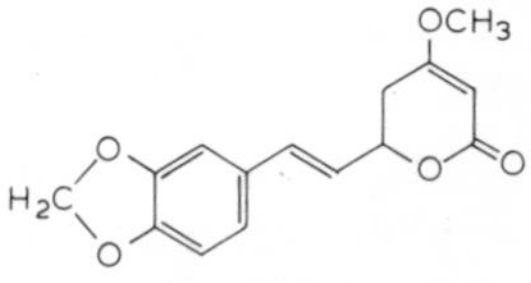

(LXV) Methysticin

neo-flavonoids respectively[423]. Again, compounds in these classes vary in the same way as for the flavonoids proper. Finally, there are the biflavones (LXI) in which two flavone nuclei are linked, usually by C–C bonds, but sometimes through one of the hydroxyl groups. The leucoanthocyanins (LXII) also form compounds in which the flavonoid nuclei are linked in this way, but in this case oligomers and polymers are formed containing two to over ten residues: these compounds are the condensed tannins[424–426].

The orientation of hydroxyl groups in flavonoid compounds is not random. In the aromatic ring of the chromone nucleus, the A ring, the vast majority of flavonoids have hydroxyl groups at C-5 and C-7, as to be expected from the polyketide origin of this ring[38,427] whereas in the side phenyl group, the B ring, hydroxyl groups are mainly found at C-4′, C-3′ and C-5′, reflecting the prephenate origin of this ring and the C-atoms of the heterocyclic nucleus[38,421,422,427]. Almost all the more highly oxidised flavonoids, that is excluding the catechins and leucoanthocyanins (LXII), occur in the leaves, flowers and fruit of tracheophytes

References p. 287

as *O*- or more rarely *C*-glycosides. In other tissues, especially in wood, the compounds occur in uncombined form[424,428].

In addition to the flavonoids and their congeners, plants contain a number of other phenolic phenylpropanoid compounds whose biosynthetic origin is identical to that of the B-ring in the former group[429,430]: indeed one class, the cinnamic acids, have been postulated as obligate precursors of this ring in the flavonoids[427]. Besides the cinnamic acids (XIII–XVI), there are coumarins, lignans which are dimeric C-18 compounds, and a number of more reduced phenylpropanoids of limited occurrence[422]. Individual compounds in all these groups vary in their hydroxylation or alkylation pattern in the same way as the flavonoids. The most common compounds are those with either a 4-, a 3,4- or a 3,4,5-oxygen substitution pattern relative to the side chain. The cinnamic acids are usually present in leaves as esters of sugars or alcoholic acids. Other compounds occur often as *O*-glycosides.

One final class of phenolic compounds which are related to the phenylpropanoids are the hydroxybenzoic acids. Again these have the same type of hydroxylation pattern as the cinnamic acids and originate from the shikimic pathway. From the evolutionary point of view, the most interesting is gallic acid (LXIII) which occurs also with its dimer hexahydroxydiphenic acid (HHDP) and related congeners in the hydrolysable tannins[426,431]. The latter group of tannins are all esters of glucose and related sugars and are readily cleaved by hot mineral acid. HHDP, when released by hydrolysis under acid conditions immediately lactonises to give the double lactone, ellagic acid (LXIV). Congeners of HHDP which are also present in hydrolysable tannins are formed by oxidation of one of the aromatic rings. These give rise to various other lactones on hydrolysis such as chebulic acid and brevifolin[424,426,431].

(b) Biosynthesis of flavonoids

As mentioned above, the A ring of the flavonoids (and of the iso- and neo-flavonoids) arises from acetate or malonate *via* the polyketide route[38,427]. The starter unit is the phenylpropanoid precursor of the B-ring and carbon atoms -2, -3 and -4 of the flavonoid nucleus (LX). It is presumed that this precursor is a cinnamic acid, possibly having the same hydroxy substitution pattern as the final flavonoid molecule, although since hydroxylation can occur after the primary C-15 nucleus has been

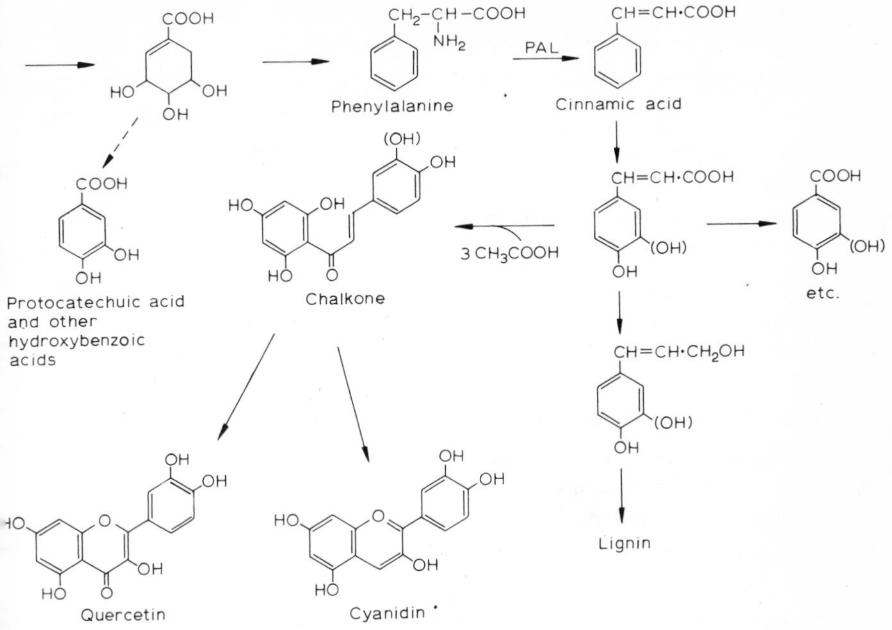

Fig. 27. Biosynthesis of flavonoids and related compounds.

formed, and most flavonoids have at least a 4′-hydroxy-group, it is often tacitly assumed that *p*-coumaric acid (XIII) is the main substance involved[427]. The cinnamic acids are assumed to arise from L-phenylalanine by the action of phenylalanine ammonia lyase (PAL) (Fig. 27). Enzymes for the hydroxylation of the cinnamic acid formed in this way to yield *p*-coumaric acid, have been demonstrated in a wide variety of vascular plants and in some higher fungi[432,433]. The addition of the three acetate units to the cinnamic acid moiety followed by aromatisation gives rise to a hydroxychalkone which is then presumed to undergo further elaboration to give the wide range of other flavonoids encountered. Alkylation, prenylation and glycosylation of hydroxy groups are postulated to be late steps in biosynthesis[421,434].

Intermediates between the cinnamic acid moieties and the chalkones have not been isolated. It seems likely therefore that as in the case of several other classes of compound (*e.g.* fatty acids), the addition of the three acetate units takes place sequentially on one enzyme matrix without

References p. 287

release of intermediates. A number of simple styryl γ-pyrones are known in which only two acetate units have been combined with the phenylpropanoid including hispidin (XII) from the fungus *Polyporus hispidus* and methysticin (LXV) from *Piper methysticum.* These have been referred to as 'failed' flavonoids, but none have been detected during investigations on the biosynthesis of the flavonoids proper.

Although there is general agreement about the role of *p*-coumaric acid in flavonoid synthesis, there is some uncertainty about the stage at which further hydroxylation of the B ring occurs. There is evidence that the obligate pathway to 3′,4′,5-trihydroxy flavonoids might not be *via* L-phenylalanine[427,435], and it has been suggested it might involve a prephenic acid-like precursor having the same hydroxylation pattern as shikimic acid[435]. It is, of course, possible that the 4′-mono- and 3′,4′-di-hydroxy flavonoids also arise from similarly substituted precursors rather than *via* the unsubstituted phenylalanine–cinnamic acid route, but the evidence is scanty. In any case, it is certain that this latter route can operate in all plants examined.

The pathway from phenylalanine to caffeic acid in the Basidiomycetes is a degradative pathway[436]. The dihydroxy acid is rapidly β-oxidised to protocatechuic acid which then undergoes fission as described earlier (see Fig. 10). Nevertheless it is probable that caffeic acid is utilised for synthetic purposes by some fungi *e.g.* by *Polyporus hispidus* in the synthesis of hispidin[293], and by other fungi which accumulate methyl isoferulate (XVII) and similar compounds[307,308].

Although the formation of alkyl or glycosyl derivatives of flavonoid hydroxy groups are presumably late steps in biosynthesis, the formation of *C*-alkyl or *C*-glycosyl groups may occur prior to the final elaboration of the flavonoid nucleus[437]. In almost all cases, *C*-substitution occurs on the A-ring of flavones and it is known that the presence of the carbonyl group at C-4 reduces the activity of this ring. It is probable that *C*-substitution, including the formation of biflavones, takes place with an intermediate in which the activity of the 4-keto group is reduced by enzymic combination. In the case of the commonly occurring leucoanthocyanin precursors (flavan-3,4-diols) no such restriction applies and polymerisation occurs readily *via* the easily formed C-4 carbonium ion[425]. In this case, the process is reversible under acidic conditions, although the majority of the ions released then undergo an irreversible acid-catalysed polymerisation which involves fission of the pyran ring. Part, however, may be further oxidised without polymerisation to yield anthocyanidins.

It should also be noted that lignin is also formed from phenylpropanoid precursors[299,300]. The cinnamyl alcohols corresponding to *p*-coumaric, ferulic and sinapic acids (XIII–XV) are all readily oxidatively polymerised and are presumed to be the primary precursors of the heteropolymer (Fig. 27). The control mechanisms which divert the phenylpropanoid precursors to cinnamic acids, coumarins, lignans, and lignin and flavonoids are, at present, unknown. There is evidence of multiple forms of PAL which may control the ultimate fate of the C-9 compounds[432,438] but there are probably more complex mechanisms involved in the overall control of the multitudinous different classes of end product which are synthesised by this pathway. The suggestion that all flavonoid bio ,nthesis is controlled solely by one PAL is too simplistic as evidenced by the fact that at low temperatures, light-controlled (phytochrome) reactions influence flavonol and leucoanthocyanin production but not the synthesis of anthocyanins[439]. Similarly the fact that phytochrome apparently controls both flavonoid and betalain production indicates that PAL activity[440] cannot only be involved, otherwise there would be too little phenylalanine for the synthesis of betalains (see p. 281).

The hydroxybenzoic acids have been shown to arise by β-oxidation of the corresponding cinnamic acids[441] as in the Basidiomycetes[436]. It now appears, however, that they are more readily formed in higher plants as in other fungi[259] by the direct aromatisation of shikimic acid (see Fig. 27)[441]. The simple benzoic acids may thus be regarded as the most primitive of all aromatic compounds derived from the shikimic pathway.

(c) The function of flavonoids

Since the flavonoids are ubiquitous in higher plants, occurring in all tracheophytes[421,442–444], bryophytes[445], and even some of the higher green algae[446], is it right to assume that their role is the same in each? Unfortunately it is not possible to answer such a question, since much of the data about the possible function of flavonoids relates only to the angiosperms. Here, some flavonoid compounds, including flavonols and cinnamic acids, have been implicated in growth, since the *ortho*-dihydroxy compounds, like quercetin (LXVI) and its glycosides, inhibit the destruction of indole acetic acid by the corresponding oxidase and related peroxidases, while monohydroxy compounds, like kaempferol (LXVII), stimulate the destruction of the hormone[447]. Another class of compounds, the coumestanes, are known to act as potent specific anti-fungal agents (phytoalexins) which are only produced when the plant suffers infection[318,448]. It is highly probable that many other classes of flavonoid compound possess similar activity, though to a lesser degree, since there are many reports of a general increase in their concentration when plants are attacked by pathogens[449]. The widely distributed polymeric leucoanthocyanins are known to inhibit a wide variety of enzymes[450] and undoubtedly confer some resistance to fungal attack in the plants in which they occur by inhibiting extracellular hydrolases. It is especially noticeable that certain woody angiosperms which lack leucoanthocyanins in their heartwoods, are more susceptible to fungal decay. A more obvious role of the coloured flavonoids in angiosperms is as attractants for flower pollinators. This is borne out, for example, by the fact that humming bird-pollinated tropical species have a higher proportion of flowers having flavonoids imparting orange hues, which attracts such birds, than do closely related temperate species[421]. More recently it has been shown that what to the human eye are colourless flavonols play a role in pollination because many insects can detect variations in the wavelength of near ultraviolet light which such compounds absorb[451]. Examination of many apparently uniformly white or pale-coloured flowers under UV light shows they exhibit a marked patterning which probably acts as a guide to pollinating insects[451].

The antifungal activities of flavonoids might be sufficient to explain the selective advantage of these compounds in non-flowering land plants, but it seems unlikely that this could be the reason for their original evolution in the Algae. It has been suggested that here, and perhaps in all plants, their role is to act as a screen against ultraviolet light in the nucleic acid-sensitive 250–270 nm region of the spectrum and perhaps in the area between 320–350 nm where NAD and other co-enzymes can be photo-oxidised[452]. Many flavonoids show very high absorbancies (log $\varepsilon = 4$ to 5) in both areas. The flavonoids are also potent anti-oxidants and metal chelators and can protect autoxidisable lipids[452]. This may partly account for their general fungistatic activity. The flavonoids, therefore, would be expected to confer an evolutionary advantage on plants which were emerging from the aquatic environment and invading the land. It should be noted that the hydroxybenzoic acids also possess most of these attributes (λ_{max} 250–270 nm log $\varepsilon \sim 4$; good antifungal and antibacterial activity[453]) and may be regarded as the primitive forerunners of the flavonoids. It has recently been shown that the primary acceptor complex for photosystem I (see p. 191) in spinach chloroplasts contains two *p*-coumaric acid esters and a mixture of flavonoid glycosides, the main aglycone being the hitherto unknown 3-*O*-methoxy-6,7-methylenedioxyquercetagetin (LXVIII)[454a]. Etioplasts and chloroplasts of barley have also been found to contain the congener of vitexin (LXIX), saponarin, in which the glucose is attached at C-6 rather than at C-8[454b]. Since these compounds are not found in any procaryote so far examined, their presence in the chloroplasts is contrary to the hypothesis of an endosymbiotic origin for these organelles[78].

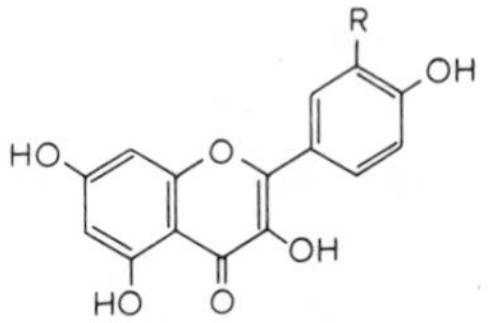

(LXVI) R=OH Quercetin
(LXVII) R=H Kaempferol

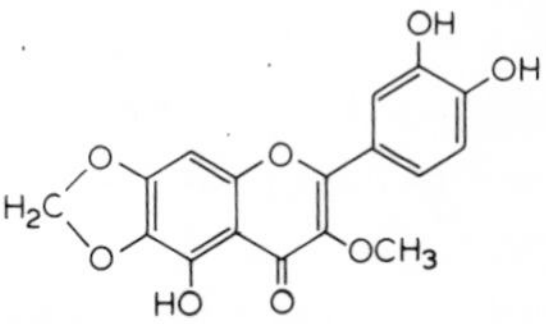

(LXVIII)

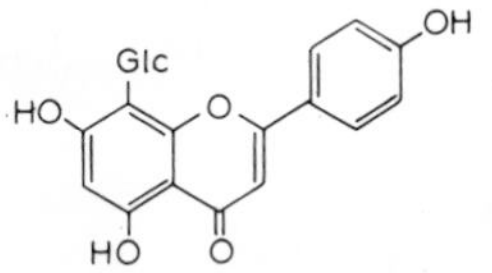

(LXIX) Vitexin

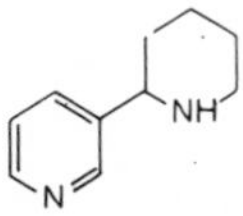

(LXX) Anabasine

However, it seems entirely possible that the ability to synthesise the pigments may have been an evolutionary step in the development of higher plant chloroplasts and are present, as suggested above, to protect the DNA and RNA against UV radiation. Further development of flavonoids as possible co-factors in growth[447,454], more potent fungicides[448] and as flower colours[421], on both the visible and the ultraviolet, were undoubtedly late features in their evolution. It is noticeable, for example, that those features of flavonoids which increase their anti-oxidant or anti-fungal properties such as extra B ring hydroxylation or polymerisation of the leucoanthocyanidins must have occurred relatively early in evolution[226,227], since such compounds predominate even in rather primitive ferns[444]. Features such as extra *O*-methylation or A ring hydroxylation which increases the range of visible colours of the flavonoids, on the other hand, are found in more advanced phyla[421,441,442]. The latter plants have, at the same time, developed more potent methods of fungal protection than the simple accumulation of tannins (both condensed and hydrolysable) which, if produced in too large a quantity, are themselves likely to have deleterious effects on the host plant itself. We can thus regard evolution in this class of compounds as the mosaic succession of one type of flavonoid for another in some areas of metabolism while other more primitive features might well be still retained in others.

(d) Distribution of flavonoids and related compounds

Flavonoids and related compounds are ubiquitous in higher plants[421–426]. Overall, the most widespread are those with a B-ring having a 3′,4′-dihydroxy substitution pattern, (Table XIX). However, there are marked disparities in the distribution of the different classes. Relatively advanced angiosperm families like the Leguminosae contain representatives of all types of flavonoids, whereas more primitive taxa usually have a restricted number of such compounds[421]. The most primitive flavonoid-containing Division of the Plant Kingdom are the Chlorophyta, the more advanced members of which contain *C*-glycosyl derivatives of the flavone apigenin (*e.g.* LXIX)[446]. *C*-Glycosides of this type are found in mosses, liverworts, ferns

TABLE XIX

Common Flavonoid Compounds[a]

Compound	*Substitution pattern*[b]		
	4-OH	*3,4-diOH*	*3,4,5-triOH*
Cinnamic acid	*p*-Coumaric acid (49)	Caffeic acid (63) Ferulic acid[c] (48)	Sinapic acid[c] (32)
Flavonol	Kaempferol (48)	Quercetin (56)	Myricetin (10)
Anthocyanidin	Pelargonidin (17)	Cyanidin (67)	Delphinidin (25) Malvidin[c] (27)
Leucoanthocyanidin[d]		Leucocyanidin (35)	Leucodelphinidin (9)

[a] Numbers in brackets give frequency of occurrence of derivatives in angiosperm leaves or, in the case of anthocyanidins, in flowers.

[b] Of B-ring in flavonoids, see formulae XIII–XVI, LX, LXII, LXVII and LXVIII.

[c] *O*-methylated in 3 or 3,5 groups.

[d] In terms of anthocyanidin produced.

and many of the relatively less advanced angiosperms[421,444]. On this basis and on the distribution of biflavonyls which are mainly found in gymnosperms, it has been suggested that *C*-substitution is a primitive feature in flavonoids.

Based mainly on their distribution in angiosperms, Harborne had adduced a number of other primitive and advanced features in flavonoids which are shown in Table XX[421]. The evolutionary status of many of these characters is self evident from their mode of synthesis while others, which appear to be plainly contradictory at first glance, can generally be satisfactorily reconciled when the overall distribution of the compounds in question is examined. Of particular interest is the suggestion put forward both by Kubitski[226] and Bate-Smith[227], on the basis of the distribution of the trihydroxy tannins, isoquinoline alkaloids and the iridoids (cyclopentane monoterpenes), that the Dilleniidae and the Rosidae (*sensu* Cronquist)[56], may be co-equal to the Magnoliidae as ancestral to the angiosperms. The last-named Subclass is more or less devoid of either ellagitannins or of leucodelphinidin-based condensed tannins. Both are regarded as primitive markers and would be expected to be richly present in the magnolian stock if indeed it were ancient. Bate-Smith has also shown that in several angiosperm taxa there is a good correlation between the presumed geographical centre of origin and the occurrence of the more primitive flavonoid features listed in Table XX[455]. However, leucoanthocyanins are absent from the most primitive tracheophytes[444] (Psilopsida, Lycopsida and most Sphenopsida), and their presence in all members of the more advanced ferns probably represents an important step in the evolution of land plants. It seems probable that they only arose when the more complex organisation of the

TABLE XX

Primitive and Advanced Characters in the Flavonoids

Primitive	*Advanced*
C-glycosylation	Complex *O*-glycosidation
C-alkylation	*O*-alkylation
Leucoanthocyanidins	2′-, 6- and 8-hydroxylation
Flavonols	Aurones
Trihydroxytannins[a] in leaves	Flavones
3-Deoxyanthocyanins	Trihydroxy-anthocyanins in flowers

[a] Leucodelphinidin and ellagic acid.

References p. 287

vascular system was developed in the ferns because this required a greater degree of protection against invading fungal hyphae. From the point of view of overall plant evolution, the ellagitannins are even more advanced than the leucoanthocyanins, for they are only found in a limited array of the more primitive angiosperms[227]. It might be expected that they would be found in several lower plant Divisions since gallic acid derivatives are relatively widespread, being found in several algae and fungi and the phenolic coupling reaction, which gives rise to the ellagitannins and many other classes of compound like the biflavonyls, appears to be relatively ubiquitous[38,456]. It is likely, therefore, that as with many other classes of secondary product the range of flavonoids found in any given plant represents a spectrum of compounds from primitive to advanced, which, as shown in Fig. 28, depends on the overall phylogenetic history of the taxa in question.

One character, however, probably does represent a primitive feature in flavonoids and that is the formation of *C*-glycosides, *C*-alkyl derivatives and biflavonyls, in which the reactive phloroglucinol A ring is substituted. It has been shown that few bryophytes and neither fungi nor algae can form *O*-glycosides from foreign phenols[457], and since non-glycosidic forms are usually toxic and not readily transported within the cell, it seems likely that the development of *C*-glycosidation was the answer. The ability to readily activate the A ring of the flavonoid molecule for this particular *C*-substitution reaction could then have been developed in the primitive tracheophytes, ferns and gymnosperms for the production of *C*-alkyl flavones and flavanones and the biflavonyls which presumably structurally are more difficult for any invading microorganism to detoxify than are the parent unsubstituted flavonoids. This procedure was quite early adapted to the formation of the antifungal polymeric leucoanthocyanins. The ability to form *C*-substituted derivatives is superceded in the more advanced angiosperms by the power to synthesise the equivalent *O*-derived compounds, and to hydroxylate the A ring. The overall increase in synthetic power has meant that a greater array of compounds is available. From this pool, the more advanced angiosperms have been able to reject some of the simpler solutions to fungal attack by the use of leucoanthocyanins, and produce new fungicides like the coumestanes which can be synthesised on demand[448]. Much of our knowledge about the present distribution of flavonoids supports these conclusions which are given in outline in Fig. 28.

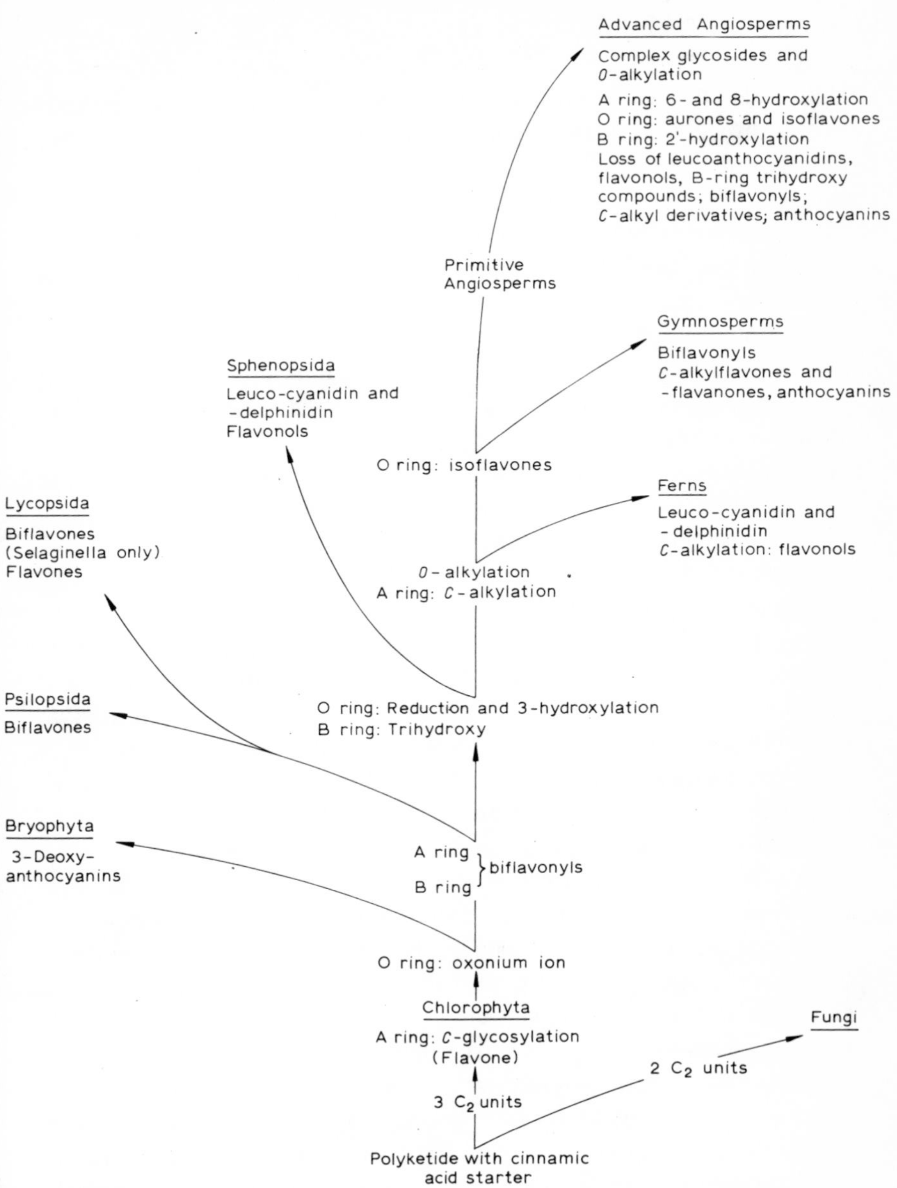

Fig. 28. Evolution of flavonoids in higher plants.

14. The alkaloids

(a) Introduction

Unlike all other classes of naturally occurring substances, the alkaloids have a diverse biosynthetic origin, and therefore, form a group of structurally varied compounds which are difficult to specify in a simple manner[30–42,458,459]. The true alkaloids[460] can be defined as nitrogen-containing compounds most of which have a relatively complex heterocyclic ring structure and which are synthesized almost exclusively by plants from a restricted group of normal protein amino acids, and exhibit marked pharmacological activity. The major groups of true alkaloids are formed from one of the following five amino acids: ornithine, lysine, phenylalanine, tyrosine or tryptophan. These compounds give rise to alkaloids by two distinct, but related routes[22,30,38] (Fig. 19). In both cases the amino acids are first decarboxylated: the diamino aliphatic amino acids then lose their α-amino group and cyclisation takes place on the terminal $-NH_2$. The aromatic amino acids, on the other hand, retain their α-amino group, and condensation to give the heterocyclic ring usually involves the incorporation of a suitable carbonyl-containing compound. In the case of phenylalanine and tyrosine, these are mainly derived from a second molecule of the amino acid itself, while carbonyl moieties arising from cyclopentane monoterpenes are incorporated into the widespread complex indole alkaloids derived from tryptophan. A third route to alkaloids starts from anthranilic acid and here both the amino and the carboxyl group are retained, cyclisation being affected by the addition of acetyl units in a modified polyketide pathway.

Two other groups of alkaloids may be distinguished[460]; the first includes compounds derived from either a preformed di- or tri-terpene or a steroid by the incorporation of nitrogen at a late stage in biosynthesis, sometimes in heterocyclic form. This group also contains peptide alkaloids and those derived from polyketides. All such compounds are obviously unrelated biosynthetically to the true alkaloids, and are usually referred to as pseudoalkaloids. The third class of alkaloids are called protoalkaloids. These are compounds which do not contain heterocyclic nitrogen and are derived from the aromatic amines, tryptamine and tyramine, derived by decarboxylation of the corresponding amino acids or from histidine by *N*-methylations and, where appropriate, ring hydroxylations. Also included in this class of alkaloids are *N*-methyl and other derivatives of the normal purines found in nucleic acids. It should be noted that both ornithine and lysine give rise to naturally occurring amines by decarboxylation. These and their congeners act as growth promoters, perhaps because of their interaction with RNA, and as cations restoring ionic imbalance[461].

(b) The biosynthesis of alkaloids

The alkaloids which arise from ornithine all contain a five-membered heterocyclic ring in which, as mentioned earlier, the γ-amino group of the amino acid is retained[22,30]. In many cases, this amino group is methylated as a first step on the biosynthetic route presumably to ensure that it does not undergo oxidative deamination. Subsequent decarboxylation and oxidation of the α-amino group leads, after ring closure, to the *N*-methyl-Δ'-pyrrolinium cation. There is a formal resemblance between the conversion of glutamic acid semi-aldehyde to proline and that of the 4-methylaminobutyraldehyde to the pyrrolinium ion (Fig. 29). It seems possible, therefore, that the enzyme catalysing the latter step on the alkaloid route may have originated by a slight modification of that responsible for the ring-closure reaction in proline biosynthesis. The reaction mechanism is very similar in both cases and presumably would thus require the same type of active centre in the enzymes responsible. All that would be needed is a change in enzyme specificity which might have been brought about by a single amino acid substitution to alter the folding of the protein chain in the region either where the substrate is attached or where access is gained to the active site. If this were true, it can be postulated that the ability to form simple pyrrolidine alkaloids is a relatively primitive feature, since it is merely an extension of normal amino acid biosynthesis. Furthermore, it might be expected that animals who can degrade proline might readily detoxify ornithine derived alkaloids in the same way, and that such compounds would be rather ineffective feeding deterrents against herbivores (see below, p. 277). However, it should be noted that in most of the simple alkaloids of this type, like hygrine (Fig. 29), the *N*-methyl-Δ'-pyrrolinium ion first formed condenses with acetoacetate at C-2, and this is usually followed by decarboxylation. This condensation reaction has no formal equivalent in the biosynthesis of either proline or ornithine. Nor does the further ring closure of the acetoacetyl moiety at C-5 of the pyrrole ring to give the tropane alkaloids such as cocaine and hyoscyamine and such compounds are obviously less easily detoxified by animals or micro-organisms[462].

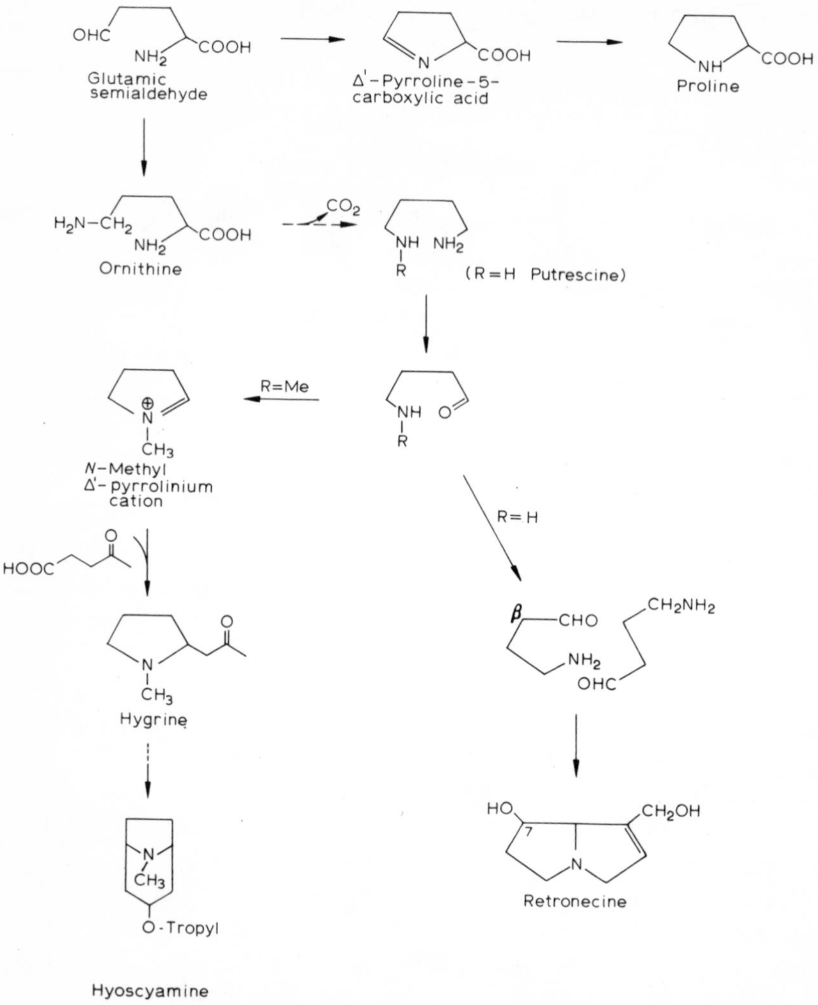

Fig. 29. Biosynthesis of ornithine-derived alkaloids.

It should be noted that neither the diamine, putrescine, which is produced by the decarboxylation of ornithine, nor any other symmetrical intermediates, appears to be involved in the biosynthesis of either hygrine or tropine, since when [2-^{14}C] ornithine is incorporated it labels only one of the carbons adjacent to the nitrogen in these alkaloids[22,30,38]. However, putrescine can be incorporated into several of the pyrrolidine alkaloids including hygrine and tropine and this demonstrates the lack of specificity of the enzymes involved. It should also be noted that the diamine may be involved in the biogenesis of pyrrolidine ring in nicotine (XX), since double labelling experiments with [2-^{14}C, δ-^{15}N] ornithine showed that here again a symmetrical intermediate was involved[30].

In addition to the relatively simple pyrrolidine alkaloids described above, ornithine is the precursor of the bicyclic necine alkaloids, which occur in certain advanced angiosperm families (Fig. 29). Here, two molecules of the amino acid are involved[22,30,38]. Specific labelling experiments demonstrate that after decarboxylation, one of these gives rise to an unmethylated pyrrolinium cation and the second loses the δ-amino group and is attached to the pyrrolinium ring nitrogen as shown. Subsequent ring closure gives the pyrrolizidine ring system. This pathway is obviously more advanced than those described above. It should be noted that most of the necine alkaloids have a hydroxyl group at C-7 as in retronecine. This position arises from the C-3 of one of the ornithines by the scheme shown in Fig. 29, and it is this β-position in the second molecule of the amino acid which is involved in the ring closure. It seems possible, therefore, that the necine alkaloids may be formed from a route involving β-hydroxyornithine. It is interesting to note that the compound, 3-hydroxyproline, which may be regarded as a congener, is known in the Leguminosae where the necines are mainly found[463]. It may be concluded that such advanced alkaloids are more potent deterrents to animals than the simple pyrrolidines since their detoxification must require the evolution of new specific enzyme systems by the herbivore.

The piperidine alkaloids, derived from lysine, are produced in an analogous manner to the pyrrolidine compounds described above[22,30,38]. Again *N*-methylation often precedes ring closure, and acetoacetate is condensed with the *N*-methyl piperidinium ring to give compounds like *N*-methylpelletierine (Fig. 30). It should be noted that the biosynthesis of lysine itself, whether by the DAP or AAA pathways (Fig. 9, p. 180), involves piperidine intermediates. This suggests that, as with the pyrrolidines, the enzymes responsible for the cyclisation step in the lysine-derived alkaloids may be little altered from that involved in the equivalent reaction in the biogenesis of the amino acid itself. However, it should be noted that not all piperidine- or pyridine-containing compounds arise from the cyclisation of lysine. The ubiquitous nicotinamide co-enzymes have quite a different origin and so do nicotine and coniine. In animals, nicotinamide is formed

References p. 287

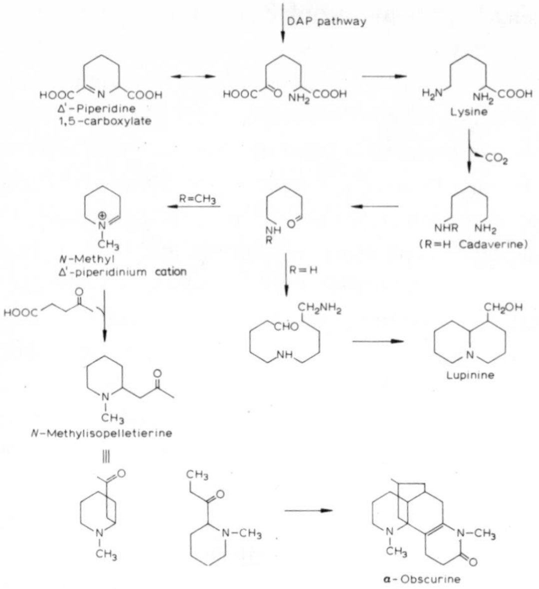

Fig. 30. Biosynthesis of lysine-derived alkaloids.

from tryptophan *via* 3-hydroxyanthranilic acid. This pathway is also used by some fungi (*e.g. Neurospora crassa*), and a few bacteria[37,38]. However, most bacteria and all higher plants which have been examined synthesise the vitamin by the condensation of glycerol and aspartic acid to yield quinolinic acid (pyridine-2,3-dicarboxylic acid), the same compound which is formed on the tryptophan route, which is then decarboxylated[38]. Plants which synthesise nicotine and anabasine (LXX) use the glycerol–aspartate route for the production of the pyridine ring in these alkaloids. The pyrrolidine and piperidine moieties of these compounds arise from ornithine and lysine respectively. Yet another way to the piperidine ring is utilised by *Conium maculatum* in the biosynthesis of coniine where a tetrapolyketide, formed from acetate, is involved and subsequent transamination with the NH_2 group from alanine yields the alkaloid[22,464].

The relative evolutionary significance of these four pathways to the pyridine or reduced pyridine ring is difficult to decide. It must be assumed that the nicotinamide co-enzymes are of a very primitive character and were necessary to the earliest procaryotic organisms. The biosynthesis of lysine may be less ancient, so that it is reasonable to assume that a route to the piperidine/pyridine ring system other than that involving the diamino acid was required. If tryptophan is regarded as another "late" amino acid, the most ancient biogenesis of nicotinic acid would involve the route from glycerol and aspartate. The condensation step between the two is formally similar but simpler, to that between anthranilic acid and phosphoribosyl pyrophosphate to give the precursor of tryptophan, and the latter reaction may well have been based on this pattern. The polyketide route to the piperidine ring is probably much more advanced. It leads to a heterocyclic ring with extra carbons in the α- rather than the β-position as is required for nicotinamide. It is surprising that this route is not found also in the fungi since these organisms are apparently better equipped to form polyketide derivatives than those from any other Division of plants[127]. It is noteworthy to find that in *Eusarium oxysporum* the biosynthesis of fusaric acid (LXXI) does involve acetate but that again aspartic acid provides the three carbons of the pyridine ring and the carboxyl group. In this organism nicotinic acid biosynthesis involves glycerol as described above[127].

Like ornithine, lysine is the precursor of a number of complex polycyclic alkaloids which contain two, three or four lysine-derived fragments. The mode of formation of these lupin alkaloids appears to be similar to that of the necines, but there is no evidence from the substitution pattern of the final products that hydroxy intermediates are involved. It is possible, however, that the corresponding diamine, cadaverine, might be an intermediate[22,37,38]. Lysine is also the starting point for the biosynthesis of the complex lycopodium alkaloids. Specific labelling experiments suggest that these may arise from *N*-methylpelletierine as shown in Fig. 30.

The biosynthesis of true alkaloids from the two aromatic acids phenylalanine and tyrosine, naturally differs from the routes described for the diamino acids, although there are analogies with the pathway to the necines and the lupin alkaloids[22,37,38]. In most cases, the alkaloids are isoquinolines which arise directly from the extra carbon atom required to form the heterocyclic ring usually coming from an activated carbonyl group (Fig. 31). In the simplest case, anhalamine, ring closure is effected on the methyl group of an *N*-methylmethoxytyramine, whereas with pellotine and many other alkaloids of this type, pyruvate provides the two-carbon fragment required (it should be noted that acetate is not incorporated into pellotine in a specific manner). In more complex cases, condensation of phenylethylamine or tryptamine takes place with the corresponding aromatic

References p. 287

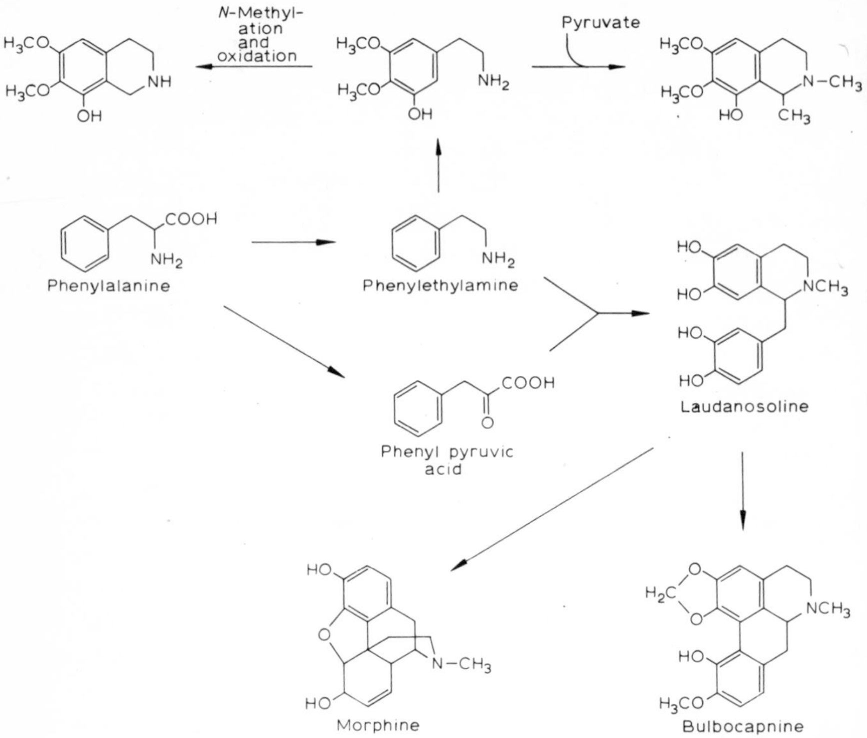

Fig. 31. Biosynthesis of phenylalanine-derived alkaloids

keto acid, such as phenylpyruvic acid. After decarboxylation and ring closure, this yields a benzylisoquinoline such as norlaudanosoline (XXV), hydroxylation occurring before or after cyclisation[22,37,38]. Norlaudanosoline and its derivatives are the precursors of a large number of different isoquinoline alkaloids. These include the morphine (XXII), aporphine (LXXII) and erythrina (LXXIII) bases[30,38], which are derived directly from the benzylisoquinoline nucleus by oxidative phenolic coupling reactions followed by further hydroxylations or reductions. A second series of alkaloids which includes the protoberberine (LXXIV), phthalide-isoquinoline (LXXV) and protopine (LXXVI) bases, are derived in a similar manner but all have one extra carbon atom derived from methionine. The biosynthetic routes to all types of isoquinoline alkaloids involve no steps which are analogous in any way to those which lead to the biosynthesis of the precursor amino acids themselves or to their degradation. The

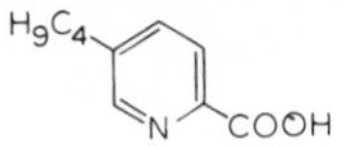

(LXXI) Fusaric acid

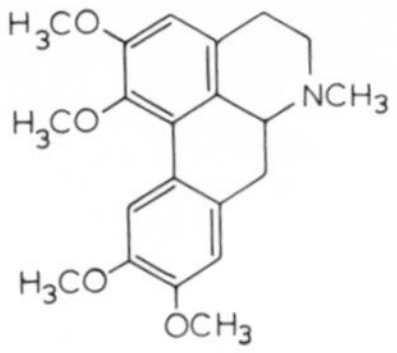

(LXXII) Glaucine

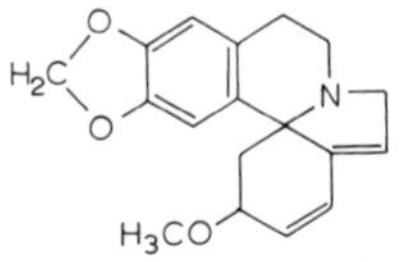

(LXXIII) Erythraline

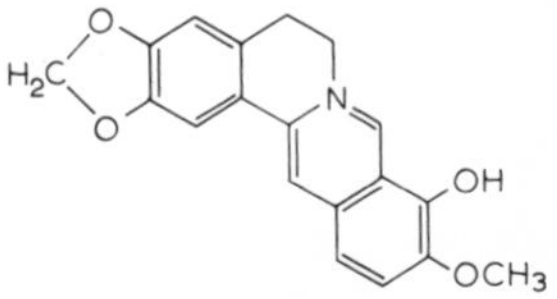

(LXXIV) Berberine

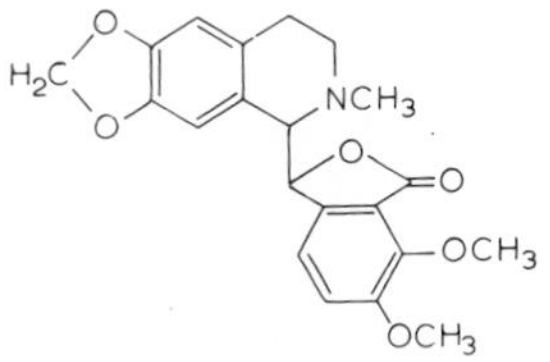

(LXXV) Hydrastine

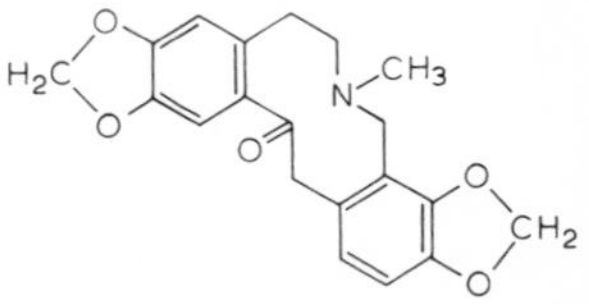

(LXXVI) Protopine

pathways can be thus regarded as being much more advanced than those starting from ornithine and lysine.

In all the alkaloids described above, the first steps include the decarboxylation of both the aromatic amino acid and its corresponding keto analogue. In colchicine (LXXVII) and its congeners, however, one of the carboxyl groups is retained, the first formed heterocyclic intermediate being a phenylethyl- rather than a benzyl-isoquinoline[38].

Like phenylalanine and tyrosine, tryptophan forms a few alkaloids which have an additional six-membered heterocyclic ring, analogous to that found in the isoquinolines (LXXVIII), and these are formed by similar routes[22,30,38]. Cognate compounds having two five- rather than a five- and a six-membered heterocyclic ring are also known together with corresponding dimers[30]. The majority of the indole alkaloids, however, are derived by the combination of tryptamine with a C-9 or C-10 moiety derived from the terpene iridoid compound, loganin (LXXIX)[30]. This undergoes a series of modifications before incorporation into three different classes of complex indole alkaloid, for example, ibogaine (LXXX). As with the isoquinolines, the pathways of biosynthesis of the complex

References p. 287

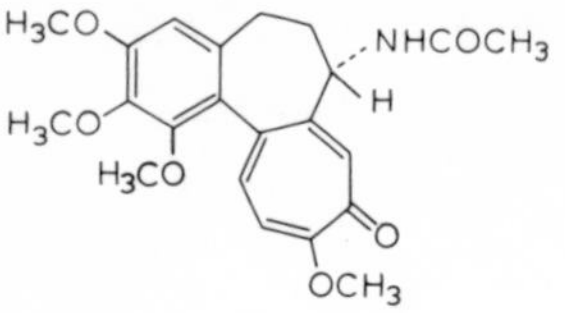

(LXXVII) Colchicine

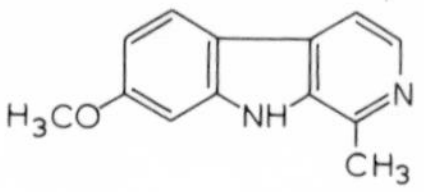

(LXXVIII) Harmine

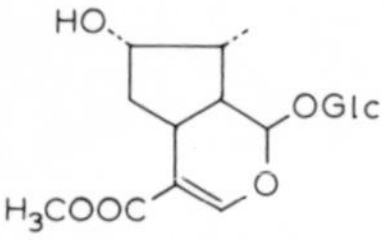

(LXXIX) Loganin

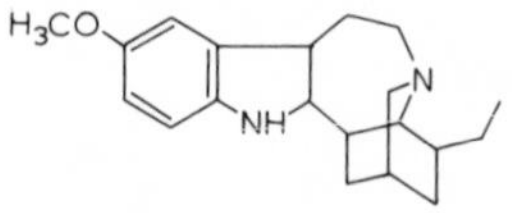

(LXXX) Ibogaine

indole alkaloids bear no relation to that leading to the parent amino acid. Indeed they also involve a number of novel steps in the metabolism of the monoterpenes which are themselves considered as advanced (see p. 238). Such pathways must be considered as being very advanced indeed, since they produce compounds which are even more difficult to degrade than the isoquinolines. A complex pathway involving tryptophan and a single mevalonic acid unit occurs in *Claviceps purpurea* and certain members of the Convolvulaceae, giving rise to the ergoline alkaloids[30]. In the fungi these compounds arise from 4-dimethylallyl tryptophan which is also an intermediate in the biosynthesis of cyclopiazonic acid (LXXXI) from *Penicillium cyclopium*[127].

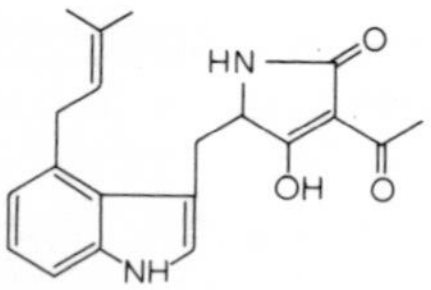

(LXXXI) β-Cyclopiazonic acid

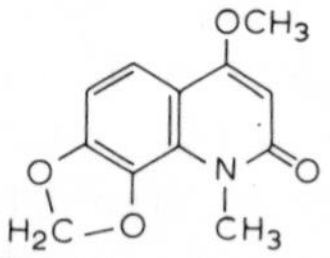

(LXXXII) Casimiroine

Alkaloids from anthranilic acid all contain a quinoline or acridine ring system[30]. The extra carbon atoms required for the elaboration of these rings arise from acetate or malonate by a modification of the well-known polyketide route (for example, casimiroine, LXXXII). Although this class of alkaloids has a relatively restricted distribution in higher plants, the mode of biosynthesis seems to be relatively primitive since polyketides are found in bacteria. It is notable that the quinoline compounds derived from anthranilic acid found in fungi derive the extra carbon atoms from the side chain of phenylalanine giving rise to 4-phenyl derivatives such as viridicatin (LXXXIII)[127].

The biosynthesis of the terpenoid protoalkaloids is touched on briefly elsewhere (p. 227). Other alkaloids of this class are of minor importance and need not concern us here.

(c) The function of alkaloids

Almost all alkaloids show some degree of toxicity to mammals and at lower doses most exert disturbing effects on metabolism[462,465]. They are probably the most effective protectants against herbivores that exist, and where present largely determine the selection of crop plants by ruminants[466]. For example, sheep avoid eating only those strains of the legume *Lupinus angustifolius*, which contain the alkaloid lupanine

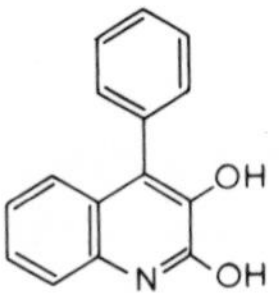

(LXXXIII) Viridicatin

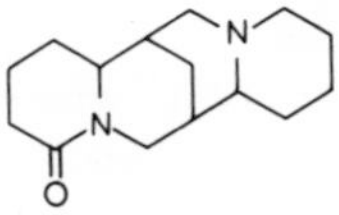

(LXXXIV) Lupanine

(LXXXIV). Even protoalkaloids such as gramine and *N*,*N*-dimethyltryptamine, can affect the acceptability of plant foods by sheep as shown by studies on the grass *Phalaris tuberosa*[466]. As in most other cases, animals have developed a sensitive means of detecting alkaloids, usually by their bitter taste. Indeed to ourselves, alkaloids are amongst the bitterest substances known, even those such as quinine which are pharmocologically useful. Certain phytophagous insects have taken advantage of this fact and have evolved ways of accumulating alkaloids and other bitter substances from plants to protect them against bird and animal predators[467]. There are several reports, however, of the role of alkaloids as feeding deterrents to insects[321] and as inhibitors of certain infecting micro-organisms[322]. It seems likely, however, that their ability to act as feeding deterrents to mammals is their most important role. As is pointed out later, the true alkaloids are absent from the ferns and gymnosperms both of which antedate the mammals by 200 or so million years. One aspect of angiosperm-mammal co-evolution is thus the development of pathways of metabolism leading to the alkaloids. Unfortunately there is little evidence

References p. 287

to show how the alkaloids act on animal systems or what detoxification mechanisms the latter have produced to deal with these compounds[462]. It has been shown that both quinine and berberine form complexes with DNA[468], and that this alone might account for their antibacterial and antifungal activity. The stereochemistry of the more complex alkaloids like berberine and yohimbine is very similar to that of certain active steroids, and it is possible that such alkaloids may act as competitive inhibitors of hormonal action. Whatever the reason for their pharmacological activity, it may be concluded that the alkaloids confer a clear cut advantage on the plants that produce them. The fact that they have a relatively limited distribution merely reflects the fact that non-alkaloidal plants have evolved other ways to deter potential predators.

(d) Distribution of alkaloids

The alkaloids are found mainly in the fungi and in the higher vascular plants[31]. Although the other Divisions of the Plant Kingdom, including the bacteria and blue-green algae, can synthesise a wide variety of amines including those involved in the synthesis of the different classes of alkaloids, they apparently do not possess the necessary enzyme systems to transform these compounds further. A few alkaloids are known in insects[469] and amphibia[362].

The fungi contain a wide variety of alkaloids, most of which are derived from tryptophan[127]. These include the ergoline alkaloids mentioned above, and sulphur-containing analogues of physostigmine, the sporodesmins (LXXXV). Tryptophan is also incorporated into the poisonous peptides of *Amanita phalloides* which also contain proline and hydroxyproline. Many fungal alkaloids contain mevalonic acid residues like the ergolenes, echinulin (LXXXVI) and cyclopiazonic acid, and a number of protoalkaloids derived from tryptamine also occur. The various alkaloids show the extreme complexity and versatility of the biosynthetic apparatus present in the fungi[127]. Many of the alkaloids and their modes of synthesis are unique, and taken with the vast range of other novel compounds certainly justify the fungi being treated as a separate Kingdom. Unfortunately the distribution of alkaloids in the fungi does not throw much light on the evolution of the compounds in plants as a whole. The co-occurrence of bufotenin (LXXXVII) in the common toad, *Bufo vulgaris*, in *Piptadenia peregrina* (Leguminosae) and the fungus *Amanita mappa* is hardly surprising, since the compound is a metabolite of tryptophan and its unmethylated analogue, serotonin, is a powerful vasoconstrictor occurring in all animals. On the other hand, the ability to hydroxylate tryptophan in the 4-position to give psilocybin (LXXXVIII) is apparently unique to fungi.

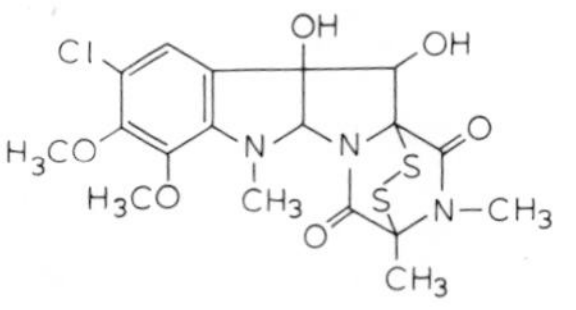

(LXXXV) Sporodesmin

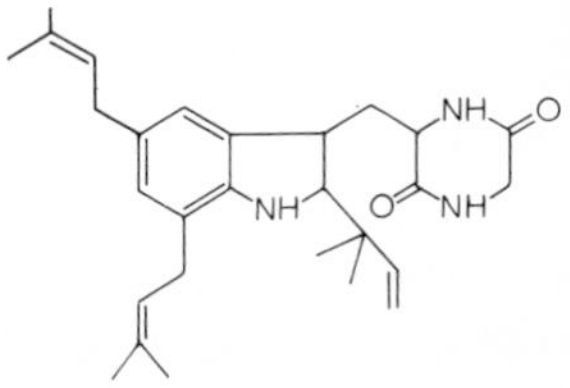

(LXXXVI) Echinulin

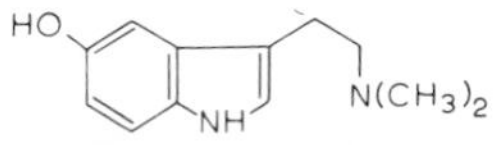

(LXXXVII) Bufotenin

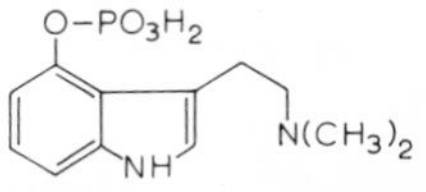

(LXXXVIII) Psilocybin

In the vascular plants the distribution of the true alkaloids appears to have a more interesting evolutionary significance[31,32,470]. None are found in the Psilopsida, and only one type of true alkaloid is found in the Lycopsida. As has been discussed above, these lycopodium alkaloids are derived directly from lysine, and it seems possible that the possession of such compounds might well have helped the lycopods to survive predation by early amphibian herbivores and perhaps accounts for their being fossilised in such enormous numbers in the coal measures of the Carboniferous period. Nevertheless, the ability to synthesise such compounds can only be regarded as an elementary step in the evolution of alkaloids, for it probably involves only a relatively slight modification of normal biosynthetic pathways.

The related Sphenopsida have devised a different type of solution to produce alkaloid-like compounds. *Equisetum palustre* synthesises the macrocyclic compound palustrine (LXXXIX) which is a derivative of *N*-methylspermidine[461]. This diamine is found in a wide variety of plants including bacteria, and is synthesised from putrescine and methionine[1]. The origin of the alicyclic ring portion of palustrine is unknown. One must class such a compound as even more primitive in the alkaloidal sense than the lycopodium alkaloids since it does not contain a ring system which is derived from one of the precursors of the true alkaloids.

References p. 287

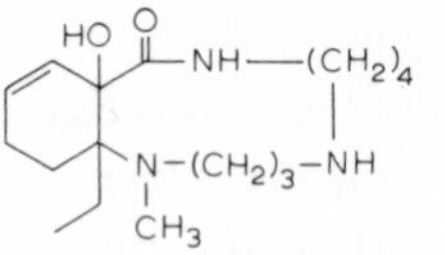

(LXXXIX) Palustrine

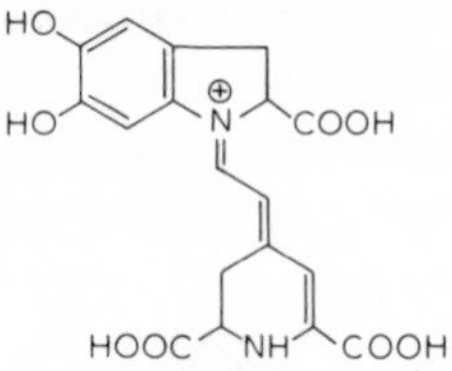

(XC) Betanidin

The ferns contain no alkaloids. Some contain cyanogenic glycosides[471] which may represent a modified alkaloid pathway, since the amino nitrogen of the parent amino acids is retained. The gymnosperms are also devoid of true alkaloids. *Ephedra* species contain relatively simple derivatives of phenylanaline, and *Taxus* has a number of toxic pseudo-alkaloidal compounds from diterpenes. In either case, the compounds, although showing marked pharmacological activity, can hardly be regarded as advances in the evolution of the alkaloids proper.

It is with the presumed most primitive angiosperms that the full flowering of alkaloid biosynthesis is achieved[31,32,472]. The Magnoliales, which are suggested by most authorities to be the most primitive of the angiosperms (see p. 151) have developed the ability to synthesise an array of advanced isoquinoline and simple indole alkaloids. Some workers, however, have suggested that this group of plants is not the most primitive of angiosperms, and that the Hamamelidae and Dileniidae (Fig. 3) are at least collateral with the magnolioid stock[224–227]. From the alkaloid point of view, this suggestion is interesting, since it has been postulated that the iridoid glycosides (*e.g.* LXXIX) are relatively primitive characters in the evolution of the flowering plants, and it thus seems probable that the complex indoles for which the iridoids are precursors, may be equally ancient.

However, the Hamamelidae are almost devoid of true alkaloids. Indeed, only one family, the Moraceae, contains any alkaloids derived from aromatic acids, and these are isoquinolines. The percentage of alkaloidal species for the Subclass as a whole is only 0.55%, which contrasts strongly with the proportion of such species in the Magnoliidae (4.7%). The Dilleniidae are like the Hamamelidae in having 0.6% alkaloidal species and complex indole alkaloids only occur in relatively advanced families. It appears, therefore, that the isoquinoline alkaloids may have been the first of the more advanced alkaloids to evolve.

The calculations on the proportion of the alkaloidal species have been made on the basis of reported occurrences of alkaloids, and, therefore, are subject to quite large sampling errors. Furthermore, no differentiation has been made between true alkaloids and the other classes, but bearing these drawbacks in mind, the distribution of the alkaloids in the other sub-classes of the angiosperms shows no consistent trend between the ability to accumulate these compounds and presumed evolutionary advancement[229]. Distribution is uneven and even closely related families contain widely different proportions of alkaloid-containing genera and quite often widely different alkaloid types.

Further analysis is therefore needed before any more detailed conclusions can be drawn, but it is obvious that alkaloids from the aromatic amino acids are characteristic of the angiosperms, as are those alkaloids from ornithine and lysine which possess the more complex features such as the polycyclic necine and lupine derivatives.

(e) The betalains

One other class of alkaloid-like compounds are the betalains[251,472]. These have a very limited distribution in the Plant Kingdom, being only present in nine of the eleven families of the angiosperm order Caryophyllales (*sensu* Cronquist). All members of these betalain families examined contain the red-coloured betacyanins (glycosides or acylated glycosides of betanidin, XC) and seven also contain yellow betaxanthins in which the indole moiety of betanidin is replaced by other amino acids including proline, aspartic and glutamic acids and methionine. None of the betalain families contain anthocyanins, the orange to blue flavonoid pigments, although they do synthesise other flavonoid compounds including flavones, flavonols and leucoanthocyanins (see p. 256). Conversely, the other two members of the Caryophyllales, the Caryophyllaceae and Molluginaceae contain anthocyanins but no betalains. On these grounds it has been suggested that these latter two families should be excluded from the Order, although morphological criteria are against such a split. It appears better for the moment to group all 11 families together, but more note should be taken of the betalains as a character when advancing phylogenetic schemes involving the angiosperms. It seems highly likely for example, that the Caryophyllaceae and Molluginaceae are more closely linked than is often proposed, and probably diverged from the rest of the Order. Indeed, the betalain families may have evolved separately from the other angiosperms as suggested by Kubitski[226] and others[251].

The biosynthesis of betalains involves the formation of betalamic acid (Fig. 8) by the oxidative cleavage of the ring of 3,4-dihydroxyphenylalanine (dopa) between C-4 and C-5. As mentioned earlier, this oxidation is obviously related to the unusual *meta* oxidation of protocatechuic acid by certain *Pseudomonas* (p. 179).

Unfortunately, we do not know whether this mode of degradation of the aromatic ring is rare in higher plants or not. Nor do we know whether it is an ancient character in the bacteria. The only other product arising from the ring opening of an aromatic amino acid in angiosperms is the cyanogenic glycoside (Fig. 8) from *Thalictrum aquilegifolium*[252]. This arises from dopa by the more usual oxidative cleavage between the hydroxyl groups. A similar compound triglochinin (XCI) has been isolated from *Triglochin maritinum*[473]. These plants are regarded as primitive members of the dicots and monocots respectively. In evolutionary terms, then, it appears that the ability to utilise the products obtained by cleavage of the aromatic ring may indeed be primitive, but more investigation is needed on the problem of aromatic degradation in higher plants, before definite conclusions can be drawn.

15. Cyanogenic compounds

Plants which yield hydrocyanic acid on crushing or on treatment with dilute mineral acids are widespread in nature[471]. Over 600 angiosperms, about half a dozen gymnosperms and nearly 30 ferns which have this property have been reported. In addition, cyanogenesis occurs in a few bacteria, over two dozen fungi and several millipedes and centipedes. The mechanism of cyanogenesis has been established in a very few cases, mainly in flowering plants where, with one exception, the precursors of hydrogen cyanide have been shown to be glycosides of α-hydroxynitriles (XCII). These compounds give rise to the unstable aldehyde cyanohydrins on hydrolysis which decompose to yield the aldehyde and HCN[471].

All the bacteria which produce hydrogen cyanide, do so continuously. There is no evidence for the existence of a combined form. In *Chromobacterium violaceum*, the gas originates from glycine, and it seems possible that this same mechanism operates in some of the cyanogenic fungi. However, this is not necessarily true for all fungi, since two higher plant cyanogenic glucosides, linamarin (XCII) and lotaustralin (XCIII) have been isolated from an unidentified psychrophilic basidiomycete known to be responsible

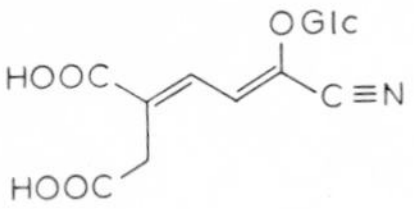

(XCI) Triglochinin

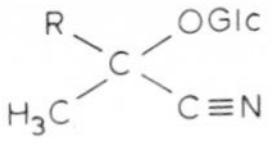

(XCII) R = CH_3 Linamarin
(XCIII) R = C_2H_5 Lotaustralin

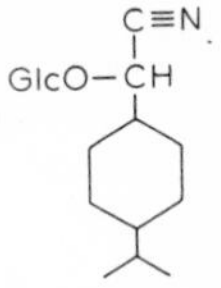

(XCIV) Polydesmus glycoside

for a rot in alfalfa (*Medicago sativa*), a plant known to be weakly cyanogenic. The fungus synthesised the two compounds in an analogous way to that in higher plants, and also contained specific hydrolases to split the glucosides[474].

With the sole exception of gynocardin from *Gynocardia odorata* and *Pangium edule*, both Flacourtiaceae, all the higher plant cyanogenic glycosides are formed from the corresponding L-amino acids by a route involving decarboxylation and oxidation of the C–N bond. The individual steps have not been worked out in all cases, but it seems likely that the common route is analogous to that shown for prunasin from L-phenylanine (Fig. 32). It is not certain whether the hydroxylation of the β-carbon takes place at the aldoxime or nitrile stage. The other cyanogenic glycosides whose structure is known arise from L-leucine (acacipetalin), L-isoleucine (lotaustralin), L-valine (linamarin), L-phenylalanine (amygdalin, lucumin, prunasin, sambunigrin and vicianin, the difference between these compounds lying in their stereochemistry and the nature of the sugar group), L-tyrosine (dhurrin, *Nandina* glucoside, and taxiphyllin), and *m*-tyrosine (zierin). In addition, the glucoside from the millipede *Polydesmus vicinus* has been shown to have the structure (XCIV), but it seems unlikely that it is formed from the corresponding 4-isopropylphenylalanine; it may well be formed by the direct addition of HCN to the mevalonate-engendered cuminaldehyde[475]. It is interesting to note that another polydesmoid millipede *Apheloria corrugata*, as a defence mechanism, has glands containing mandelonitrile and a hydrolase which are mixed if the insects are attacked leading to the emission of HCN and benzaldehyde, both of which are repellents to predators[315].

Cyanogenetic glycosides are obviously related to the alkaloids, and also to the glucosinolates (Fig. 32)[474]. Since cyanogenetic compounds are found in ferns they may play a defensive role equivalent to that of alkaloids in higher plants[475]. The glucosinolates are restricted to the families of the Capparidales[474] and have been shown to be anti-biting factors. None of the plants in this Order are alkaloidal.

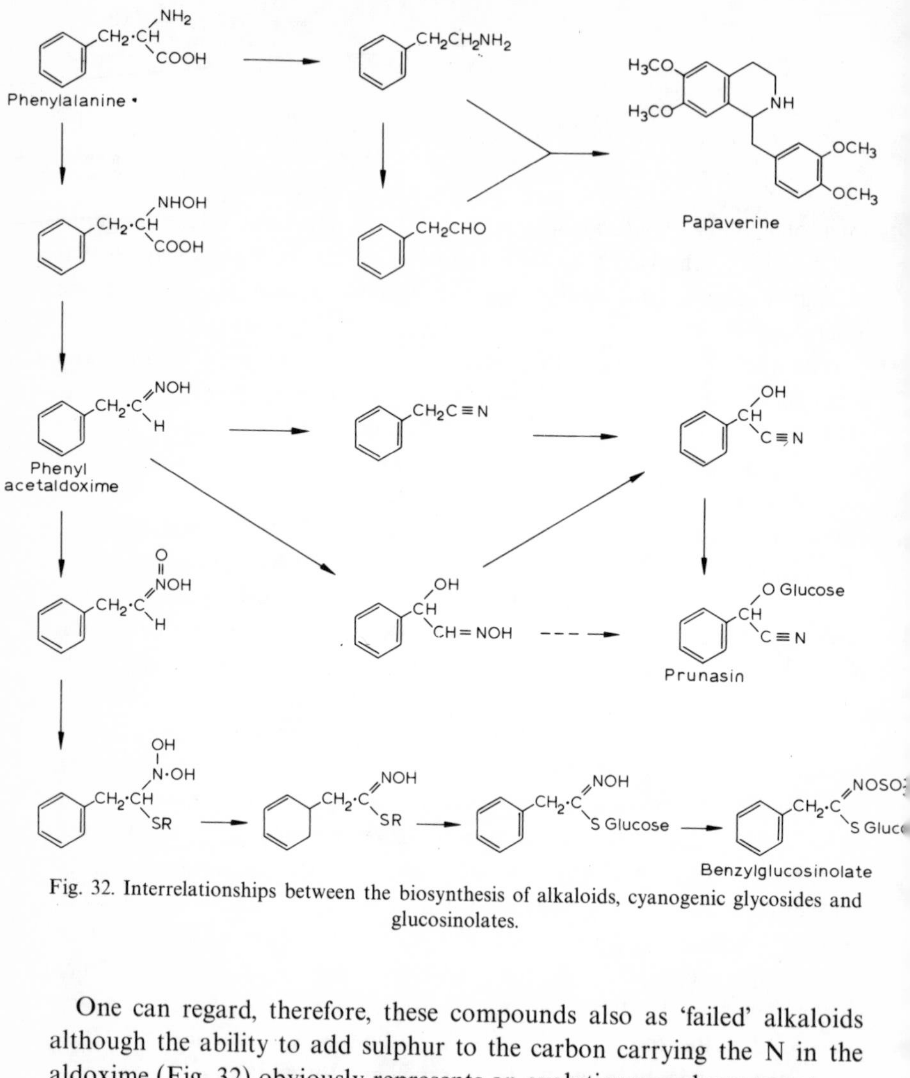

Phenylalanine
Phenyl acetaldoxime
Papaverine
Prunasin
Benzylglucosinolate

Fig. 32. Interrelationships between the biosynthesis of alkaloids, cyanogenic glycosides and glucosinolates.

One can regard, therefore, these compounds also as 'failed' alkaloids although the ability to add sulphur to the carbon carrying the N in the aldoxime (Fig. 32) obviously represents an evolutionary advancement.

Taking all the above information together the evolution of alkaloids can be summed up as shown in Fig. 33. The polyamines derived from putrescine and cadaverine are undoubtedly the most primitive since they occur in bacteria and in ribosomes, besides being associated with histones in the control of DNA. Palustrine (LXXXIX), which is derived from cadaverine may thus be regarded as the most ancient alkaloid and

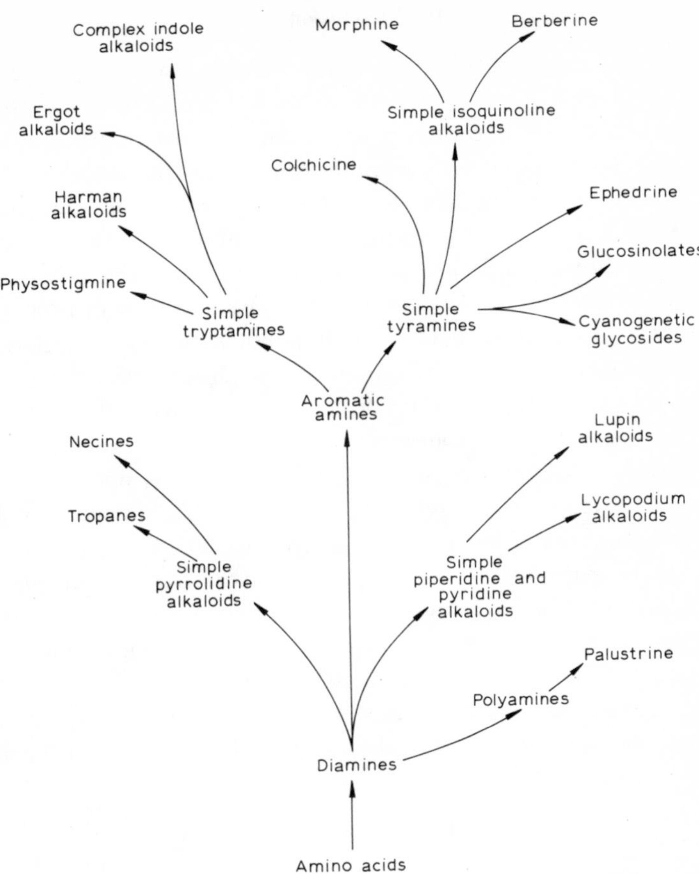

Fig. 33. The evolution of alkaloids.

it is not surprising to find it only occurs in the Equisetales. The next step was the ability to use the diamines, presumably as *N*-methyl derivatives for the elaboration of the simple pyrrolidine and piperidine alkaloids. From these the more complex necines, lupinines and *Lycopodium* alkaloids arose later. The next breakthrough was the ability to form and use the aromatic amines and its diversion to ephedrine, and the cyanogenic glycosides and glucosinolates. Finally the ability to form benzyliso-quinolines and complex indoles marks the final chapter in alkaloidal evolution.

References p. 287

16. Conclusions

The underlying theme of this Chapter has been to display the versatility of plant biochemistry and to stress that the end products of the diverse secondary biosynthetic pathways of any given plant determine, to a large extent, the niche that plant occupies in the total ecosystem. Since evolution, in the broadest sense, can be conceived as a succession of continuously changing ecosystems, it is obvious that a plant will survive only if it has, or can develop through selection or mutation, the necessary biochemical elaboration to face new challenges from variations in climate or from other invading organisms, plant or animal. It is not surprising, therefore, to find that the main problem in attempting to chart biochemical evolution in the Plant Kingdom is the enormous diversity of solutions which apparently work! One can see some evolutionary advantage in the development of more advanced energy-trapping systems in photosynthesis, in the development of cell walls that offer flexibility, strength and impenetrability, and in the utilisation of flavonoids as specific antifungal agents in higher plants. But in many other cases there is no clear picture and one can only argue advance on the rather dubious grounds of increasing biochemical complexity. Nevertheless, it is hoped that the suggestions which have been made will stimulate further thought and form a basis for future experiment and argument. For diversity is the very stuff of evolution and woe betide anyone who tries to elaborate a unified biochemical theorem.

ACKNOWLEDGEMENTS

Many colleagues have helped me through correspondence and in discussion in the preparation of this Chapter and I would like to thank especially E. C. Bate-Smith, Don Boulter, John Bu'Lock, Patrick Echlin, Jeffrey Harborne, Dan Janzen, L. Schoonhoven, Rainer Scora, R. H. Whittaker and Carol Williams. I am also most grateful to Marcel Florkin for his forbearance and continued support. Finally, I am delighted to acknowledge the help given in the preparation of this Chapter by Miss M. E. Wilkinson.

REFERENCES

1 G. G. Simpson and W. S. Beck, *Life, An Introduction to Biology*, 2nd edn., Harcourt, Brace and World, New York, 1965.
2 G. G. Simpson, *The Meaning of Evolution*, Revised edn., Yale University Press, New Haven, 1967.
3 A. Lee McAlester, *The History of Life*, Prentice Hall, Englewood Cliffs, New Jersey, 1968.
4 M. Calvin, *Chemical Evolution*, Oxford University Press, London, 1969.
5 G. de Beer, *A Handbook on Evolution*, 4th edn., British Museum, London, 1970.
6 J. B. Harborne (Ed.), *Phytochemical Phylogeny*, Academic Press, London, 1970.
7 P. Echlin, in ref. 6. pp. 1–19.
8 J. W. Schopf, *Biol. Rev. Cambridge Phil. Soc.*, 45 (1970) 317.
9 E. D. Hanson, *Animal Diversity*, 2nd edn., Prentice Hall, Englewood Cliffs, New Jersey, 1964.
10 M. Florkin (Ed.), *Aspects of the Origin of Life*, Pergamon, Oxford, 1960.
11 I. G. Gass, P. J. Smith, R. C. L. Wilson (Eds.), *Understanding the Earth*, Artemis Press, Sussex, U.K. 1971.
12 R. Buvet, C. Ponnamperuma and E. Schoffeniels (Eds.), *Molecular Evolution*, Vol. 1, *Chemical Evolution and the Origin of Life*, North Holland, Amsterdam, 1971.
13 E. Schoffeniels (Ed.), *Molecular Evolution*, Vol. 2, *Biochemical Evolution and the Origin of Life*, North Holland, Amsterdam, 1971.
14 K. P. Oakley and R. M. Muir-Wood, *The Succession of Life through Geological Time*, British Museum, London, 1967.
15 R. H. Whittaker, *Science*, 163 (1969) 150.
16 T. D. Brock, *Biology of Micro-organisms*, Prentice-Hall, Englewood Cliffs, N.J., 1970.
17 J. R. A. Pollock and R. Stevens (Eds.), *Dictionary of Organic Compounds*, 4th edn., with Supplements, Eyre and Spottiswood, London, 1965–1971.
18 *The Merck Index*, 8th edn., Merck, Rahway, New Jersey, 1971.
19 M. Florkin and E. Stotz (Eds.), *Comprehensive Biochemistry*, Vols. 1–30, Elsevier, Amsterdam, 1961–1975.
20 T. A. Geissman, in Ref. 19, Vol. 9, p. 215.
21 H. J. Nicholas, in Ref. 19, Vol. 20, p. 1.
22 I. D. Spencer, in Ref. 19, Vol. 20, p. 231.
23 R. Bentley and I. A. Campbell, in Ref. 19, Vol. 20, p. 415.
24 S. M. Siegel, in Ref. 19, Vol. 26A, p. 1.
25 J. M. Ghuysen, J. L. Strominger and D. J. Tipper, in Ref. 19, Vol. 26A, p. 53.
26 S. B. Hendricks and H. W. Siegelman, in Ref. 19, Vol. 27, p. 211.
27 L. N. M. Duysens and J. Amesz, in Ref. 19, Vol. 27, p. 237.
28 R. Ikan, *Natural Products*, Academic Press, London, 1969.
29 N. R. Farnsworth (Ed.), *Pharmacognosy Titles*, Vol. V, *Cumulative Genus Index*, Dept. of Pharmacognosy and Physiology, Univ. Illinois, Chicago, Ill.
30 S. W. Pelletier (Ed.), *Chemistry of the Alkaloids*, Van Nostrand/Reinhold. New York, 1970.
31 R. F. Raffauf, *A Handbook of Alkaloids and Alkaloid Containing Plants*, Wiley-Interscience, New York, 1970.
32 R. H. F. Manske and H. L. Holmes (Eds.), *The Alkaloids*, Vols. 1–12, Academic Press, New York, 1950–1969.
33 J. B. Pridham (Ed.), *Terpenoids in Plants*, Academic Press, London, 1967.
34 T. W. Goodwin (Ed.), *Aspects of Terpenoid Chemistry and Biochemistry*, Academic Press, London, 1971.

35 R. H. Thomson, *Naturally Occurring Quinones*, 2nd edn. Academic Press, London, 1971.
36 C. Hitchcock and B. W. Nichols, *Plant Lipid Biochemistry*, Academic Press, London, 1971.
37 P. Bernfeld (Ed.), *Biogenesis of Natural Compounds*, 2nd edn., Pergamon, Oxford, 1967.
38 T. A. Geissman and D. H. G. Crout, *Organic Chemistry of Secondary Metabolism*, Freeman Cooper, San Francisco, 1969.
39 J. B. Pridham and T. Swain (Eds.), *Biosynthetic Pathways in Higher Plants*, Academic Press, London, 1965.
40 J. B. Hendrickson, *The Molecules of Nature*, Benjamin, New York, 1965.
41 J. D. Bu'Lock, *The Biosynthesis of Natural Products*, McGraw-Hill, London, 1965.
42 A. H. Der Marderosian, in H. W. Youngken Jr. (Ed.), *Food-Drugs from the Sea*, Marine Technology Society, Washington, 1970.
43 A. H. Der Marderosian, *J. Pharm. Sci.*, 58 (1969) 1–33.
44 M. Florkin and H. Mason, *Comparative Biochemistry*, Vols. 1–8, Academic Press, New York, 1960–1964.
45 G. A. D. Haslewood, *Bile Salts*, Methuen, London, 1967.
46 M. O. Dayhoff, *Atlas of Protein Sequence and Structure*, Vol. 4, National Biomedical Research Foundation, Silver Springs, Md., 1970.
47 V. G. Dethier and E. Stellar, *Animal Behaviour*, 2nd edn., Prentice-Hall, Englewood Cliffs, N.J., 1964.
48 A. Roe and G. G. Simpson (Eds.), *Behaviour and Evolution*, Yale Univ. Press, New Haven, 1958.
49 G. L. Stebbins, *Processes of Organic Evolution*, Prentice-Hall, Englewood Cliffs, N.J., 1966.
50 G. L. Stebbins, *Taxon*, 20 (1971) 3.
51 W. H. Wagner Jr., in V. H. Heywood (Ed.), *Modern Methods in Plant Taxonomy*, Academic Press, London, 1968.
52 F. Ehrendorfer, *Taxon*, 19 (1970) 185.
53 G. G. Simpson, *Principles of Animal Taxonomy*, Columbia Univ. Press, New York, 1961.
54 E. Mayr, *Animal Species and Evolution*, Harvard Univ. Press, Cambridge, Mass., 1963.
55 D. Briggs and S. M. Walters, *Plant Variation and Evolution*, McGraw-Hill, New York, 1969.
56 A. Cronquist, *The Evolution and Classification of Flowering Plants*, Houghton, Mifflin, Boston, 1968.
57 P. H. Davis and V. H. Heywood, *Principles of Angiosperm Taxonomy*, Oliver and Boyd, Edinburgh, 1963.
58 A. Takhtajan, *Flowering Plants—Origin and Dispersal*, (C. Jeffrey, Translator), Oliver and Boyd, Edinburgh, 1969.
59 L. F. Laporte, *Ancient Environments*, Prentice-Hall, Englewood Cliffs, N.J., 1968.
60 R. A. Phinney, Ed., *The History of the Earth's Crust*, Princeton Univ. Press, Princeton, N.J., 1968.
61 H. N. Andrews Jr., *Studies in Paleobotany*, Wiley, New York, 1961.
62 H. P. Banks, *Evolution and Plants of the Past*, Wadsworth, Belmont, Calif., 1970.
63 P. Bell and C. Woodcock, *The Diversity of Green Plants*, 2nd edn., Arnold, London, 1971.
64 A. Cronquist, *Introductory Botany*, 2nd edn. Harper, Row, New York, 1971.
65 N. S. Parihar, *Bryophyta*, 4th edn., Central Book Depot, Allahabad, 1962.
66 N. S. Parihar, *Pteridophytes*, 5th edn., Central Book Depot, Allahabad, 1965.
67 E. V. Watson, *The Structure and Life of Bryophytes*, 2nd edn., Hutchinson, London, 1967.
68 K. R. Sporne, *The Morphology of Pteridophytes*, 3rd edn., Hutchinson, London, 1970.

69 J. E. Gottlieb, *Plants; Adaptation through Evolution*, Reinhold, New York, 1968.
70 C. J. Alexopoulos and H. C. Bold, *Algae and Fungi*, McMillan, New York, 1967.
71 M. E. Hale Jr., *The Biology of Lichens*, Arnold, London, 1967.
72 W. T. Doyle, *The Biology of Higher Cryptogams*, McMillan, New York, 1970.
73 J. Hogg, *Edinburgh New Phil. J.*, 12 (1860) 216.
74 E. Haeckel, *Systematische Phylogenie*, Vol. 1, Reimer, Berlin, 1894.
75 H. F. Copeland, *The Classification of Lower Organisms*, Pacific Books, Palo Alto, Calif., 1956.
76 R. H. Whittaker, *Quart. Rev. Biol.*, 34 (1959) 210.
77 L. Margulis, *Evolution*, 25 (1971) 242.
78 L. Margulis, *Origin of Eukaryotic Cells*, Yale Univ., New Haven, 1970.
79 L. Margulis, *Science*, 161 (1968) 1020.
80 P. Echlin and I. Morris, *Biol. Revs. Cambridge Phil. Soc.*, 40 (1965) 143.
81 R. M. Klein and A. Cronquist, *Quart. Rev. Biol.*, 42 (1967) 105.
82 R. Y. Stanier, M. Doudoroff and E. A. Adelberg, *The Microbial World*, 2nd edn., Prentice-Hall, Englewood Cliffs, N.J., 1963.
83 C. Linnaeus, *Systema Naturae*, Haak, Leyden, 1735.
84 B. D. Davis, R. Dulbecco, H. N. Eisen, H. S. Ginsberg and W. B. Woud, *Principles of Microbiology and Immunology*, Harper and Row, New York, 1968.
85 G. E. Hutchinson, *A Treatise on Limnology*, Vol. 2, Wiley, New York, 1967.
86 J. D. Ebert, and I. M. Sussex, *Interacting Systems in Development*, 2nd edn., Holt, Rinehart and Winston, New York, 1970.
87 C. P. Swanson, *The Cell*, 3rd edn., Prentice-Hall, Englewood Cliffs, N.J., 1969.
88 L. E. Mettler and T. G. Gregg, *Population Genetics and Evolution*, Prentice-Hall, Englewood Cliffs, N.J., 1969.
89 R. P. Levine, *Genetics*, Holt, Rinehart and Winston, New York, 1968.
90 P. E. Hartman and S. R. Suskind, *Gene Action*, 2nd edn., Prentice-Hall, Englewood Cliffs, N.J., 1969.
91 V. Grant, *The Origin of Adaptations*, Columbia Univ. Press, New York, 1963.
92 P. Echlin, in H. P. Charles and B. C. J. G. Knights (Eds.), *Organisation and Control in Procaryotic and Eucaryotic Cells*, Cambridge Univ. Press, London, 1971.
93 P. Echlin, *Advan. Organic Geochem.*, Pergamon, Oxford, 1971.
94 N. G. Carr and I. W. Craig, in ref. 6, pp. 119–143.
95 S. Bartnicki-Garcia, in ref. 6, pp. 81–103.
96 S. Bartnicki-Garcia, *Ann. Rev. Microbiol.*, 22 (1968).
97 V. H. Heywood, *Plant Taxonomy*, Arnold, London, 1967.
98 J. Heslop-Harrison, *New Concepts in Flowering Plant Taxonomy*, Heinemann, London, 1953.
99 L. Benson, *Plant Classification*, Heath, Boston, 1957.
100 V. H. Heywood (Ed.), *Modern Methods in Plant Taxonomy*, Academic Press, London, 1968.
101 E. Mayr, *Principles of Systematic Zoology*, McGraw-Hill, New York, 1969.
102 E. Mayr (Ed.), *The Species Problem*, Amer. Assoc. Adv. Sci. Publication No. 50, Washington, 1957.
103 J. Heslop-Harrison, in T. Swain (Ed.), *Chemical Plant Taxonomy*, Academic Press, London, 1963.
104 D. L. Eicher, *Geologic Time*, Prentice-Hall, Englewood Cliffs, N.J., 1968.
105 P. Harris, in ref. 11, pp. 53–69.
106 L. V. Berkner and L. C. Marshall, *Proc. Natl. Acad. Sci. (U.S.)*, 53 (1965) 1169.
107 P. Cloud, in ref. 11, pp. 151–155.
108 P. C. Sylvester-Bradley, in ref. 11, pp. 123–141.

109 S. W. Fox (Ed.), *The Evolution of Prebiological Systems*, Academic Press, New York, 1965.
110 A. G. Smith, in ref. 11, pp. 213–231.
111 F. J. Vine, in ref. 11, pp. 233–249.
112 J. G. Hawkes and P. Smith, *Nature*, 207 (1965) 48.
113 M. R. House, in ref. 11, pp. 193–211.
114 S. A. Tyler and E. S. Barghoorn, *Science*, 119 (1954) 606.
114a G. Eglington, *Pure Appl. Chem.*, 34 (1973) 611.
115 E. S. Barghoorn and J. W. Schopf, *Science*, 152 (1966) 758.
116 J. W. Schopf, E. S. Barghoorn, M. D. Maser and R. O. Gordon, *Science*, 149 (1965) 1365.
117 J. W. Schopf and E. S. Barghoorn, *Science*, 156 (1967) 508.
118 E. S. Barghoorn and J. W. Schopf, *Science*, 150 (1965) 337.
119 A. Allsop, *New Phytologist*, 68 (1969) 591.
120 W. G. Chaloner, *Biol. Rev. Cambridge Phil. Soc.*, 45 (1970) 353.
121 W. G. Chaloner and K. Allen, in ref. 6, pp. 21–30.
122 S. Moorbath, in ref. 11, pp. 41–51.
123 A. Cronquist, *Botan. Rev.*, 26 (1960) 426.
124 T. W. Goodwin, in T. W. Goodwin (Ed.), *Biochemistry of Plant Pigments*, Academic Press, London, 1965.
125 J. H. Burnett, *Fundamentals of Mycology*, Arnold, London, 1968.
126 J. H. Birkinshaw, in G. C. Ainsworth and A. S. Sussman (Eds.), *The Fungi*, Vol. 1, Academic Press, New York, 1965.
127 W. B. Turner, *Fungal Metabolites*, Academic Press, London, 1971.
128 J. Muller, *Biol. Rev. Cambridge Phil. Soc.*, 45 (1970) 417.
129 Hsen-Hsu Hu, *Science Record*, 3 (1950) 221.
130 E. J. H. Corner, *The Life of Plants*, Weidenfeld and Nicolson, London, 1964.
131 J. G. Hawkes (Ed.), *Chemotaxonomy and Serotaxonomy*, Academic Press, London, 1968.
132 M. Nei, *Nature*, 221 (1969) 40.
133 T. Ohta and M. Kimura, *Nature*, 233 (1971) 118.
134 F. Crick, *Nature*, 234 (1971) 25.
135 R. J. Britten and E. H. Davidson, *Science*, 165 (1969) 349.
136 R. L. Watts and D. C. Watts, *Nature*, 217 (1968) 1125.
137 R. L. Watts and D. C. Watts, *J. Theoret. Biol.*, 20 (1968) 227.
137a H. Rees and R. N. Jones, *Intern. Rev. Cytol.*, 32 (1972) 53.
138 E. Freese, *Angew. Chem., Internat. Edn.*, 8 (1969) 12.
139 M. Kimura and T. Ohta, *J. Molec. Evolution*, 1 (1971) 1.
140 R. E. Dickerson, *J. Molec. Evolution*, 1 (1971) 26.
141 T. Ohta, *J. Molec. Evolution*, 1 (1971) 150.
142 R. Holmquist, *J. Molec. Evolution*, 1 (1971) 211.
143 W. M. Fitch, *J. Molec. Evolution*, 1 (1971) 84.
144 R. F. Lyndon, in J. B. Pridham (Ed.), *Plant Cell Organelles*, Academic Press, London, 1968, p.16.
145 D. E. Kohne, in ref. 131, p. 135.
146 R. J. Britten and D. E. Kohne, *Science*, 161 (1968) 529.
147 P. M. B. Walker, *Nature*, 219 (1968) 228.
148 E. M. Southern, *Nature*, 227 (1970) 794.
149 F. Crick, *Nature*, 234 (1971) 25.
150 J. Paul, *Nature*, 238 (1972) 444.
150a G. M. Evans, H. Reis, C. L. Snell and S. Sun, *Proc. 3rd Oxford Chromosome Conf.*, 1973.

151 G. M. Tomkins, T. D. Gelehrter, D. Granner, D. Martin Jr., H. H. Samuels and E. B. Thompson, *Science*, 166 (1969) 1474.
152 B. H. Hoyer, B. J. McCarthy and E. T. Bolton, *Science*, 144 (1964) 959.
153 A. J. Bendich and E. T. Bolton, *Plant Physiol.*, 42 (1967) 959.
154 Z. A. Medvedev, *J. Molec. Evolution*, 1 (1972) 270.
155 W. Hennig and P. M. B. Walker, *Nature*, 225 (1970) 915.
155a J. A. Mazrimas and F. T. Hatch, *Nature New Biol.*, 240 (1972) 102.
156 I. Gibson and G. Hewitt, *Nature*, 225 (1970) 67.
157 M. Stroun, P. Charles, P. Anker and S. R. Pelc, *Nature*, 216 (1967) 716.
158 M. Sampson and D. D. Davies, *Exptl. Cell Res.*, 43 (1966) 199.
159 B. L. Turner, *Taxon*, 18 (1969) 134.
160 J. de Ley, in T. Dohzhansky, M. K. Hecht and W. C. Steere (Eds.), *Evolutionary Biology*, Vol. 2, Appleton Century Crofts, New York, 1968, p. 102.
161 J. de Ley, in ref. 13, p. 297.
162 M. Mandel, *Ann. Rev. Microbiol.*, 23 (1969) 239.
163 R. Hall, *Botan. Rev.*, 35 (1969) 285.
164 U. E. Loening, *Pure Appl. Chem.*, 34 (1973) 579.
165 U. E. Loening, *J. Mol. Biol.*, 38 (1968) 355.
166 U. E. Loening and J. Ingle, *Nature*, 215 (1967) 363.
167 H. J. Gould, in ref. 131, p. 149.
168 B. V. Milborrow (Ed.), *Biosynthesis and its Control in Plants*, Academic Press, London, 1973.
169 J. Ingle, in ref. 168, p. 70.
170 D. Boulter, *Ann. Rev. Plant Physiol.*, 21 (1970) 91.
171 P. J. Lea and R. D. Norris, *Phytochem.*, 11 (1972) 2897.
172 H. G. Zachau, *Angew. Chem., Intern. Edn.*, 8 (1969) 711.
173 M. Staehelin, *Experientia*, 27 (1971) 1.
174 T. M. Jukes and L. Gatlin, *Prog. Nucleic Acid Res.*, 11 (1971) 303.
175 J. Hindley and D. H. Staples, *Nature* 224 (1969) 964.
176 P. I. Payne and T. A. Dyer, *Nature New Biol.*, 235 (1972) 145.
177 R. M. Dowben, *General Physiology*, Harper and Row, Boston, Mass., 1969.
178 R. A. Raff and H. R. Mahler, *Science*, 177 (1972) 575.
179 B. E. H. Maden, *Prog. Biophys. Mol. Biol.*, 22 (1971) 129.
180 R. J. Ellis, *Planta*, 91 (1970) 329.
181 G. H. Pigott and N. G. Carr, *Science*, 175 (1972) 1259.
182 G. H. Pigott and N. G. Carr, in ref. 6, p. 119.
183 O. Öpik, in J. B. Pridham (Ed.), *Plant Cell Organelles*, Academic Press, London, 1968, p. 47.
184 D. E. Fairbrothers, *The Serological Museum Bulletin*, No. 41 (1969).
185 D. E. Fairbrothers, in V. H. Heywood (Ed.), *Modern Methods in Plant Taxonomy*, Academic Press, London, 1968, p. 141.
186 J. G. Vaughn, *Sci. Prog. Oxford*, 56 (1968) 205.
187 D. Boulter, D. A. Thurman and B. L. Turner, *Taxon*, 15 (1966) 135.
188 G. H. Dixon, *Essays in Biochemistry*, 2 (1966) 147.
189 W. M. Fitch and E. Margoliash, *Taxon*, 20 (1971) 51.
190 C. Nolan and E. Margoliash, *Ann Rev. Biochem.*, 37 (1968) 727.
191 E. Zuckerkandl and L. Pauling, in V. Bryson and H. J. Vogel (Eds.), *Evolving Genes and Proteins*, Academic Press, New York, 1965, p. 97.
192 E. Margoliash, W. M. Fitch and R. E. Dickerson, in ref. 13, p. 52.
193 R. L. Watts, in ref. 13, p. 14.
194 R. Holmquist, *J. Molec. Evolution*, 1 (1972) 134.

195 R. E. Dickerson, *J. Molec. Evolution*, 1 (1971) 26.
196 J. Buettner-Janusch and R. L. Hill, in V. Bryson and H. J. Vogel, (Eds.), *Evolving Genes and Proteins*, Academic Press, New York, 1965, p. 167.
197 A. Hafleigh and C. A. Williams, *Science*, 151 (1966) 1534.
198 G. Pfleiderer, R. Linke and G. Reinhardt, *Comp. Biochem. Physiol.*, 33 (1970) 955.
199 C. L. Oakley, in J. R. Norris and D. W. Ribbons (Eds.), *Methods in Microbiology*, Academic Press, London, 1971, p. 173.
200a C. Manwell and C. M. Ann Baker, *Molecular Biology and the Origin of Species*, Sidgwick and Jackson, London, 1970.
200b S. H. Boger, *Nature*, 239 (1972) 453.
201 B. L. Turner, *Taxon*, 18 (1969) 134.
202 F. G. Young, *Proc. Roy. Soc., (London)*, B, 157 (1962) 1.
203 E. Margoliash and W. M. Fitch, *Ann. N.Y. Acad. Sci.*, 151 (1968) 359.
204 H. Harris, *Brit. Med. Bull.*, 25 (1969) 5.
205 R. C. Richmond, *Nature*, 255 (1970) 1029.
206 H. Neurath, K. A. Walsh and W. P. Winter, *Science*, 158 (1967) 1638.
207 J. A. Adelson, *Nature*, 229 (1971) 321.
208 E. J. W. Barrington, in ref. 13, p. 174.
209 D. C. Watts, in ref. 13, p. 150.
210 J. Yourno, T. Kohno and J. R. Roth, *Nature*, 228 (1970) 820.
211 H. B. White and N. O. Kaplan, *J. Mol. Evolution*, 1 (1972) 158.
212 A. C. Wilson and N. O. Kaplan, in C. A. Leone (Ed.), *Taxonomic Biochemistry and Serology*, Ronald Press, New York, 1964, p. 321.
213 S. H. Boyer, E. F. Crosby, T. F. Thurman, A. N. Noyes, G. F. Fuller, S. E. Leslie, M. K. Shepard and C. N. Herndon, *Science*, 166 (1969) 1428.
214 M. Goodman, J. Barabas, G. Matsuda and G. W. Moore, *Nature*, 233 (1971) 604.
215 T. H. Jukes and R. Holmquist, *Science*, 177 (1972) 530.
216 D. Boulter, *Pure and Appl. Chem.*, 34 (1973) 539.
217 D. Boulter, J. A. M. Ramshaw, E W. Thomson, M. Richardson and R. H. Brown, *Proc. Roy. Soc., (London), Ser. B*, 181 (1972) 441.
218 R. E. Dickerson, *J. Mol. Biol.*, 57 (1971) 1.
219 R. E. Dickerson, T. Takano, D. Eisenberg, O. B. Kallai, L. Samson, A Cooper and E. Margoliash, *J. Biol. Chem.*, 246 (1971) 1511.
220 D. Boulter and J. A. M. Ramshaw, *Phytochem.*, 11 (1972) 279.
221 W. M. Fitch and E. Margoliash, *Science*, 155 (1967) 279.
222 G. N. Lance and W. T. Williams, *Computer J.*, 9 (1967) 373.
223 W. M. Fitch and E. Margoliash, *Brookhaven Symp. Biol.*, 21 (1969) 217.
224 A. D. J. Meeuse, *Acta Bot. Neerl.*, 19 (1970) 61.
225 A. D. J. Meeuse, *Acta Bot. Neerl.*, 19 (1970) 133.
226 K. Kubitski, *Taxon*, 18 (1969) 360.
227 E. C. Bate-Smith, *Nature*, 236 (1972) 353.
228 A. J. Eames, *Morphology of the Angiosperms*, Harper & Row, New York, 1961.
229 K. Sporne, *New Phytologist*, 69 (1970) 1161.
230 R. H. Tschudy and R. A. Scott (Eds.), *Aspects of Palynology*, Wiley-Interscience, New York, 1969.
230a J. A. M. Ramshaw, L. R. Richardson, B. J. Meatyard, R. H. Brown, M. Richardson, E. W. Thompson and D. Boulter, *New Phytologist*, 71 (1972) 733.
231 H. Matsubara, T. H. Jukes and C. R. Cantor, *Brookhaven Symp. Biol.*, 21 (1968) 201.
232 D. O. Hall, R. Cammack and K. K. Rao, *J. Pure and Applied Chem.*, (1973) (in the press).
233 D. O. Hall, R. Cammack and K. K. Rao, *Nature*, 233 (1971) 136.

234 K. K. Rao, R. Cammack, D. O. Hall and R. Johnson, *Biochem. J.*, 122 (1971) 257.
235 J. F. Lavering, R. W. Le Maitre and B. W. Chappel, *Nature*, 230 (1971) 18.
236 D. H. Kenyon and G. Steinman, *Biochemical Predestination*, McGraw Hill, New York, 1969.
237 N. O. Kaplan, in V. H. Bryson and H. J. Vogel (Eds.), *Evolving Genes and Proteins*, Academic Press, New York, 1965.
238 A. Baich and M. Johnson, *Nature*, 218 (1968) 464.
239 S. Zamenhof and H. H. Eichorn, *Nature*, 216 (1967) 456.
240 R. Y. Stanier, in J. G. Hawkes (Ed.), *Chemotaxonomy and Serotaxonomy*, Academic Press, London, 1968.
241 M. H. Richmond, *Essays in Biochemistry*, 4 (1968) 108.
242 H. Tristram, *Sci. Progr. (Oxford)*, 56 (1968) 449.
243 P. J. Casselton, *Sci. Progr. (Oxford)*, 57 (1969) 207.
244 P. Datta, *Science*, 165 (1969) 556.
245 J. L. Cánovas, L. N. Ornston and R. Y. Stanier, *Science*, 156 (1967) 1696.
246 B. J. Miflin, in ref. 168, p. 49.
247 P. D. J. Weitzman and D. Jones, *Nature*, 219 (1968) 270.
248 D. R. Thatcher and R. B. Cain, *Biochem. J.*, 120 (1970) 291.
249 D. L. Rann and R. B. Cain, *Biochem. J.*, 114 (1969) 77P.
250 R. B. Cain, R. F. Bilton and D. A. Darrah, *Biochem. J.*, 108 (1968) 799.
251 T. J. Mabry and A. S. Dreiding, *Recent Advances in Phytochemistry*, 1 (1968) 145.
252 D. Sharples, M. S. Spring and J. R. Stoker, *Phytochem.*, 11 (1972) 3069.
253 H. J. Vogel, *Amer. Nat.*, 98 (1964) 435.
254 H. Kreisler, (1969), quoted in G. C. Ainsworth, *Dictionary of the Fungi*, 6th Edn., Commonwealth Mycological Institute, Kew, Great Britain, 1971.
255 D. C. Watts, *Adv. Comp. Physiol. Biochem.*, 3 (1968) 1.
256 C. H. Doy, *Rev. Pure and Appl. Chem.*, 18 (1968) 41.
257 R. A. Jensen and D. S. Nasser, *J. Bacteriol.*, 95 (1968) 188.
258 M. B. Berlyn and N. H. Giles, *J. Bacteriol.*, 99 (1969) 222.
259 S. I. Ahmed and N. H. Giles, *J. Bacteriol.*, 99 (1969) 231.
260 M. B. Berlyn, S. I. Ahmed and N. H. Giles, *J. Bacteriol.*, 104 (1970) 768.
261 M. Luckner, *Secondary Metabolism in Plants and Animals*, Chapman and Hall, London, 1972.
262 D. G. Gilchrist, T. S. Woodin, M. S. Johnson and T. Kosuge, *Plant Physiol.*, 49 (1972) 52.
263 I. Zelitch, *Photosynthesis, Photorespiration and Plant Productivity*, Academic Press, New York, 1971.
264 N. Bishop, *Ann. Rev. Biochem.*, 40 (1971) 197.
265 H. Metzer, *Progress in Photosynthesis Research*. 3 Vols., International Union of Biological Sciences, 1969.
266 B. Kok, in M. B. Wilkins (Ed.), *Physiology of Plant Growth and Development*, McGraw Hill, London, 1969, p. 335.
267 W. M. Laetsch, *Sci. Progr. (Oxford)*, 57 (1969) 323.
268 P. Echlin, *Symp. Soc. Gen. Microbiol.*, 20 (1970) 221.
269 N. Pfenning, *Ann. Rev. Microbiol.*, 21 (1967) 285.
270 R. K. Ellsworth, in C. O. Chichester (Ed.), *The Chemistry of Plant Pigments*, Academic Press, New York, 1972, p. 85.
271 L. Bogorad and R. F. Troxler, in Ref. 37, p. 247.
272 L. Bogorad, in T. W. Goodwin (Ed.), *Chemistry and Biochemistry of Plant Pigments*, Academic Press, London, 1965, p. 29.
273 C. ÓhEocha, in T. W. Goodwin (Ed.), *Chemistry and Biochemistry of Plant Pigments*, 1965, p. 175.

274 A. Bennett, and L. Bogorad, *Biochemistry*, 10 (1971) 3625.
275 D. J. Chapman, W. J. Cole and H. W. Siegelman, *Am. J. Bot.*, 55 (1968) 314.
276 C. Brooks and D. J. Chapman, *Phytochem.*, 11 (1972) 2663.
277 H. W. Siegelman, D. J. Chapman and W. J. Cole, in T. W. Goodwin (Ed.), *Porphyrins and Related Compounds*, Academic Press, London, 1968.
278 W. R. Briggs and H. V. Rice, *Ann. Rev. Plant Physiol.*, 23 (1972) 293.
279 N. K. Boardman, *Ann. Rev. Plant Physiol.*, 21 (1970) 115.
280 D. C. Fork and J. Amesz, *Ann. Rev. Plant Physiol.*, 20 (1969) 305.
281 R. B. Park and P. V. Sane, *Ann. Rev. Plant Physiol.*, 22 (1971) 395.
282 M. D. Hatch and C. R. Slack, *Ann. Rev. Plant Physiol.*, 21 (1970) 141.
283 D. H. Northcote, *Ann. Rev. Plant Physiol.*, 23 (1972) 113.
284 D. H. Northcote, *Essays Biochem.*, 5 (1969) 89.
285 P. E. Pilet, *Les Parois Cellulaires*, Doin, Paris, 1971.
286 R. Cleland, *Ann. Rev. Plant Physiol.*, 22 (1971) 197.
287 D. T. A. Lamport, *Ann. Rev. Plant Physiol.*, 21 (1970) 235.
288 O. Kandler and J. Ghysen, *Bacteriol. Rev.*, 32 (1968) 425.
289 G. O. Aspinall, *Polysaccharides*, Pergamon, Oxford, 1970.
290 G. A. Cooper-Driver and T. Swain, *J. Linn. Soc. (Bot.)*, 67 (1973) Suppl. 1, p. 111.
291 C. Jeuniaux, in ref. 13, p. 304.
292 D. Tyrrell, *Botan. Rev.*, 35 (1969) 305.
292a Ch. Jeuniaux, *Chitine et Chitinolyse*, Masson, Paris, 1963.
293 J. Bu'Lock and H. Smith, *Experientia*, 17 (1961) 553.
294 R. H. Thompson, in T. W. Goodwin (Ed.), *Chemistry and Biochemistry of Plant Pigments*, Academic Press, New York, 1965, p. 333.
295 W. Z. Hassid, *Science*, 165 (1969) 137.
296 S. W. Fox, *Pure and Appl. Chem.*, 34 (1973) 641.
297 R. W. Kent, in ref. 44, Vol. 7, p. 93.
298 K. C. Highnam and L. Hull, *The Comparative Endocrinology of the Invertebrates*, Arnold, London, 1969.
299 I. A. Pearl, *The Chemistry of Lignin*, Dekker, New York, 1967.
300 S. A. Brown, *Ann. Rev. Plant Physiol.*, 17 (1966) 223.
301 R. K. Ibrahim, G. H. N. Towers and R. D. Gibbs, *J. Linn. Soc. (Bot.)*, 58 (1962) 223.
302 G. H. N. Towers, and W. S. G. Maass, *Phytochem.*, 4 (1965) 57.
303 S. M. Siegel, *Amer. J. Botany*, 56 (1969) 175.
304 R. F. Leo and E. S. Barghoorn, *Science*, 158 (1970) 582.
305 R. J. Bandoni, K. Moore, P. V. SubbaRao and G. H. N. Towers, *Phytochem.*, 7 (1968) 205.
306 K. Moore, P. V. SubbaRao and G. H. N. Towers, *Biochem. J.*, 106 (1968) 507.
307 H. Shimazono and F. F. Nord, *Arch. Biochem. Biophys.*, 78 (1958) 263.
308 H. Shimazono and F. F. Nord, *Arch. Biochem. Biophys.*, 87 (1960) 140.
309 K. R. Markham and L. J. Porter, *Phytochem.*, 8 (1969) 1777.
310 B. Z. Siegel and S. M. Siegel, *Plant Physiol.*, 57 (1970) 285.
311 J. D. Bu'Lock and A. J. Powell, *Experientia*, 21 (1965) 55.
312 E. D. Weinberg, *Advan. Microbiol. Physiol.*, 4 (1970) 1.
313 J. W. Foster, *Chemical Activities of Fungi*, Academic Press, New York, 1949.
314 J. R. D. McCormick, in D. Gottlieb and P. D. Shaw (Eds.), *Antibiotics*, Vol. 2, Springer, Bonn, 1967, p. 113.
315 E. Sondheimer and J. B. Simone (Eds.), *Chemical Ecology*, Academic Press, New York, 1970.
316 J. B. Harborne (Ed.), *Phytochemical Ecology*, Academic Press, London, 1972.
317 D. A. Levin, *Amer. Naturalist*, 105 (1971) 157.

318 I. A. M. Cruickshank, *World Rev. Pest Control*, 5 (1966) 161.
319 G. S. Fraenkel, *Entomol. Exptl. Appl.*, 12 (1969) 473.
320 R. H. Whittaker and P. P. Feeny, *Science*, 171 (1971) 757.
321 L. M. Schoonhoven, in: T. J. Mabry (Ed.), *Rec. Advan. Phytochem.*, 4 (1970) 128.
322 J. E. Watkin and D. S. Magrill, *4th Intern. Symp. Chem. Nat. Prod., Abstracts*, (1966) 146.
323 M. L. Rueppel and H. Rapoport, *J. Amer. Chem. Soc.*, 92 (1970) 5528.
324 E. Leete, G. B. Bodem and M. F. Manuel, *Phytochem.*, 10 (1971) 2687.
325 K. Mothes, *Lloydia*, 29 (1966) 156.
326 P. Laland, *Z. Physiol. Chem.*, 204 (1932) 112.
327 L. Fowden, *Phytochem.*, 11 (1972) 2271.
328 L. Fowden, in ref. 168, p. 323.
329 D. E. Morland, *Ann. Rev. Plant Physiol.*, 18 (1967) 365.
330 T. W. Goodwin, in ref. 39, p. 37.
331 V. E. Davis, and M. J. Walsh, *Science*, 167 (1970) 1005.
332 G. Cohen and M. Collins, *Science*, 167 (1970) 1749.
333 J. Asselineau, *Les Lipides Bactériens*, Hermann, Paris, 1962.
334 N. A. Sørensen, in *Recent Advances in Phytochemistry*, Vol. 1, Appleton Century Crofts, New York, 1968, p. 187.
335 F. Bohlmann, *Fortschr. Chem. Org. Naturstoffe*, 25 (1967) 1.
336 G. Eglinton and R. J. Hamilton, *Science*, 156 (1967) 1322.
337 A. T. James, in J. B. Harborne and T. Swain (Eds.), *Perspectives in Phytochemistry* Academic Press, London, 1969.
338 P. K. Stumpf, *Ann. Rev. Biochem.*, 38 (1969) 159.
339 D. N. Brindley, S. Matsumura and K. Bloch, *Nature*, 224 (1967) 666.
340 J. D. Bu'Lock and G. N. Smith, *J. Chem. Soc., (London) C*, (1967) 332.
341 T. Kaneda, *J. Biol. Chem.*, 238 (1963) 1229.
342 D. Branton, *Ann. Rev. Plant Physiol.*, 20 (1969) 209.
343 E. D. Korn, *Ann. Rev. Biochem.*, 38 (1969) 263.
344 P. Mazliak, *Les Membranes Protoplasmiques*, Doin, Paris, 1971.
345 I. A. Wolff, *Science*, 154 (1966) 1140.
346 F. B. Shorland, in T. Swain (Ed.), *Chemical Plant Taxonomy*, Academic Press, London, 1963, p. 253.
347 P. Mazliak, *Progr. Phytochemistry*, 1 (1968) 49.
348 P. E. Kolattukudy, *Science*, 159 (1968) 498.
349 B. W. Nichols and A. T. James, *Progr. Phytochemistry*, 1 (1968) 1.
350 G. R. Jamieson and E. H. Reid, *Phytochem.*, 11 (1972) 471.
351 R. W. Holton and H. H. Blecker, in *Properties and Products of Algae*, Plenum, New York, 1970.
352 B. W. Nichols and R. S. Appleby, *Phytochem.*, 8 (1969) 1907.
353 R. F. Lee and A. R. Loeblich, *Phytochem.*, 10 (1971) 593.
354 J. T. Martin and B. E. Juniper, *The Cuticles of Plants*, Arnold, London, 1970.
355 K. H. Overton (Ed.), *Terpenoids and Steroids*, Vol. 1, Chemical Society, London, 1971.
356 K. H. Overton (Ed.), *Terpenoids and Steroids*, Vol. 2, Chemical Society, London, 1972.
357 T. W. Goodwin, *Natural Substances Formed Biologically from Mevalonic Acid*, (Biochemical Society Symposium No. 29), Academic Press, London.
358 W. D. Ollis and I. O. Sutherland, in W. D. Ollis (Ed.), *Recent Developments in the Chemistry of Natural Phenolic Compounds*, Pergamon, Oxford, 1961.
359 T. W. Goodwin, in Ref. 33, p. 1.
360 R. B. Clayton, in Ref. 315, p. 235.
361 H. M. Fox and H. G. Vevers, *The Nature of Animal Colours*, Sidgwick and Jackson, London, 1960.

362 R. B. Clayton, in Ref. 34, p. 1.
363 G. P. Moss, in Ref. 355, p. 221.
364 G. P. Moss, in Ref. 356, p. 197.
365 R. Croteau, A. J. Burbolt and W. D. Loomis, *Phytochem.*, 11 (1972) 2459.
366 R. Croteau, A. J. Burbolt and W. D. Loomis, *Phytochem.*, 11 (1972) 2937.
367 M. J. O. Francis, in Ref. 34, p. 29.
368 W. D. Loomis, in Ref. 33, p. 59.
369 V. Herout, in Ref. 34, p. 53.
370 V. Herout and F. Šorm, in J. B. Harborne and T. Swain (Eds.), *Perspectives in Phytochemistry*, Academic Press, London, 1969, p. 139.
371 A. C. Oehlschlager and G. Ourisson, in Ref. 33, p. 83.
372 J. MacMillan, in Ref. 34, p. 183.
373 L. J. Goad, in Ref. 357, p. 45.
374 J. D. Bu'Lock, *Pure Appl. Chem.*, 34 (1973) 435.
375 F. W. Hemming, in Ref. 357, p. 105.
376 D. R. Threlfall and G. R. Whistance, in Ref. 34, p. 357.
377 W. Sanderman, in Ref. 44. Vol. 3, p. 503.
378 J. A. Rudinsky *Science*, 166 (1969) 884.
379 G. B. Pitman, *Science*, 166 (1969) 905.
380 E. C. Bate-Smith, in Ref. 6, p. 45.
381 E. von Rudloff, personal communication.
382 C. H. Muller and C.-H. Chou, in Ref. 6, p. 201.
383 B. Bartholomew, *Science*, 170 (1970) 1210.
384 B. V. Milborrow, in Ref. 34, p. 137.
385 A. W. Barksdale, *Science*, 166 (1969) 831.
386 H. H. Rees, in Ref. 34, p. 181.
387 C. M. Williams, in Ref. 315, p. 103.
388 D. Janzen, *Pure and Appl. Chem.*, 24 (1973) .
389 C. A. West, in Ref. 168, p. 143.
390 C. W. Bird, J. M. Lynch, S. J. Pirt and W. W. Reid, *Tetrahedron Letters*, (1971) 3189.
391 L. J. Goad, in Ref. 33, p. 159.
392 G. F. Gibbon, L. J. Goad and T. W. Goodwin, *Phytochem.*, 6 (1967) 677.
393 V. Herout, *Progr. Phytochem.*, 2 (1970) 143.
394 F. Boltari, A. Marsili, I. Morelli and M. Pacchiani, *Phytochem.*, 11 (1972) 2519.
395 M. J. Kulshreshta, D. K. Kulshreshtha and R. P. Raslogi, *Phytochem.*, 11 (1972) 2369.
396 J. C. Wallwork and F. L. Crane, *Progr. Phytochem.*, 2 (1970) 267.
397 T.-C. Lee, T. H. Lee and C. O. Chichester, *Phytochem.*, 11 (1972) 697.
398 O. Isler (Ed.), *Carotenoids*, Birkhäuser, Basel, 1971.
399 T. W. Goodwin, in Ref. 34, p. 315.
400 T. W. Goodwin, in Ref. 398, p. 577.
401 G. Britton, in Ref. 34, p. 255.
402 O. B. Weeks, in Ref. 34, p. 291.
403 C. P. Whittingham, in T. W. Goodwin (Ed.), *Chemistry and Biochemistry of Plant Pigments*, Academic Press, London, 1965, p. 357.
404 G. M. Curry, in M. B. Wilkins (Ed.), *Physiology of Plant Growth and Development*, McGraw-Hill, London, 1969, p. 245.
405 T. W. Goodwin, in T. Swain (Ed.), *Comparative Phytochemistry*, Academic Press, London, 1966, p. 121.
406 J. Brooks, P. R. Grant, M. D. Muir, P. van Gijzel and G. Shaw, (Eds.), *Sporopollenin*, Academic Press, London, 1971.
407 S. Liaaen-Jensen, in Ref. 34, p. 223.

408 G. P. Moss, in Ref. 355, p. 198.
409 G. P. Moss, in Ref. 366, p. 180.
410 J. Heslop-Harrison, *Pollen: Development and Physiology*, Butterworth, London, 1971.
411 G. Shaw, in Ref. 6, p. 31.
412 M. Barcier, *Progr. Phytochem.*, 1 (1970) 1.
413 W. G. Rosen, *Ann. Rev. Plant Physiol.*, 19 (1968) 435.
414 J. Muller, *Biol. Rev. Cambridge Phil. Soc.*, 45 (1970) 417.
415 W. G. Chaloner, *Biol. Rev. Cambridge Phil. Soc.*, 45 (1970) 353.
416 W. G. Chaloner and G. Orbel, in Ref. 406, p. 273.
417 G. Shaw, in Ref. 406, p. 305.
418 J. Brooks, in Ref. 406, p. 351.
419 M. J. Strohl and M. K. Seikel, *Phytochem.*, 4 (1965) 383.
420 T. Swain, unpublished work.
421 J. B. Harbone, *Comparative Biochemistry of the Flavonoids*, Academic Press, London, 1967.
422 T. Swain and E. C. Bate-Smith, in Ref. 44, Vol. 3, p. 755.
423 W. D. Ollis, *Recent Advan. Phytochem.*, 1 (1968) 329.
424 W. D. Hillis, *Phytochem.*, 11 (1972) 1207.
425 D. G. Roux, *Phytochem.*, 11 (1972) 1219.
426 T. Swain, in J. Bonner and J. Varner (Eds.), *Plant Biochemistry*, Academic Press, New York, 1969, p. 552.
427 H. Grisebach and W. Barz, *Naturwissenschaften*, 56 (1969) 538.
428 W. E. Hillis, *Wood Extractives*, Academic Press, New York, 1962.
429 P. Ribereau-Gayon, *Plant Phenolics*, Oliver and Boyd, Edinburgh, 1972.
430 J. B. Harborne, *Biochemistry of Phenolic Compounds*, Academic Press, London, 1964.
431 E. Haslam, *The Chemistry of Vegetable Tannins*, Academic Press, London, 1966.
432 K. R. Hanson and E. A. Havir, *Recent Advan. Phytochem.*, 4 (1972) 45.
433 E. L. Camms and G. H. N. Towers, *Phytochem.*, 12 (1973) 961.
434 H. Grisebach, *Pure and Appl. Chem.*, 34 (1973) 487.
435 T. Swain and C. Williams, *Phytochem.*, 9 (1970) 2115.
436 G. H. N. Towers in J. B. Harborne and T. Swain (Eds.), *Perspectives in Phytochemistry*, Academic Press, London, 1969, p. 179.
437 J. W. Wallace, T. J. Mabry and R. E. Alston, *Phytochem.*, 8 (1969) 2287.
438 S. Ahmed and T. Swain, *Phytochem.*, 9 (1970) 2287.
439 L. L. Creasy and T. Swain, *Phytochem.*, 6 (1966) 501.
440 M. G. de Nicola, M. Piatelli, V. Castro-Giovanni and V. Amico, *Phytochem.*, 11 (1972). 1027.
441 M. Zenk, in H. Wagner and L. Hörhammer (Eds.), *Pharmacognosy and Phytochemistry*, Springer, Berlin, 1971, p. 314.
442 E. C. Bate-Smith, *J. Linn. Soc. (Bot.)*, 58 (1962) 95.
443 E. C. Bate-Smith, *J. Linn. Soc. (Bot.)*, 60 (1968) 325.
444 B. Voirin and P. Lebreton, *Boissiera*, 19 (1971) 259.
445 G. Benz, O. Martensson and E. Nilsson, *Acta Chem. Scand.*, 20 (1966) 277.
446 K. R. Markham and L. J. Porter, *Phytochem.*, 8 (1969) 1771.
447 A. W. Galston, in J. B. Harborne and T. Swain (Eds.), *Perspectives in Phytochemistry*, Academic Press, London, 1969, p. 193.
448 B. J. Deverall, in Ref. 316, p. 217.
449 I. A. M. Cruickshank and D. R. Perrin, in Ref. 430, p. 511.
450 J. L. Goldstein and T. Swain, *Phytochem.*, 2 (1963) 371.
451 W. R. Thompson, J. Meinwald, D. Aneshanoley and T. Eisner, *Science*, 177 (1972) 530.
452 T. Swain, in Ref. 13, p. 274.

453 H. Schildknecht, *Angew. Chem., Internat. Edn.*, 9 (1970) 1.
454 R. J. Pryce, *Phytochem.*, 11 (1972) 1759.
454a W. Oettmeier and A. Henpel, *Z. Naturforsch.*, 27b (1972) 177.
454b J. A. Saunders and J. W. McClure, *Amer. J. Bot.*, 59 (1972) 673.
455 E. C. Bate-Smith, personal communication.
456 C. H. Hassall and A. I. Scott, in W. D. Ollis (Ed.), *Recent Developments in the Chemistry of Natural Phenolic Compounds*, Pergamon, Oxford, 1961, p. 119.
457 J. B. Pridham, *Phytochem.*, 3 (1964) 493.
458 J. E. Saxton (Ed.), *The Alkaloids*, Vol. 1, Chemical Society, London, 1971.
459 J. E. Saxton (Ed.), *The Alkaloids*, Vol. 2, Chemical Society, London, 1972.
460 R. Hegnauer, in T. Swain (Ed.), *Comparative Phytochemistry*, Academic Press, London, 1966, p. 211.
461 T. A. Smith, *Biol. Rev. Cambridge Phil. Soc.*, 46 (1971) 201.
462 D. E. Hathway (Ed.), *Foreign Compound Metabolism in Mammals*, Chemical Society, London, 1972.
463 L. Fowden, *Progr. Phytochem.*, 2 (1970) 203.
464 M. F. Roberts, *Phytochem.*, 10 (1971) 3057.
465 F. Schlitter, in Ref. 458, p. 463.
466 G. Arnold and J. L. Hill, in Ref. 6, p. 72.
467 M. Rothschild, in Ref. 6, p. 1.
468 E. J. Smith, *Science*, 168 (1970) 421.
469 K. Schildknecht, *Endeavour*, 30 (1971) 136.
470 J. J. Williams and B. G. Schubert, *USDA Tech. Bull.*, U.S. Govt. Printing Office Washington, 1961, p. 1234.
471 R. Eyjolfsson, *Fortschr. Chem. Org. Naturst.*, 28 (1970) 74.
472 L. Kimler, J. Mears, T. J. Mabry and H. Rosler, *Taxon*, 19 (1970) 875.
473 R. Eyjolfsson, *Phytochem.*, 9 (1970) 854.
474 M. G. Ettlinger and A. J. Kjaer, *Recent Advan. Phytochem.*, 1 (1968) 59.
475 D. A. Jones, in Ref. 6, p. 63.

Appendix to Chapter II

Biochemical Evolution in Plants

T. SWAIN

Royal Botanic Gardens, Kew (Great Britain)

Since this Chapter was prepared, a number of extra observations have been made which throw new light on several of the topics which have been discussed.

The possibility that the remains of ancient plants were contaminated with saprophytic fungi and bacteria before fossilisation has already been mentioned (pp. 142, 143). In most cases this occurrence would have little or no effect on the overall morphology of the fossils, but it might be expected to markedly affect their chemistry. Recently, Ourisson and his colleagues [476, 477] have presented evidence that this is so and that bacterial contamination of higher plant fossils may be more or less universal. They showed that homohopane and higher homologues of hopane (XXXVII, p. 232) were consistently present as major components of the cyclic fraction in Messel oil shale (50 million years old) and in other sediments of 15 to 300 million years of age. These compounds, which are completely absent from higher plants, have now been shown to be characteristic of *Acetobacter zinnerum* and related bacteria [477]. The results suggest that in order to use the fossil chemistry of higher plants for phylogenetic purposes, it would be better to restrict investigations to those products which are specific to such plants, like lignin, flavonoids and alkaloids. It is also apparent that the triterpenoid chemistry of bacteria has changed little over the last 300 million years, and indicates that the chemistry of modern organisms may well have been fixed since the time of their emergence.

On the nucleic acid front, it may be noted that the old dicta of 'one gene, one enzyme' or 'one gene, one polypeptide' are obviously too simplistic to account for the results of genetic analysis obtained even in simple organisms such as bacteriophages. Obviously one needs to think in more complex terms, and the recent evidence from several laboratories shows that it may not be necessary to *translate* DNA entirely to observe genetic effects. It might also be noted that new combinations of the genome which lead to genetic isolation might arise in higher plant populations, especially polyploids, by selection without the need for mutation. This points again to the fact that the *population* is the ultimate basis of evolutionary advance and the concept of the species as a collection of like populations (p. 131) is a taxonomic rather than an evolutionary one.

A further analysis of the evolutionary implications of cytochrome *c* sequence data has been carried out using results from animals, fungi, protists and plants[478]. This, by and large, in respect to higher plants agrees with the results published by Boulter *et al.*[216–218] (pp. 167–170). However, one interesting finding from the analysis is that from their cytochrome *c* sequences, the fungi appear to be much more closely related to higher animals than to higher plants. This data is thus in good accord with that from the occurrence of chitin in cell walls and related structures (p. 202) and certainly supports Whittaker's contention[15] that the fungi should be treated as a separate Kingdom (p. 129).

Recent investigations on the distribution of stizolobic acid (XCV) and the related stizolobinic acid in the angiosperms have drawn attention to the suggestion by Senoh[479] that both could arise by the *meta*-cleavage of 3, 4-dihydroxyphenylalanine (DOPA) and subsequent rearrangement of the product. This, therefore, constitutes another example of aromatic ring cleavage in higher plants (see p. 179) and reinforces the plea for more research work in this area.

Another field in which recent work points to the need for a re-evaluation of our views concerns the evolution of lignin. Observations from several laboratories point to the possibility that lignin might have originally evolved in primitive land plants in a fortuitous fashion. For example, it has been

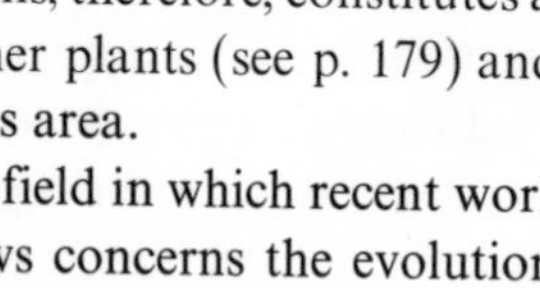

(XCV) Stizolobic acid

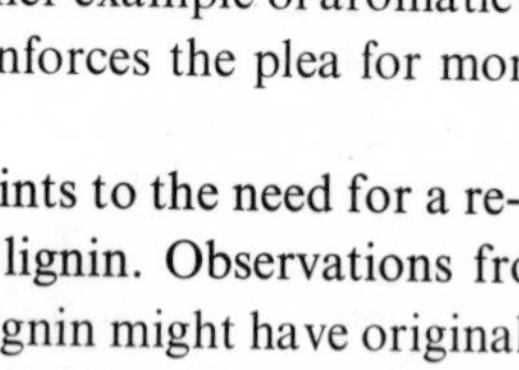

(XCVI) Lunularic acid

shown that part of the cellulose in grass cell-walls is acylated with ferulic acid (XIV, p. 206) which prevents complete hydrolysis by cellulase[480]. The same is true of barley aleurone layers[481]. Other reports indicate that the phenomenon of polysaccharide acylation by commonly occurring cinnamic and benzoic acids may be general. This could be a relic of a very ancient defence mechanism against fungal and bacterial pathogens, preventing their ingress into the cell by inhibiting their extracellular polysaccharide hydrolases[482]. If this is so, there seems to be no reason why adjacent hydroxycinnamic acids on the polysaccharide matrix should not have been cross-linked by phenolic coupling reactions, perhaps mediated by peroxidase, thus increasing the resistance of the cell to attack. An extension of such reactions, including esterification of the cinnamic acids themselves and further cross-linking, might well have led to a primitive lignin. Indeed, there are many reports of lignin–carbohydrate bonding in higher land plants[483] but the nature of the linkage between the two polymers is not known. If lignin did evolve as suggested above, it would account for the parallel occurrence of hydroxycinnamic acids and lignins with the same substitution pattern. It would also point to a need to further investigate the overall biosynthetic route to lignins in order to evaluate the role of the cinnamyl alcohols.

One other interesting observation in the area of phenolic compounds is that all the algae, including the blue-greens, and the liverworts have been shown to contain the dihydrostilbene growth inhibitor, lunularic acid (XCVI)[484]. The immediate precursor of this compound in plants is the same *p*-coumaryltriacetic acid precursor which gives rise to the chalkone (Fig. 27, p. 259); the only difference being that in lunularic acid ring closure to give the second aromatic ring involves condensation of the carbonyl group of the cinnamyl moiety with the penultimate $=CH_2$ group of the chain, the terminal carboxyl group being retained. The stilbenes are relatively rare compounds in higher plants being found mainly in gymnosperms; however, in almost every case the carboxyl group is lost. The occurrence of lunularic acid in algae shows that the primary route to the common intermediate of flavonoids is a very ancient one, and that some accident of evolution allowed the switch in ring closure to occur to change from stilbene to chalkone formation in *Chara* (p. 264). It should be noted that the absorption spectra of lunularic acid and the related stilbenes makes them less suitable than the primitive flavonoids as UV screens (p. 263). It may also be noted that the physiological activity of lunularic acid makes it unlikely that it would be accumulated in sufficient concentration to act in this way.

Finally, experiments on the taste responses of herbivorous reptiles[484] have clarified some aspects of plant–animal co-evolution. The results show that reptiles can detect and are repelled by tannins in plants at about the same concentrations as deter mammals. Their detection limit for alkaloids, on the other hand, is about one thousand times lower than that of mammals. It seems likely from this data that the advent of the early angiosperms in the mid Cretaceous with more highly astringent hydrolysable tannins, followed by the evolution of angiosperms containing new types of difficulty detoxifiable alkaloids based on phenylalanine, could have been a most important factor in the gradual disappearance of the giant ruling reptiles during the late Cretaceous. These feeding responses therefore strongly support the idea that the primitive angiosperm was more nearly related to the present-day Dilleniidae and Rosidae rather than the Magnoliidae[226, 227]. It appears, therefore, that none of the present-day sub-divisions of the angiosperms are patristically related, but that all were derived separately from proto-angiosperm stock of early Cretaceous origin.

476 A. Ensminger, P. Albrecht, G. Ourisson, B. J. Kimble, J. R. Maxwell and G. Eglinton, *Tetrahedron Letters*, (1972) 3861.
477 G. Ourisson, Paper given to the Nobel Symposium No. 25, *Chemistry in Botanical Classification*, Lidingo, Stockholm, August 1973.
478 P. J. McLaughlin and M. O. Dayhoff, *J. Mol. Evol.*, 2 (1973) 99.
479 S. Senoh, *Nippon Kagaku Zasshi*, 86 (1965) 1087.
480 R. D. Hartley, *Phytochemistry*, 12 (1973) 661.
481 R. G. Fulcher, T. P. O'Brien and J. W. Lee, *Austral. J. Biol. Sci.*, 25 (1972) 23.
482 T. Swain and J. Friend, *Soc. Exptl. Biol.*, Abstracts of the January 1974 meeting.
483 I. M. Morrison, *Phytochemistry*, 13 (1974) in the press.
484 T. Swain, *Brit. Herpetological Soc. Newsletter*, January 1974.

Subject Index